OTHER CARTOON BOOKS BY
GRADY KLEIN AND YORAM BAUMAN, Ph.D.

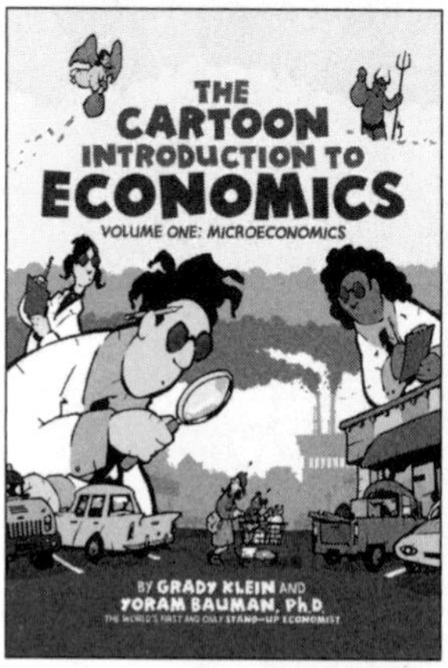
THE CARTOON INTRODUCTION TO ECONOMICS
VOLUME ONE: MICROECONOMICS
BY GRADY KLEIN AND YORAM BAUMAN, Ph.D.
THE WORLD'S FIRST AND ONLY STAND-UP ECONOMIST

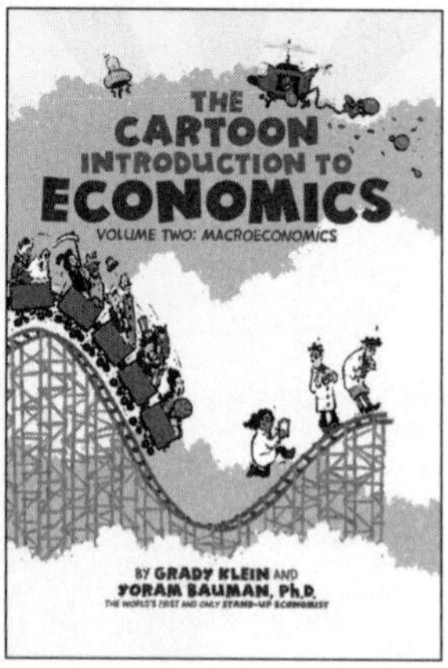
THE CARTOON INTRODUCTION TO ECONOMICS
VOLUME TWO: MACROECONOMICS
BY GRADY KLEIN AND YORAM BAUMAN, Ph.D.
THE WORLD'S FIRST AND ONLY STAND-UP ECONOMIST

AND BY GRADY KLEIN
AND ALAN DABNEY, Ph.D.

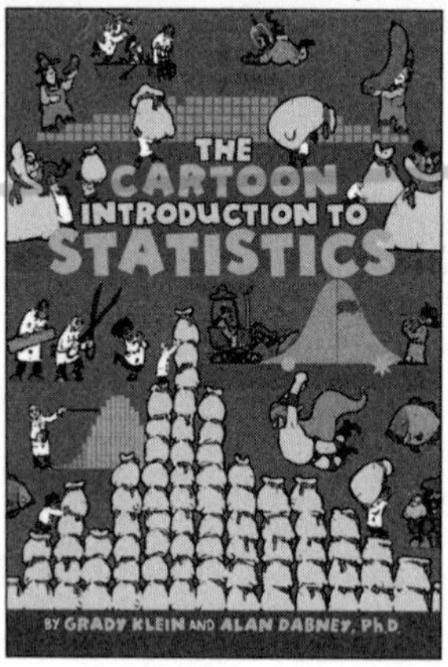
THE CARTOON INTRODUCTION TO STATISTICS
BY GRADY KLEIN AND ALAN DABNEY, Ph.D.

THE
CARTOON
INTRODUCTION TO
CALCULUS

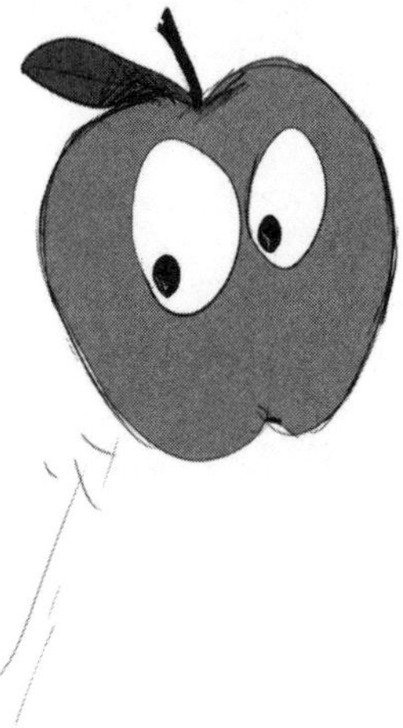

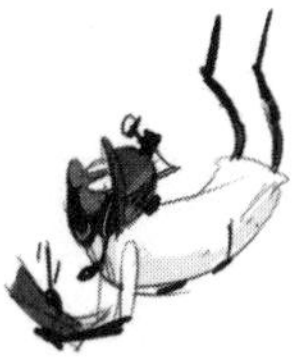

THE CARTOON INTRODUCTION TO CALCULUS

BY GRADY KLEIN AND
YORAM BAUMAN, Ph.D.
THE WORLD'S FIRST AND ONLY STAND-UP ECONOMIST

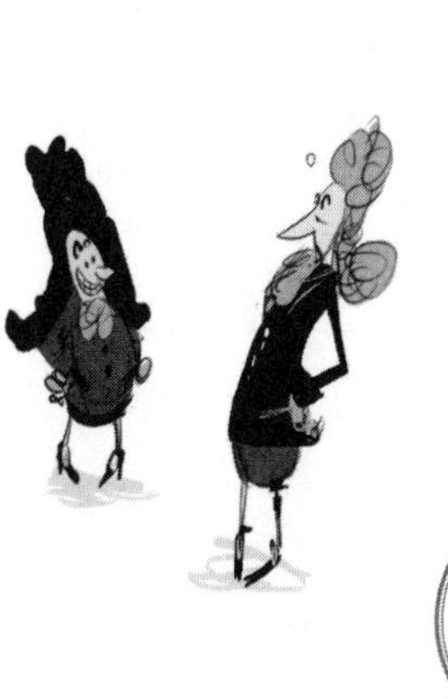

A NOVEL GRAPHIC FROM HILL AND WANG
A DIVISION OF FARRAR, STRAUS AND GIROUX
NEW YORK

HILL AND WANG
A DIVISION OF FARRAR, STRAUS AND GIROUX
120 BROADWAY, NEW YORK 10271

LIBRARY OF CONGRESS CATALOGING-IN-PUBLICATION DATA
NAMES: KLEIN, GRADY. | BAUMAN, YORAM.
TITLE: THE CARTOON INTRODUCTION TO CALCULUS / BY GRADY KLEIN AND YORAM BAUMAN, PH. D.,
   THE WORLD'S FIRST AND ONLY STAND-UP ECONOMIST.
OTHER TITLES: INTRODUCTION TO CALCULUS
DESCRIPTION: FIRST EDITION. | NEW YORK : HILL AND WANG, A DIVISION OF FARRAR, STRAUS AND
   GIROUX, 2019.
IDENTIFIERS: LCCN 2018047855 | ISBN 9780809033690 (PAPERBACK)
SUBJECTS: LCSH: CALCULUS—CARICATURES AND CARTOONS. | CALCULUS—HUMOR.
CLASSIFICATION: LCC QA303.2 .K54 2019 | DDC 515—DC23
LC RECORD AVAILABLE AT HTTPS://LCCN.LOC.GOV/2018047855

OUR BOOKS MAY BE PURCHASED IN BULK FOR PROMOTIONAL, EDUCATIONAL, OR BUSINESS USE.
PLEASE CONTACT YOUR LOCAL BOOKSELLER OR THE MACMILLAN CORPORATE
AND PREMIUM SALES DEPARTMENT AT 1-800-221-7945, EXTENSION 5442,
OR BY E-MAIL AT MACMILLANSPECIALMARKETS@MACMILLAN.COM.

WWW.FSGBOOKS.COM
WWW.TWITTER.COM/FSGBOOKS • WWW.FACEBOOK.COM/FSGBOOKS

1 3 5 7 9 10 8 6 4 2

FOR AWCK, LRCK & BLCK
—GFCK

FOR LAURA, ZADIE & TEDDY
—YB

CONTENTS

PART ONE:
OVERVIEW

1: **INTRODUCTION**...3
2: **SPEED**...15
3: **AREA**...29
4: **THE FUNDAMENTAL THEOREM OF CALCULUS**...43
5: **LIMITS**...55

PART TWO:
DERIVATIVES

6: **LIMITS AND DERIVATIVES**...71
7: **THE CALCULUS TOOLKIT**...83
8: **EXTREME VALUES**...95
9: **OPTIMIZATION**...107
10: **ECONOMICS**...119

PART THREE:
INTEGRALS

11: **INTEGRATION, THE HARD WAY**...133
12: **INTEGRATION, THE EASY WAY**...147
13: **THE FUNDAMENTAL THEOREM, REVISITED**...159
14: **PHYSICS**...173
15: **LIMITS BEYOND LIMITS**...187

GLOSSARY...201

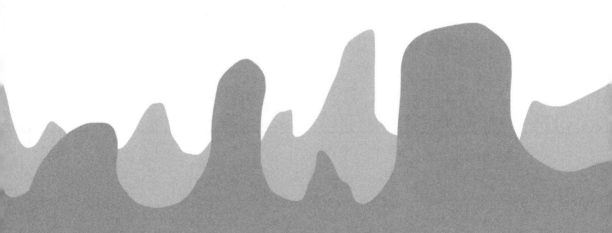

PART ONE
OVERVIEW

CHAPTER 1
INTRODUCTION

CALCULUS IS ABOUT **TWO MATHEMATICAL MOUNTAINS.**

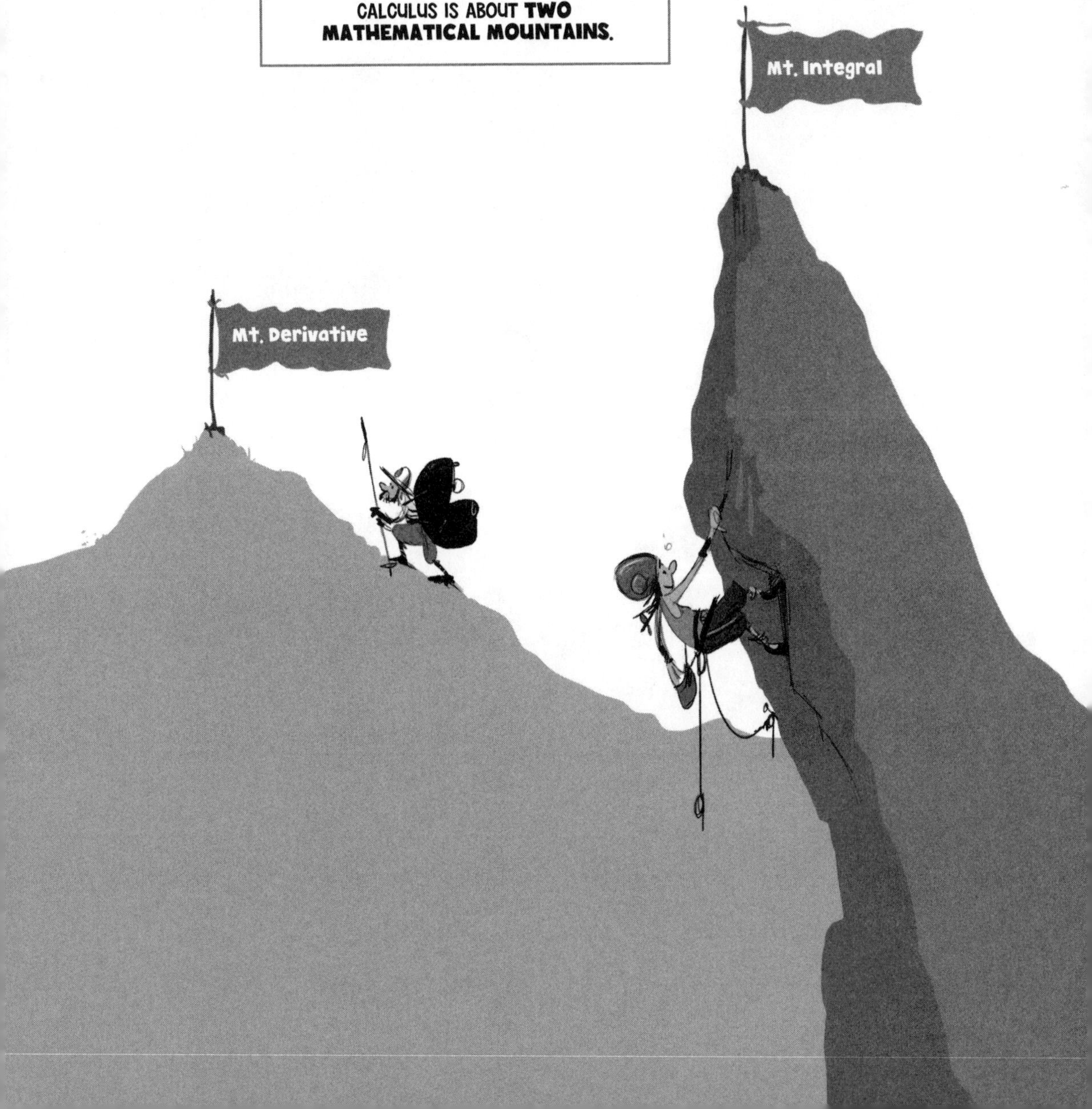
Mt. Integral
Mt. Derivative

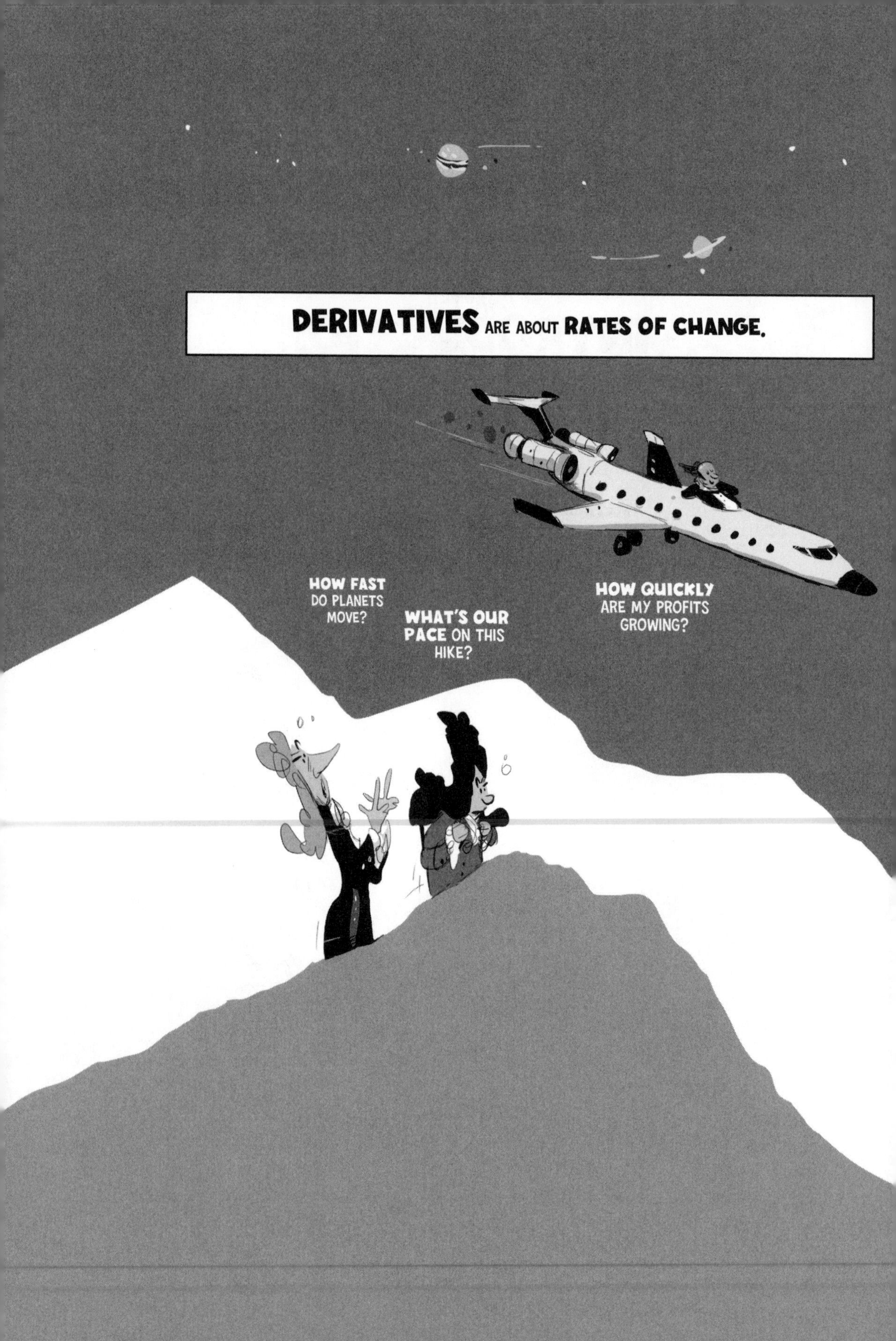

DERIVATIVES ARE ABOUT RATES OF CHANGE.
HOW FAST DO PLANETS MOVE?
WHAT'S OUR PACE ON THIS HIKE?
HOW QUICKLY ARE MY PROFITS GROWING?

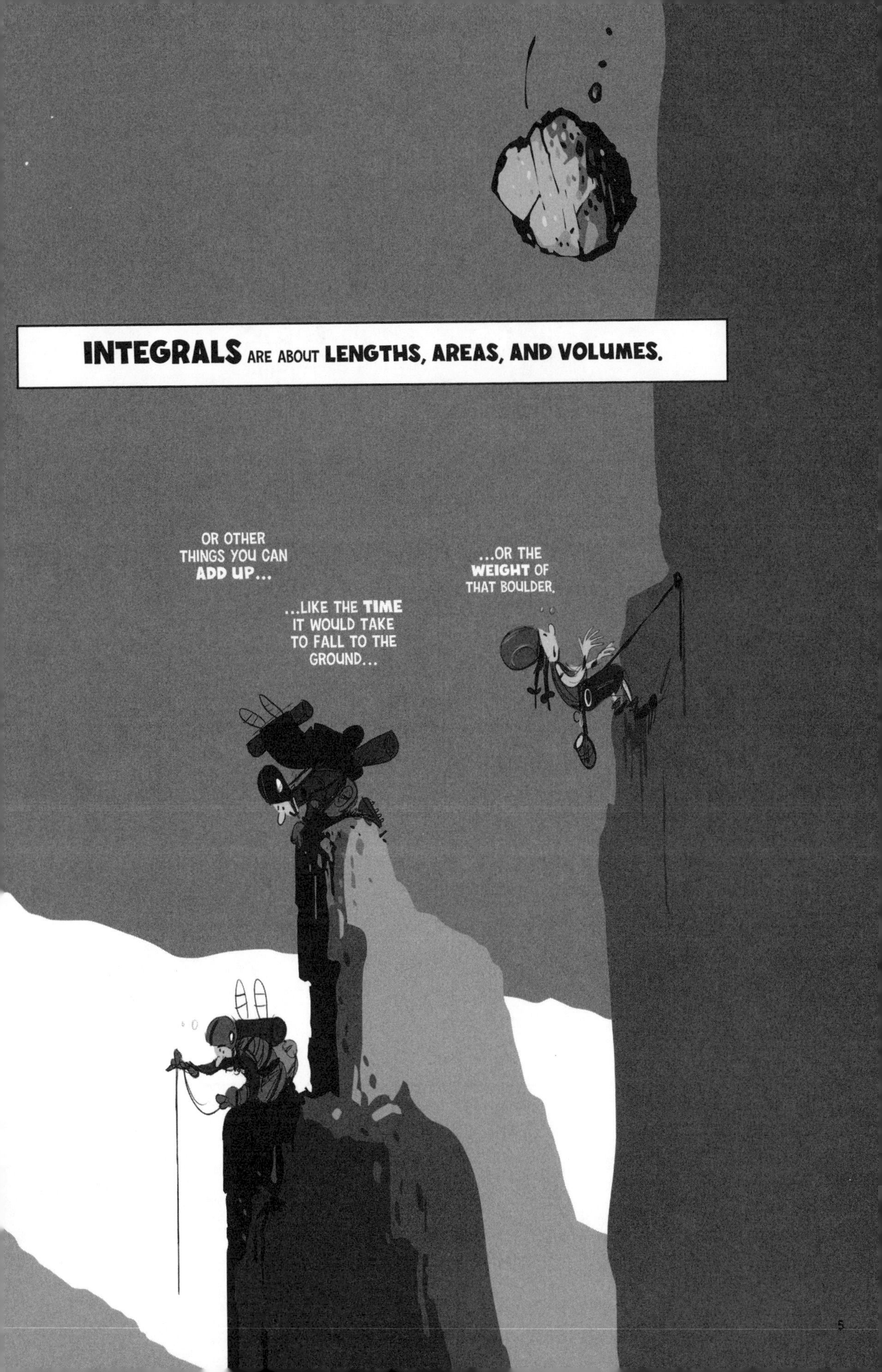
INTEGRALS ARE ABOUT LENGTHS, AREAS, AND VOLUMES.
OR OTHER THINGS YOU CAN ADD UP...
...LIKE THE TIME IT WOULD TAKE TO FALL TO THE GROUND...
...OR THE WEIGHT OF THAT BOULDER.

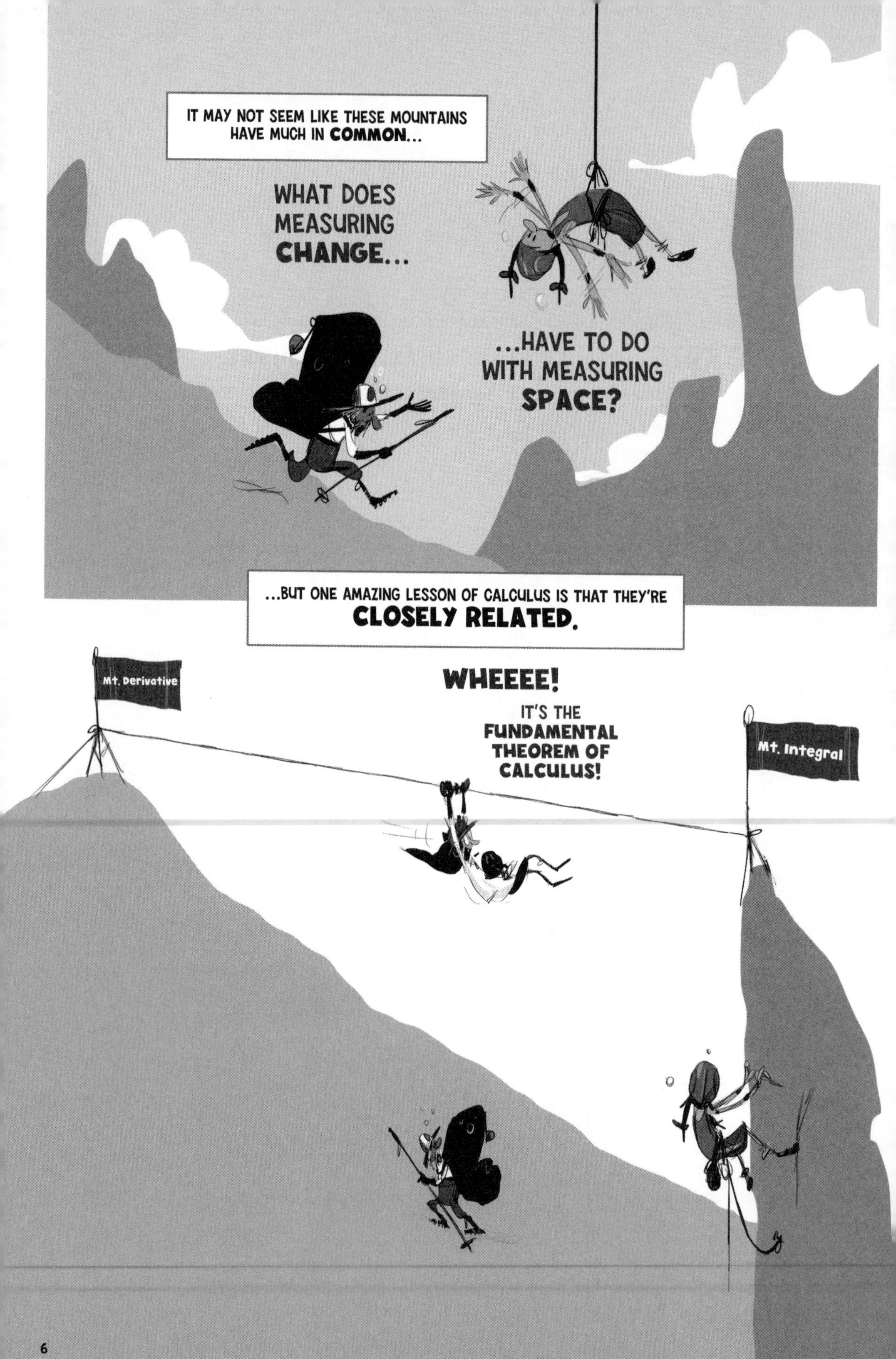
IT MAY NOT SEEM LIKE THESE MOUNTAINS HAVE MUCH IN COMMON...
WHAT DOES MEASURING CHANGE...
...HAVE TO DO WITH MEASURING SPACE?
...BUT ONE AMAZING LESSON OF CALCULUS IS THAT THEY'RE CLOSELY RELATED.
Mt. Derivative
WHEEEE!
IT'S THE FUNDAMENTAL THEOREM OF CALCULUS!
Mt. Integral

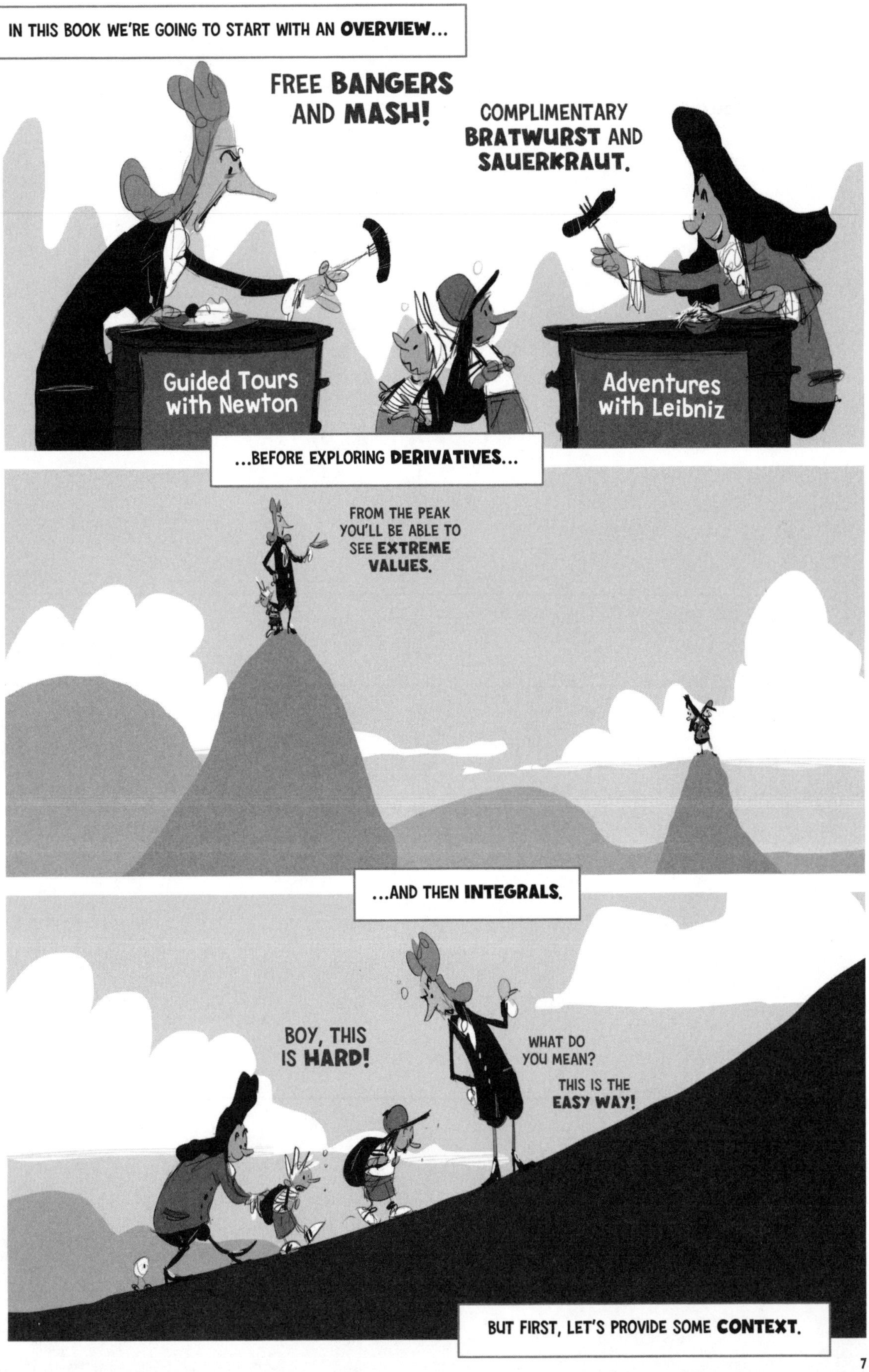
IN THIS BOOK WE'RE GOING TO START WITH AN **OVERVIEW**...
FREE **BANGERS** AND **MASH!**
COMPLIMENTARY **BRATWURST** AND **SAUERKRAUT**.
Guided Tours with Newton
Adventures with Leibniz
...BEFORE EXPLORING **DERIVATIVES**...
FROM THE PEAK YOU'LL BE ABLE TO SEE **EXTREME VALUES**.
...AND THEN **INTEGRALS**.
BOY, THIS IS **HARD!**
WHAT DO YOU MEAN?
THIS IS THE **EASY WAY!**
BUT FIRST, LET'S PROVIDE SOME **CONTEXT**.

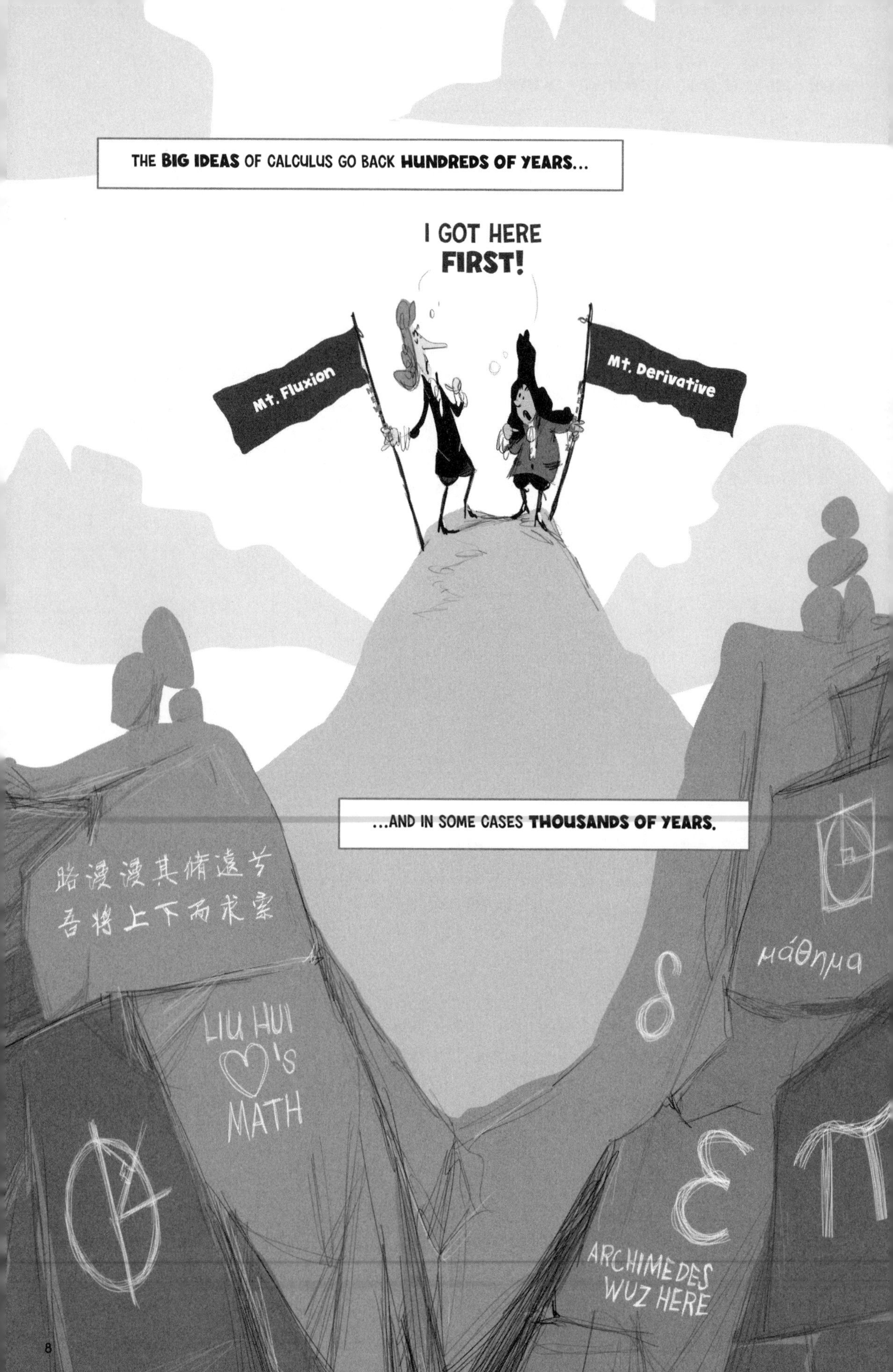

THE BIG IDEAS OF CALCULUS GO BACK HUNDREDS OF YEARS...
I GOT HERE FIRST!
Mt. Fluxion
Mt. Derivative
...AND IN SOME CASES THOUSANDS OF YEARS.
路漫漫其脩遠兮
吾將上下而求索
μάθημα
LIU HUI ♡'S MATH
ARCHIMEDES WUZ HERE

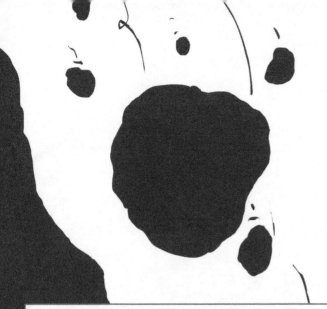

UNFORTUNATELY, THESE **BIG IDEAS** HAVE BEEN OBSCURED BY **TWO MORE RECENT DEVELOPMENTS.**

AVALANCHE!

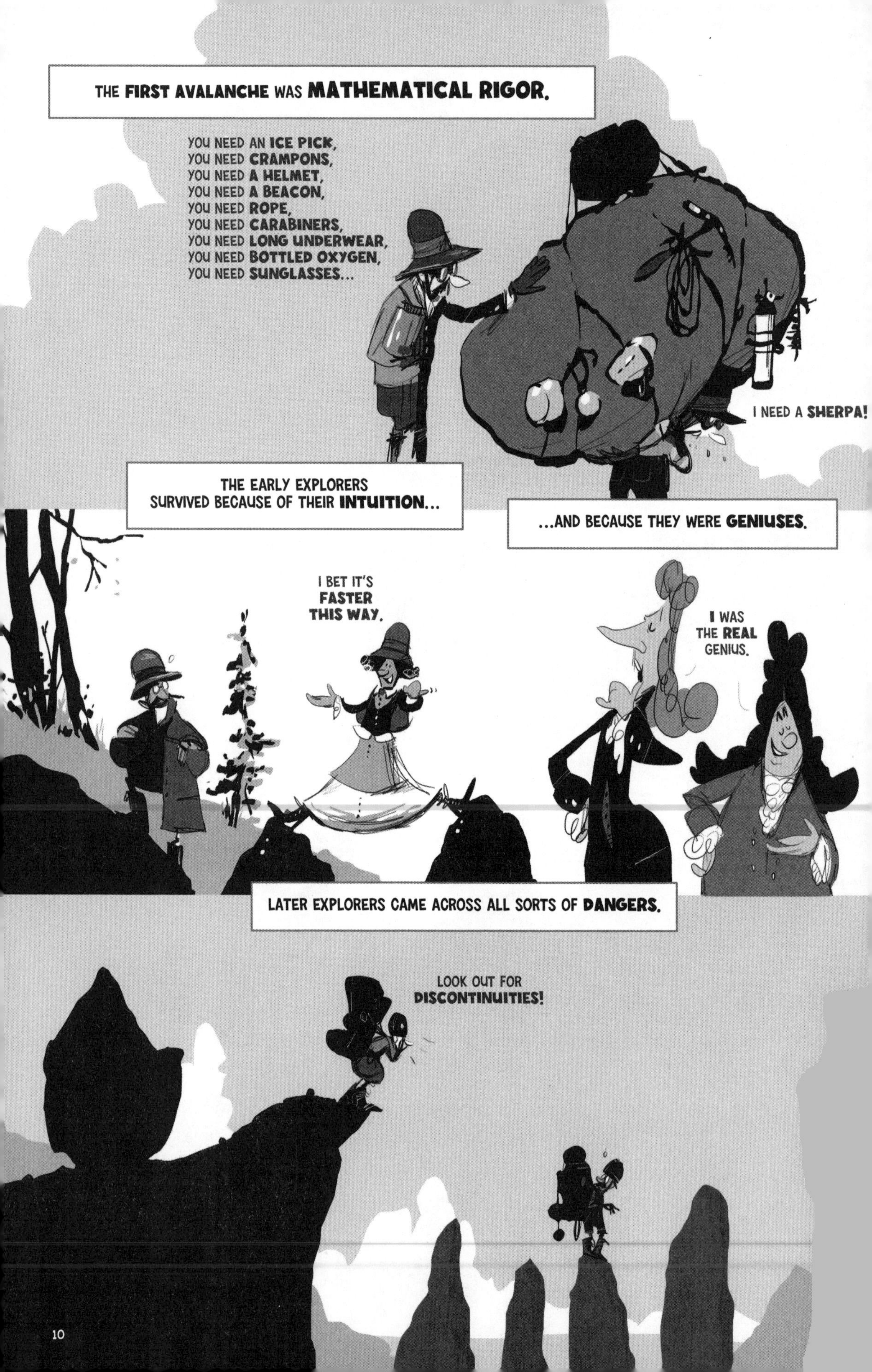

THE FIRST AVALANCHE WAS MATHEMATICAL RIGOR.
YOU NEED AN ICE PICK, YOU NEED CRAMPONS, YOU NEED A HELMET, YOU NEED A BEACON, YOU NEED ROPE, YOU NEED CARABINERS, YOU NEED LONG UNDERWEAR, YOU NEED BOTTLED OXYGEN, YOU NEED SUNGLASSES...
I NEED A SHERPA!
THE EARLY EXPLORERS SURVIVED BECAUSE OF THEIR INTUITION...
...AND BECAUSE THEY WERE GENIUSES.
I BET IT'S FASTER THIS WAY.
I WAS THE REAL GENIUS.
LATER EXPLORERS CAME ACROSS ALL SORTS OF DANGERS.
LOOK OUT FOR DISCONTINUITIES!

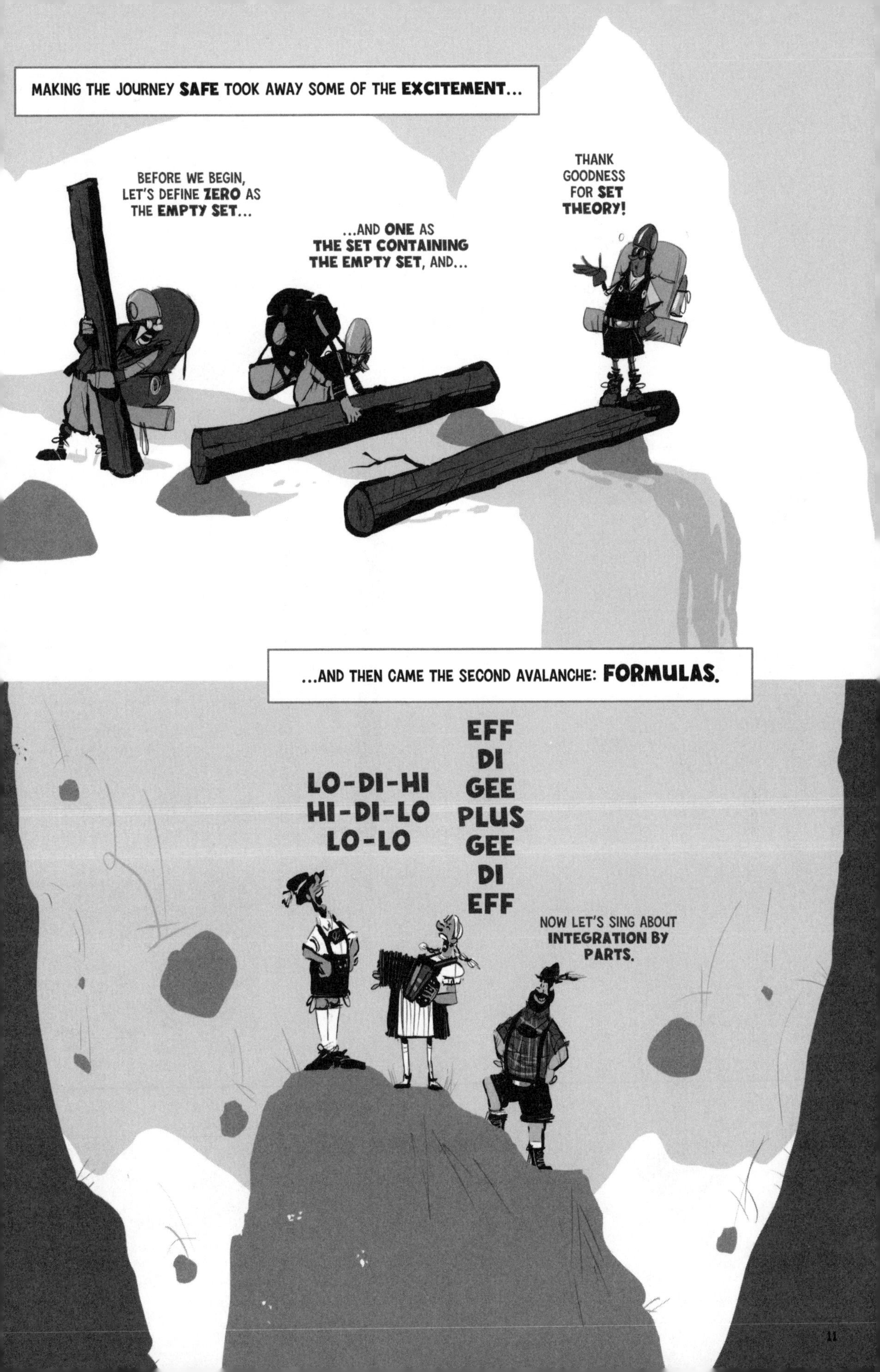
MAKING THE JOURNEY SAFE TOOK AWAY SOME OF THE EXCITEMENT...
BEFORE WE BEGIN, LET'S DEFINE ZERO AS THE EMPTY SET...
...AND ONE AS THE SET CONTAINING THE EMPTY SET, AND...
THANK GOODNESS FOR SET THEORY!
...AND THEN CAME THE SECOND AVALANCHE: FORMULAS.
LO-DI-HI HI-DI-LO LO-LO
EFF DI GEE PLUS GEE DI EFF
NOW LET'S SING ABOUT INTEGRATION BY PARTS.

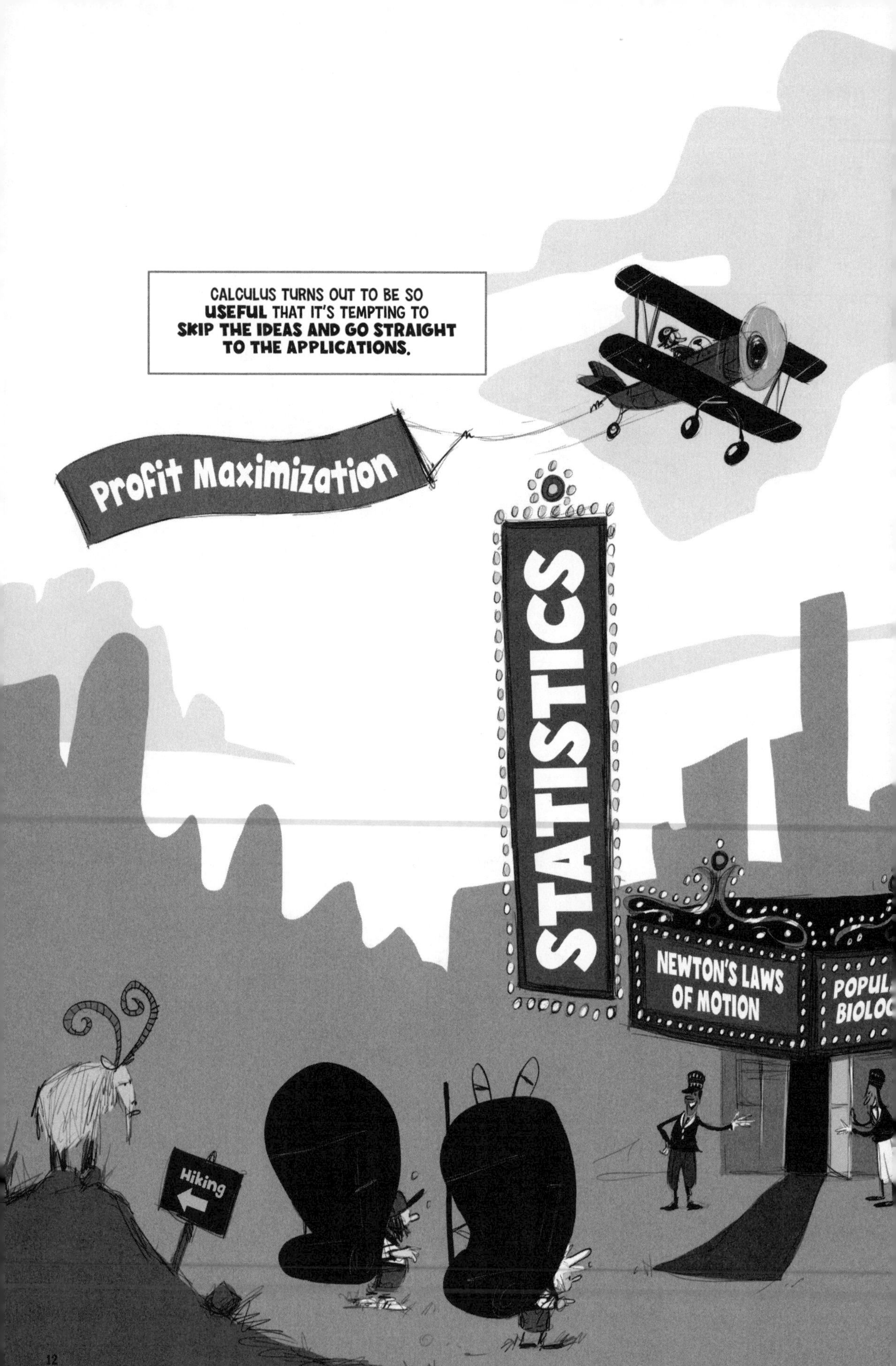
CALCULUS TURNS OUT TO BE SO USEFUL THAT IT'S TEMPTING TO SKIP THE IDEAS AND GO STRAIGHT TO THE APPLICATIONS.
Profit Maximization
STATISTICS
NEWTON'S LAWS OF MOTION
POPUL
BIOLO
Hiking

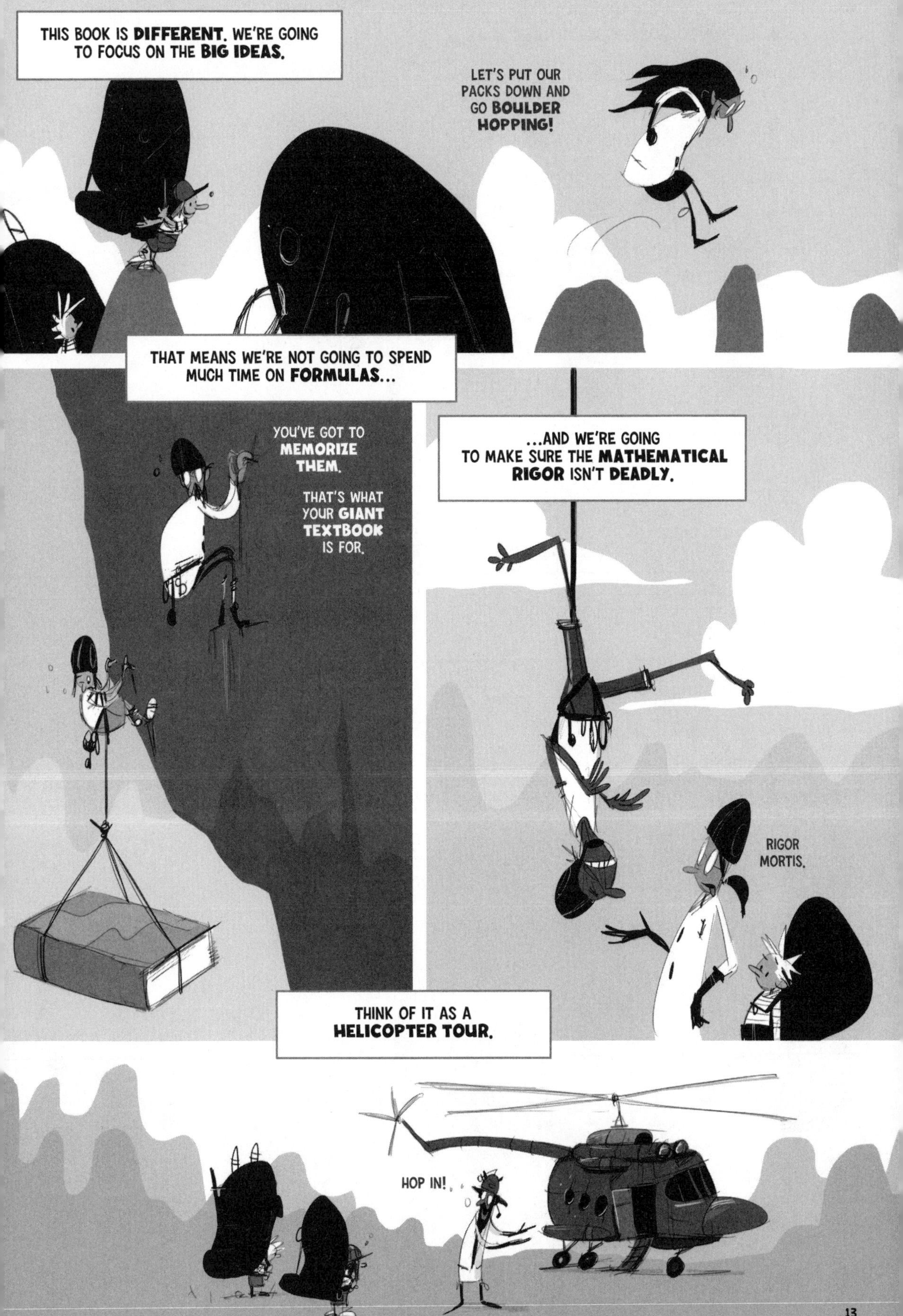
THIS BOOK IS **DIFFERENT.** WE'RE GOING TO FOCUS ON THE **BIG IDEAS.**
LET'S PUT OUR PACKS DOWN AND GO **BOULDER HOPPING!**
THAT MEANS WE'RE NOT GOING TO SPEND MUCH TIME ON **FORMULAS...**
YOU'VE GOT TO **MEMORIZE THEM.**
THAT'S WHAT YOUR **GIANT TEXTBOOK** IS FOR.
...AND WE'RE GOING TO MAKE SURE THE **MATHEMATICAL RIGOR** ISN'T **DEADLY.**
RIGOR MORTIS.
THINK OF IT AS A **HELICOPTER TOUR.**
HOP IN!

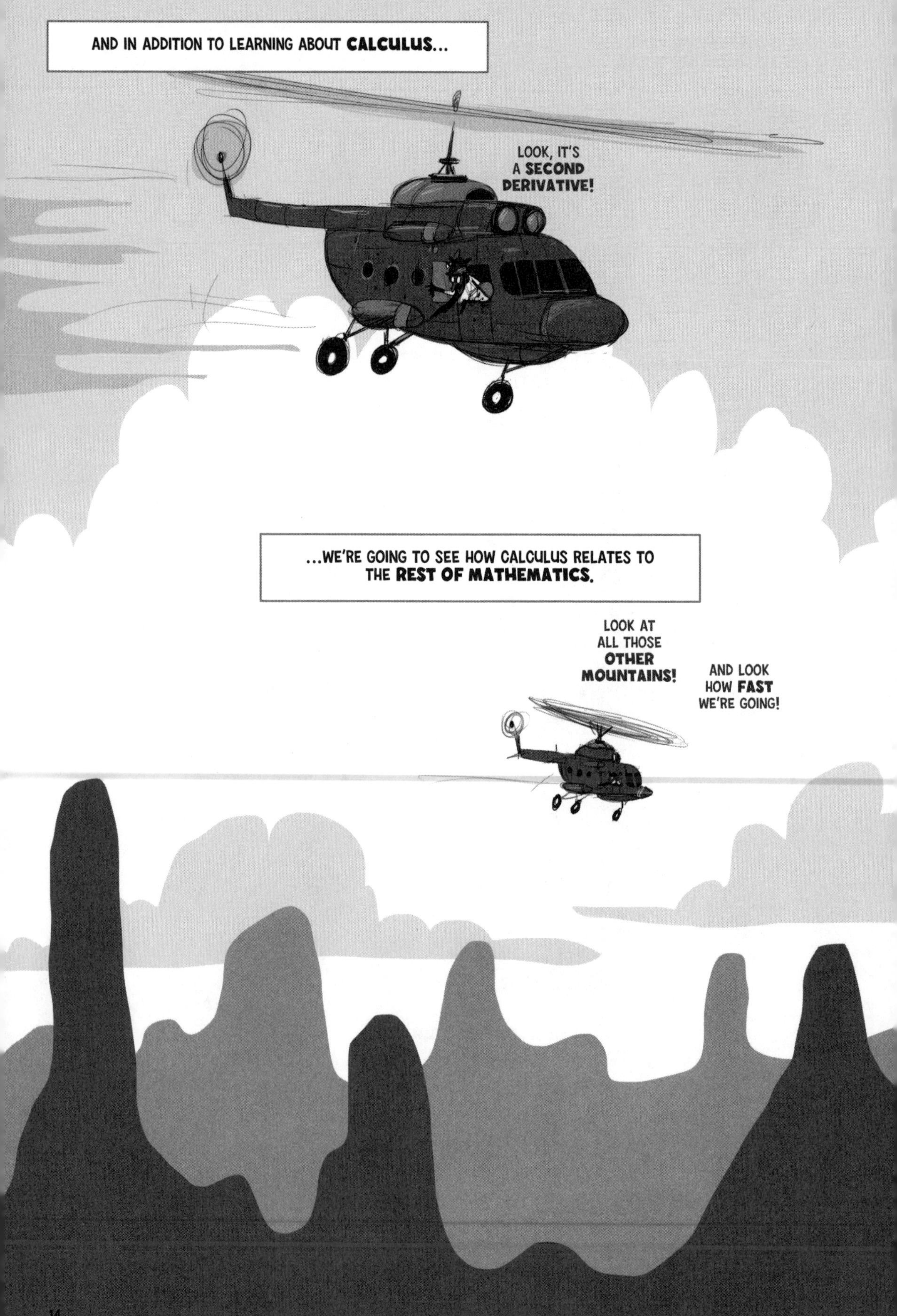
AND IN ADDITION TO LEARNING ABOUT CALCULUS...
LOOK, IT'S A SECOND DERIVATIVE!
...WE'RE GOING TO SEE HOW CALCULUS RELATES TO THE REST OF MATHEMATICS.
LOOK AT ALL THOSE OTHER MOUNTAINS!
AND LOOK HOW FAST WE'RE GOING!

CHAPTER 2
SPEED

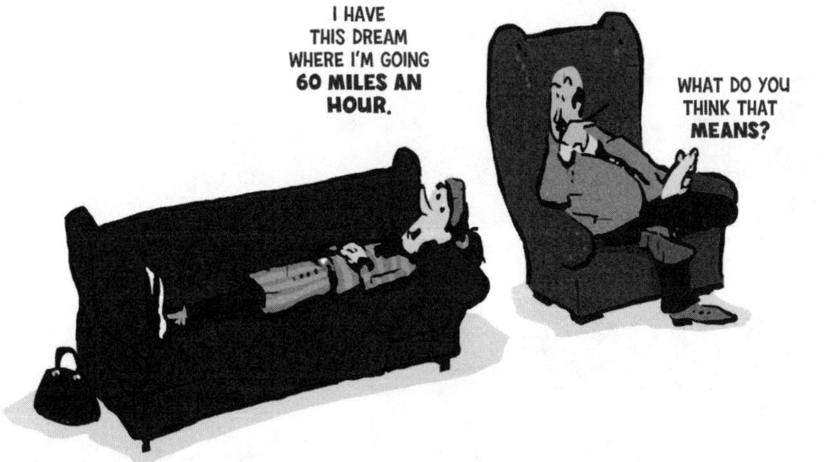
I HAVE THIS DREAM WHERE I'M GOING **60 MILES AN HOUR.**
WHAT DO YOU THINK THAT **MEANS?**

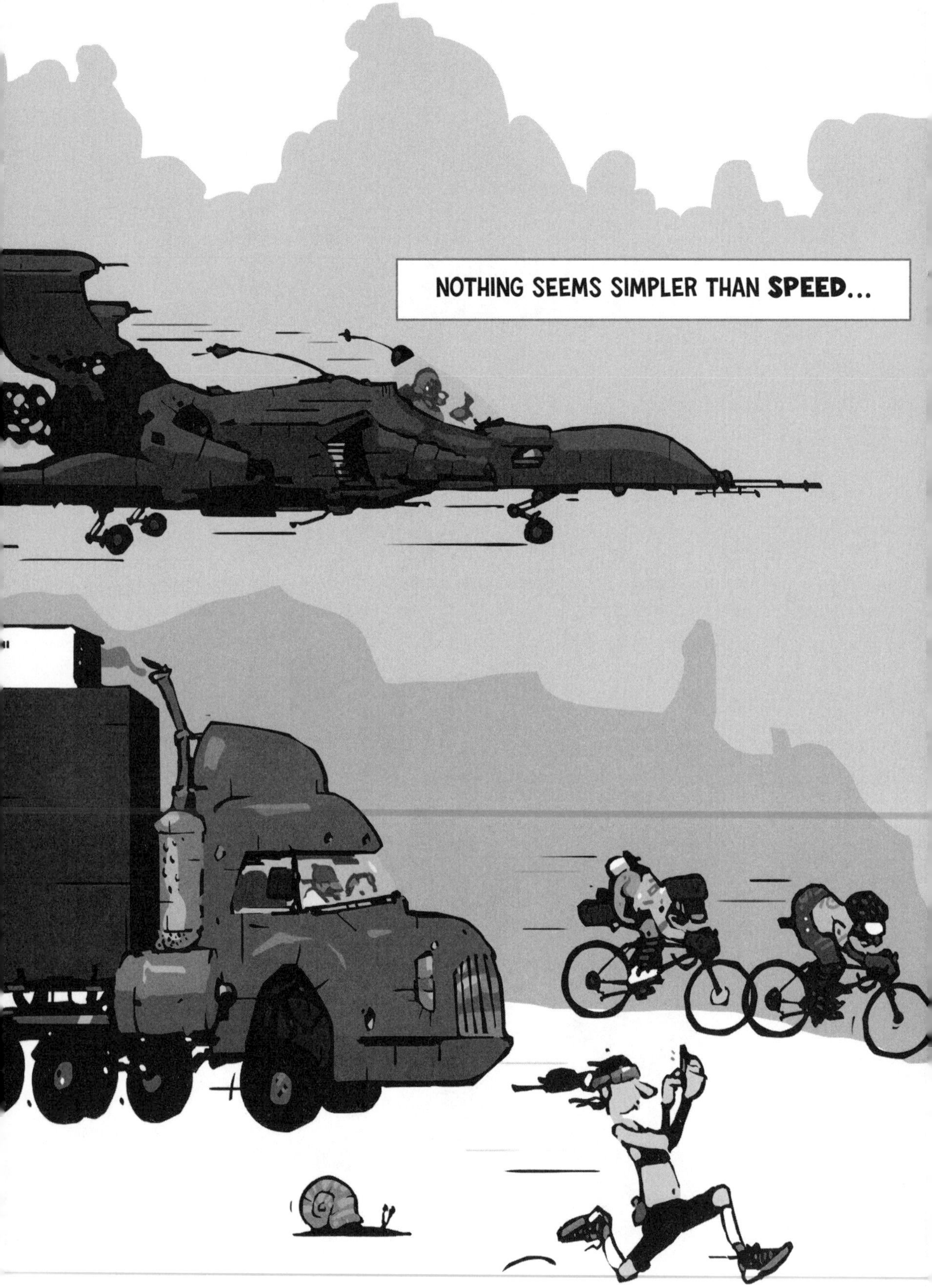
NOTHING SEEMS SIMPLER THAN SPEED...

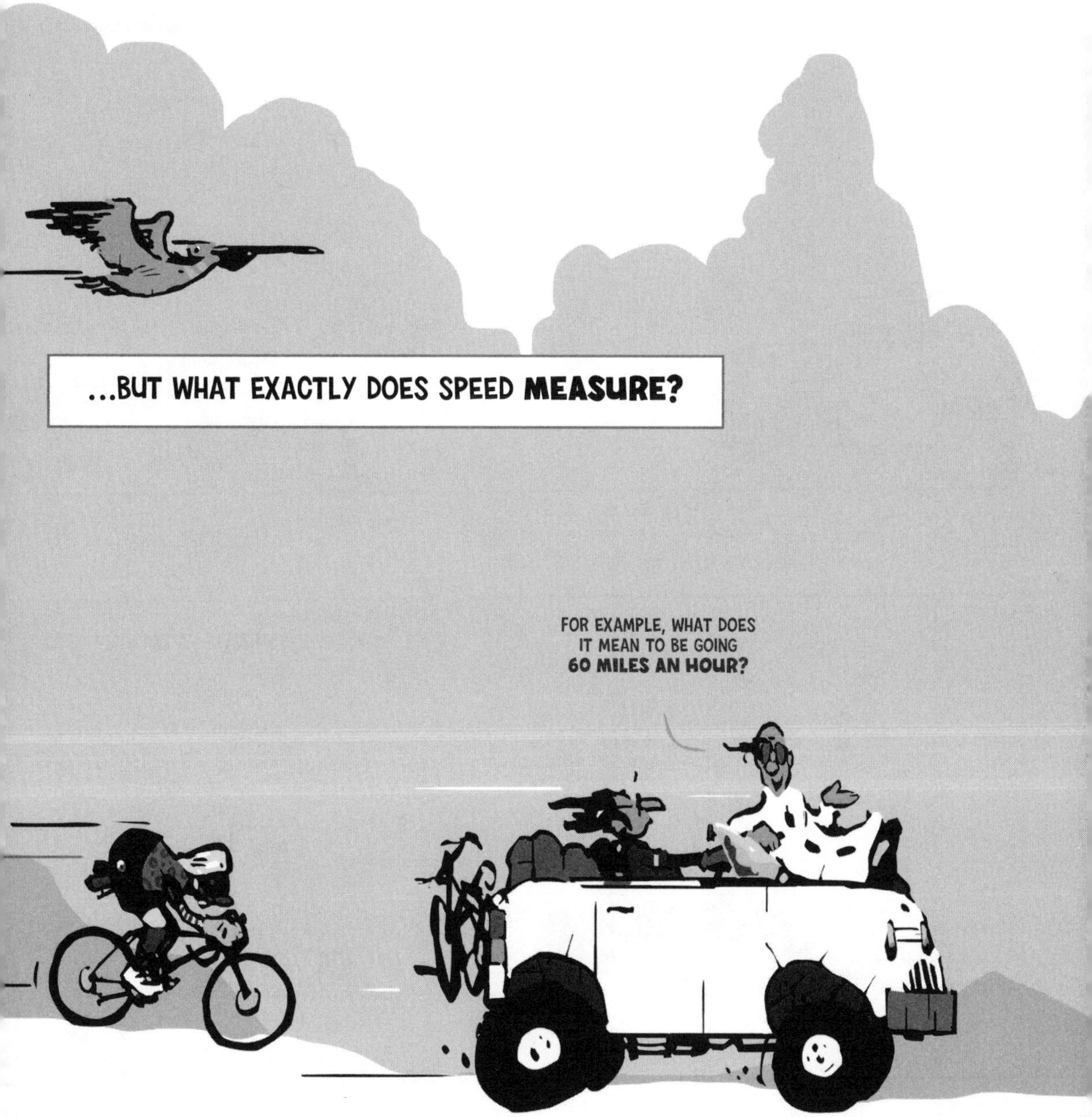
...BUT WHAT EXACTLY DOES SPEED MEASURE?
FOR EXAMPLE, WHAT DOES IT MEAN TO BE GOING 60 MILES AN HOUR?

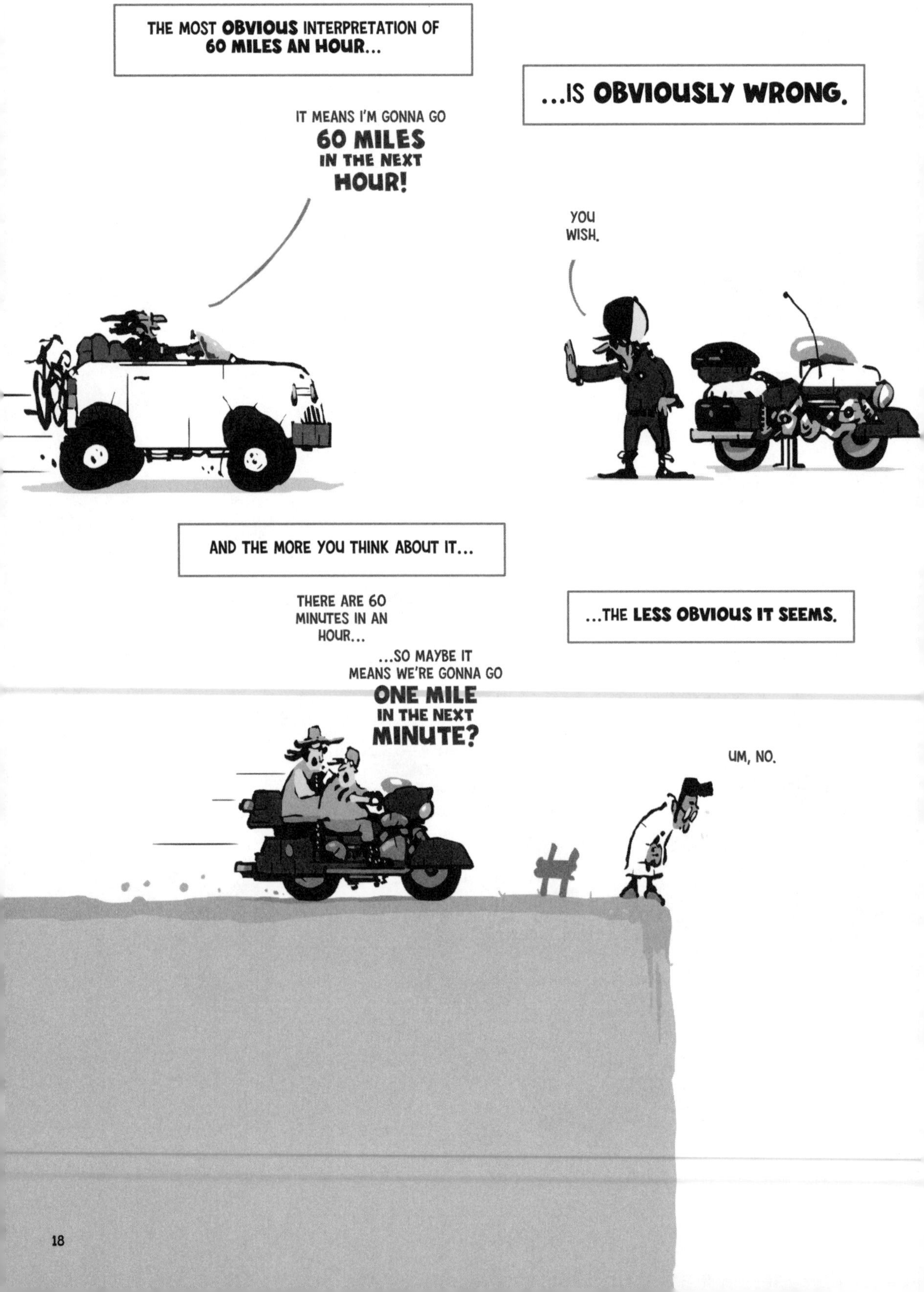
THE MOST OBVIOUS INTERPRETATION OF 60 MILES AN HOUR...
IT MEANS I'M GONNA GO 60 MILES IN THE NEXT HOUR!
...IS OBVIOUSLY WRONG.
YOU WISH.
AND THE MORE YOU THINK ABOUT IT...
THERE ARE 60 MINUTES IN AN HOUR...
...SO MAYBE IT MEANS WE'RE GONNA GO ONE MILE IN THE NEXT MINUTE?
...THE LESS OBVIOUS IT SEEMS.
UM, NO.

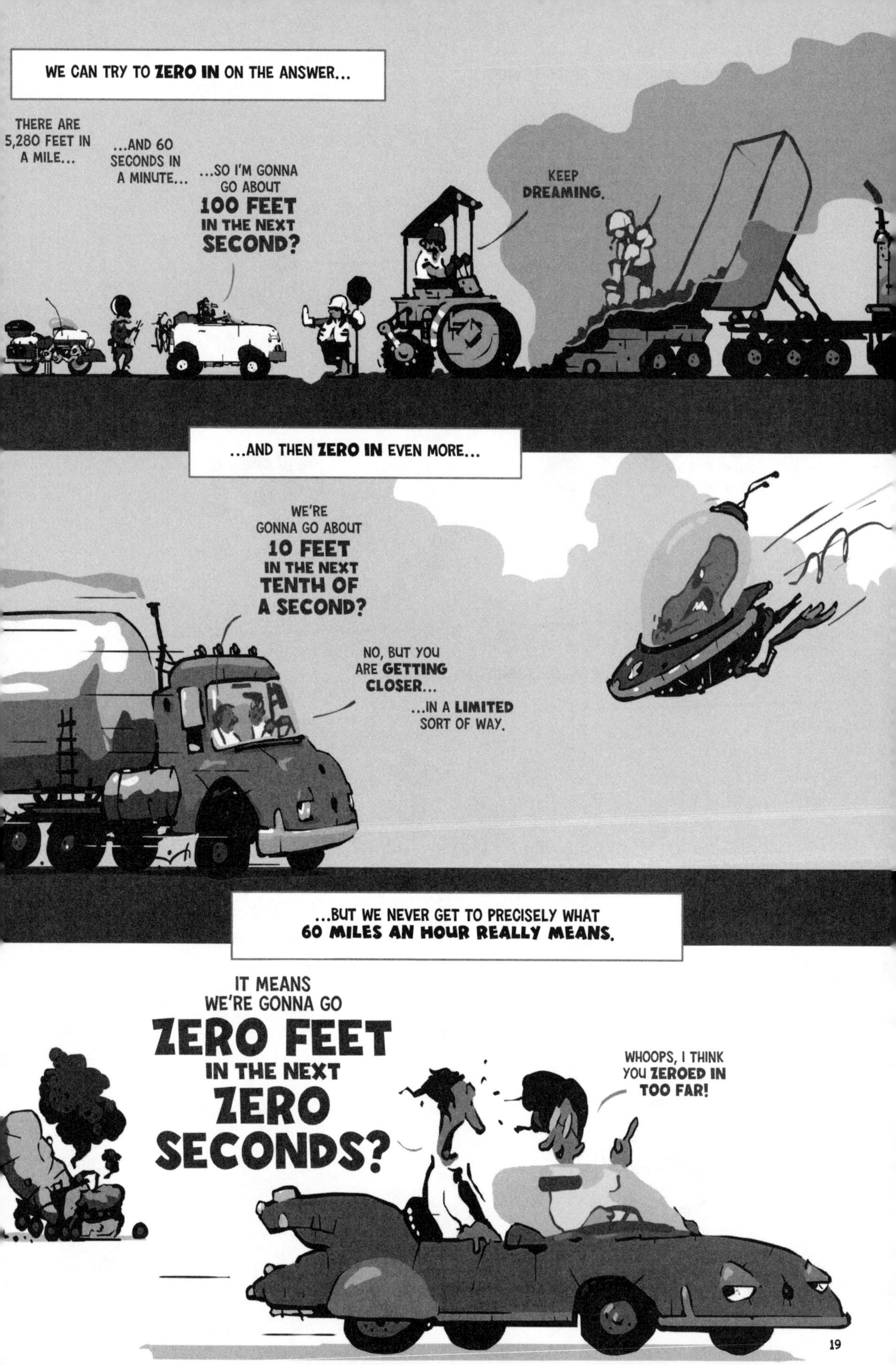
WE CAN TRY TO **ZERO IN** ON THE ANSWER...
THERE ARE 5,280 FEET IN A MILE...
...AND 60 SECONDS IN A MINUTE...
...SO I'M GONNA GO ABOUT **100 FEET IN THE NEXT SECOND?**
KEEP **DREAMING.**
...AND THEN **ZERO IN** EVEN MORE...
WE'RE GONNA GO ABOUT **10 FEET IN THE NEXT TENTH OF A SECOND?**
NO, BUT YOU ARE **GETTING CLOSER...**
...IN A **LIMITED** SORT OF WAY.
...BUT WE NEVER GET TO PRECISELY WHAT **60 MILES AN HOUR REALLY MEANS.**
IT MEANS WE'RE GONNA GO **ZERO FEET IN THE NEXT ZERO SECONDS?**
WHOOPS, I THINK YOU **ZEROED IN TOO FAR!**

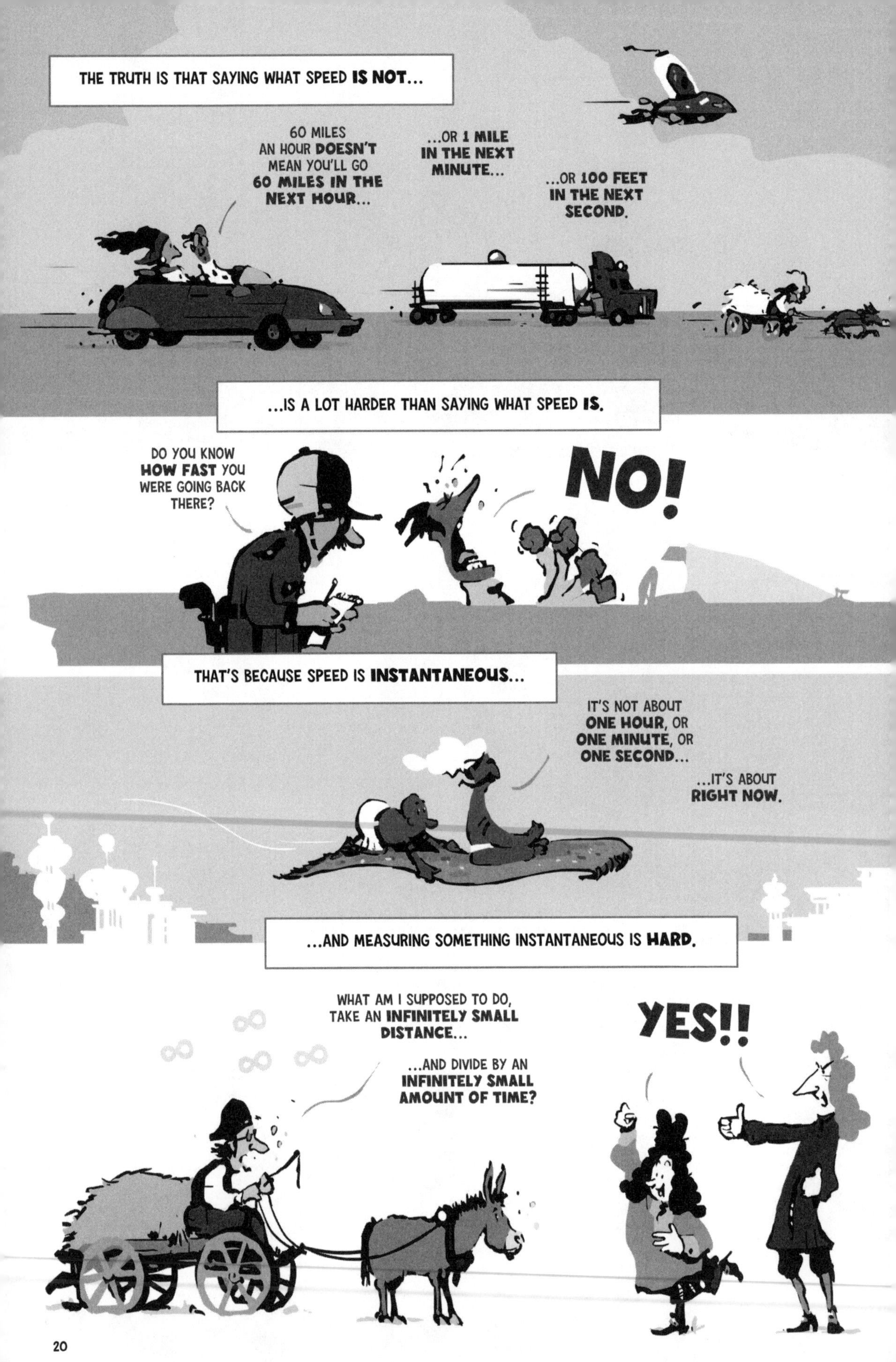

THE TRUTH IS THAT SAYING WHAT SPEED IS NOT...
60 MILES AN HOUR DOESN'T MEAN YOU'LL GO 60 MILES IN THE NEXT HOUR...
...OR 1 MILE IN THE NEXT MINUTE...
...OR 100 FEET IN THE NEXT SECOND.
...IS A LOT HARDER THAN SAYING WHAT SPEED IS.
DO YOU KNOW HOW FAST YOU WERE GOING BACK THERE?
NO!
THAT'S BECAUSE SPEED IS INSTANTANEOUS...
IT'S NOT ABOUT ONE HOUR, OR ONE MINUTE, OR ONE SECOND...
...IT'S ABOUT RIGHT NOW.
...AND MEASURING SOMETHING INSTANTANEOUS IS HARD.
WHAT AM I SUPPOSED TO DO, TAKE AN INFINITELY SMALL DISTANCE...
...AND DIVIDE BY AN INFINITELY SMALL AMOUNT OF TIME?
YES!!

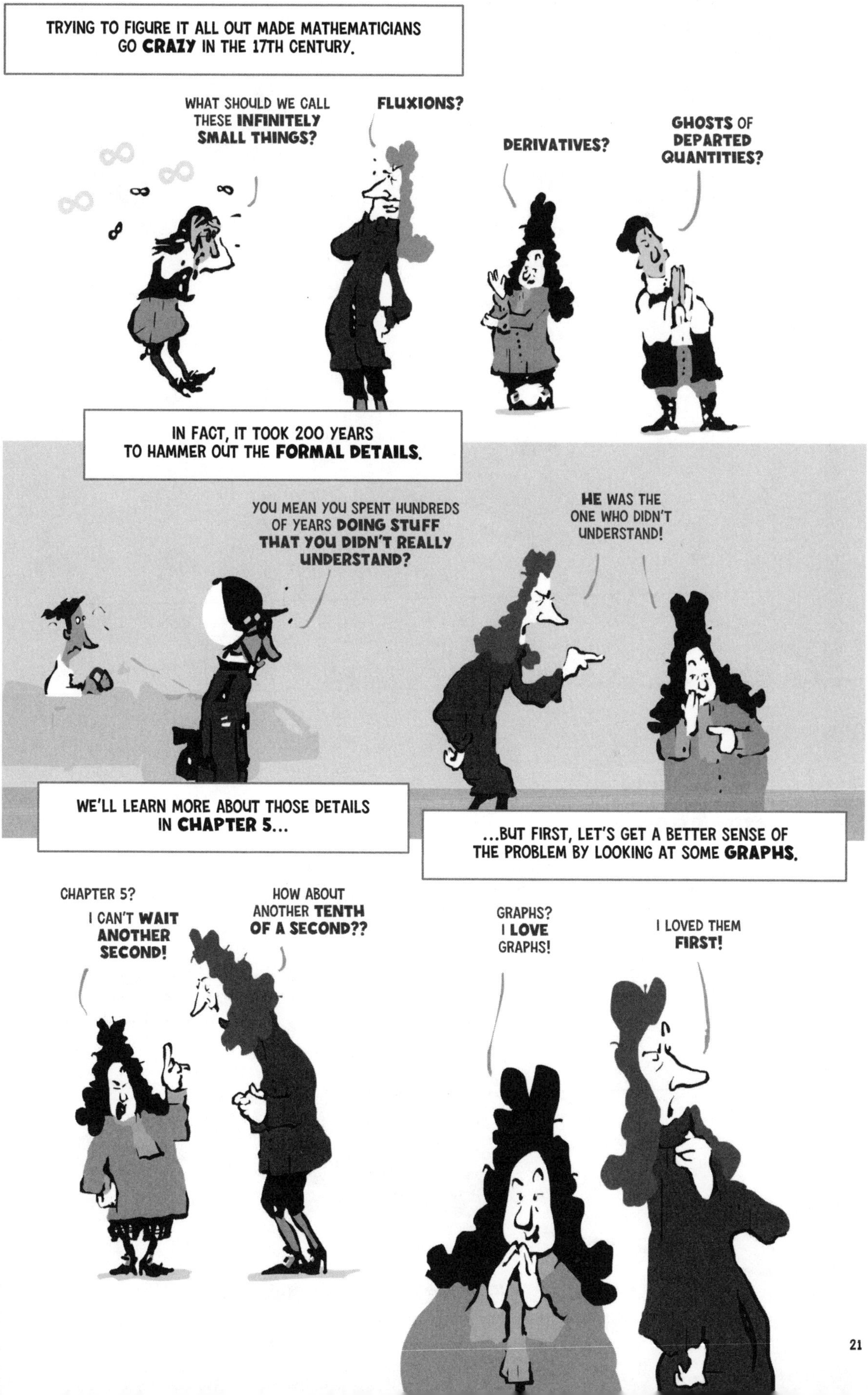

TRYING TO FIGURE IT ALL OUT MADE MATHEMATICIANS GO **CRAZY** IN THE 17TH CENTURY.
WHAT SHOULD WE CALL THESE **INFINITELY SMALL THINGS?**
FLUXIONS?
DERIVATIVES?
GHOSTS OF **DEPARTED QUANTITIES?**
IN FACT, IT TOOK 200 YEARS TO HAMMER OUT THE **FORMAL DETAILS.**
YOU MEAN YOU SPENT HUNDREDS OF YEARS **DOING STUFF THAT YOU DIDN'T REALLY UNDERSTAND?**
HE WAS THE ONE WHO DIDN'T UNDERSTAND!
WE'LL LEARN MORE ABOUT THOSE DETAILS IN **CHAPTER 5...**
...BUT FIRST, LET'S GET A BETTER SENSE OF THE PROBLEM BY LOOKING AT SOME **GRAPHS.**
CHAPTER 5? I CAN'T **WAIT ANOTHER SECOND!**
HOW ABOUT ANOTHER **TENTH OF A SECOND??**
GRAPHS? I **LOVE** GRAPHS!
I LOVED THEM **FIRST!**

CONSIDER A CAR MOVING AT A CONSTANT SPEED.

WE'RE GOING **60 MILES AN HOUR**...

...OR A **MILE A MINUTE**...

...OR ABOUT **100 FEET A SECOND**.

ON A GRAPH, TO SEE **SPEED** WE LOOK AT THE **SLOPE**.

OVER THE COURSE OF ONE HOUR, OUR POSITION CHANGES BY **60 MILES**... ...SO THIS SLOPE IS **60 mph**.

IN MATH SPEAK, THIS LINE MATCHES A **FUNCTION**:

AT ANY TIME **t**...

...THE DISTANCE TRAVELED EQUALS **F(t)**.

distance in miles F(t)

60

30

0   30 min   1 hour   t

time

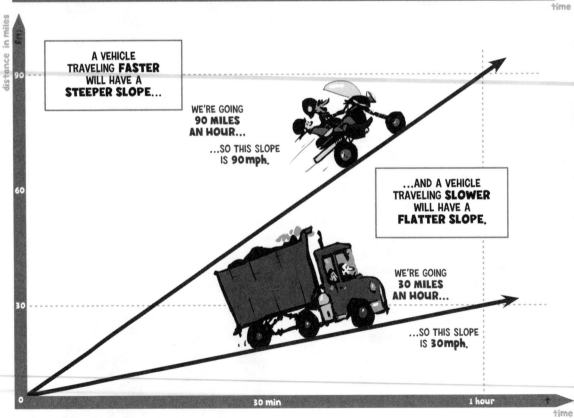

distance in miles F(t)

A VEHICLE TRAVELING **FASTER** WILL HAVE A **STEEPER SLOPE**...

90

WE'RE GOING **90 MILES AN HOUR**...

...SO THIS SLOPE IS **90 mph**.

...AND A VEHICLE TRAVELING **SLOWER** WILL HAVE A **FLATTER SLOPE**.

60

WE'RE GOING **30 MILES AN HOUR**...

30

...SO THIS SLOPE IS **30 mph**.

0   30 min   1 hour

time

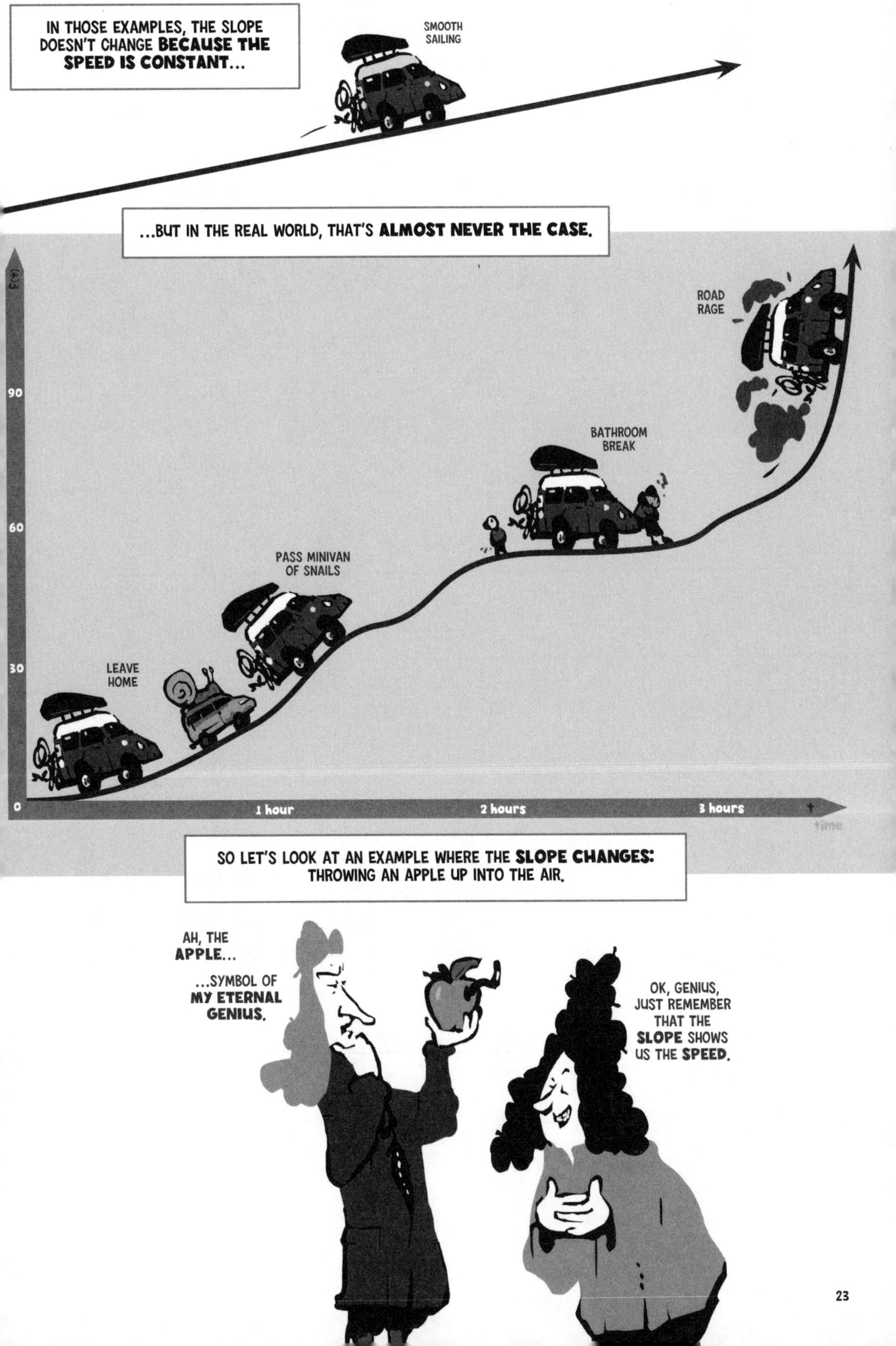
IN THOSE EXAMPLES, THE SLOPE DOESN'T CHANGE **BECAUSE THE SPEED IS CONSTANT...**
SMOOTH SAILING
...BUT IN THE REAL WORLD, THAT'S **ALMOST NEVER THE CASE.**
ROAD RAGE
90
BATHROOM BREAK
60
PASS MINIVAN OF SNAILS
30
LEAVE HOME
0
1 hour
2 hours
3 hours
t
time
SO LET'S LOOK AT AN EXAMPLE WHERE THE **SLOPE CHANGES:** THROWING AN APPLE UP INTO THE AIR.
AH, THE **APPLE...**
...SYMBOL OF **MY ETERNAL GENIUS.**
OK, GENIUS, JUST REMEMBER THAT THE **SLOPE** SHOWS US THE **SPEED.**

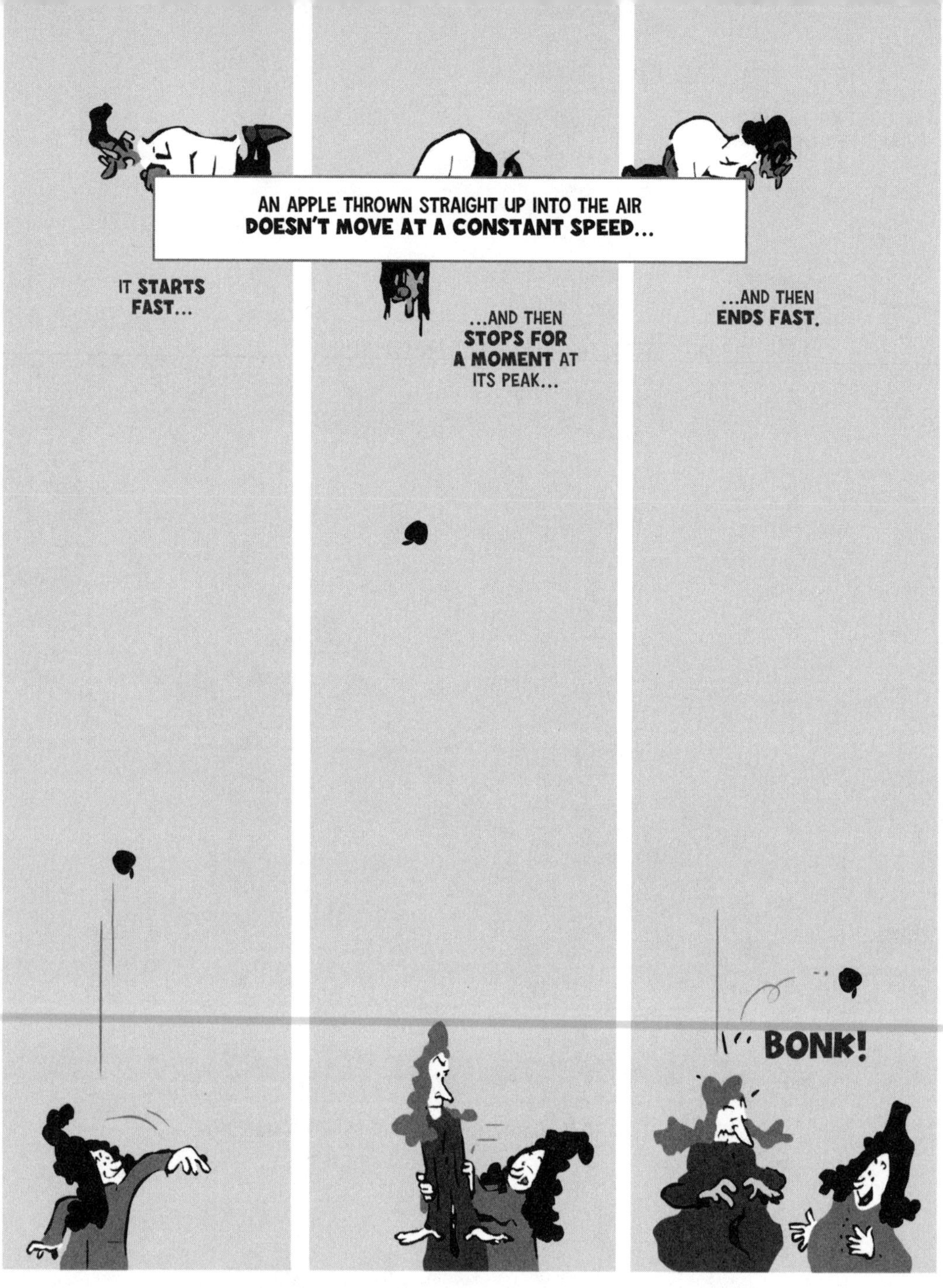
AN APPLE THROWN STRAIGHT UP INTO THE AIR
DOESN'T MOVE AT A CONSTANT SPEED...
IT **STARTS FAST...**
...AND THEN **STOPS FOR A MOMENT** AT ITS PEAK...
...AND THEN **ENDS FAST.**
BONK!

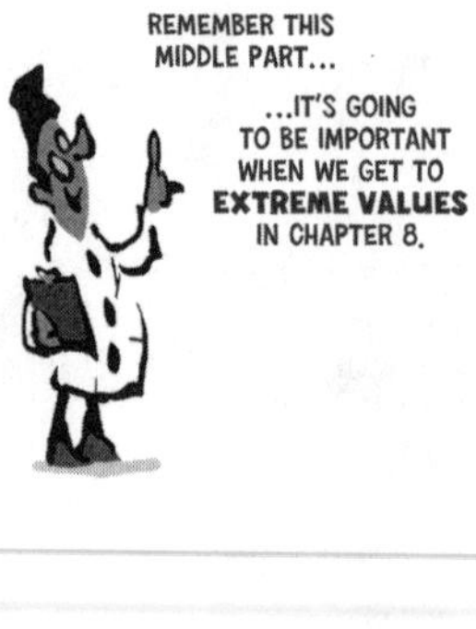
REMEMBER THIS MIDDLE PART...
...IT'S GOING TO BE IMPORTANT WHEN WE GET TO **EXTREME VALUES** IN CHAPTER 8.

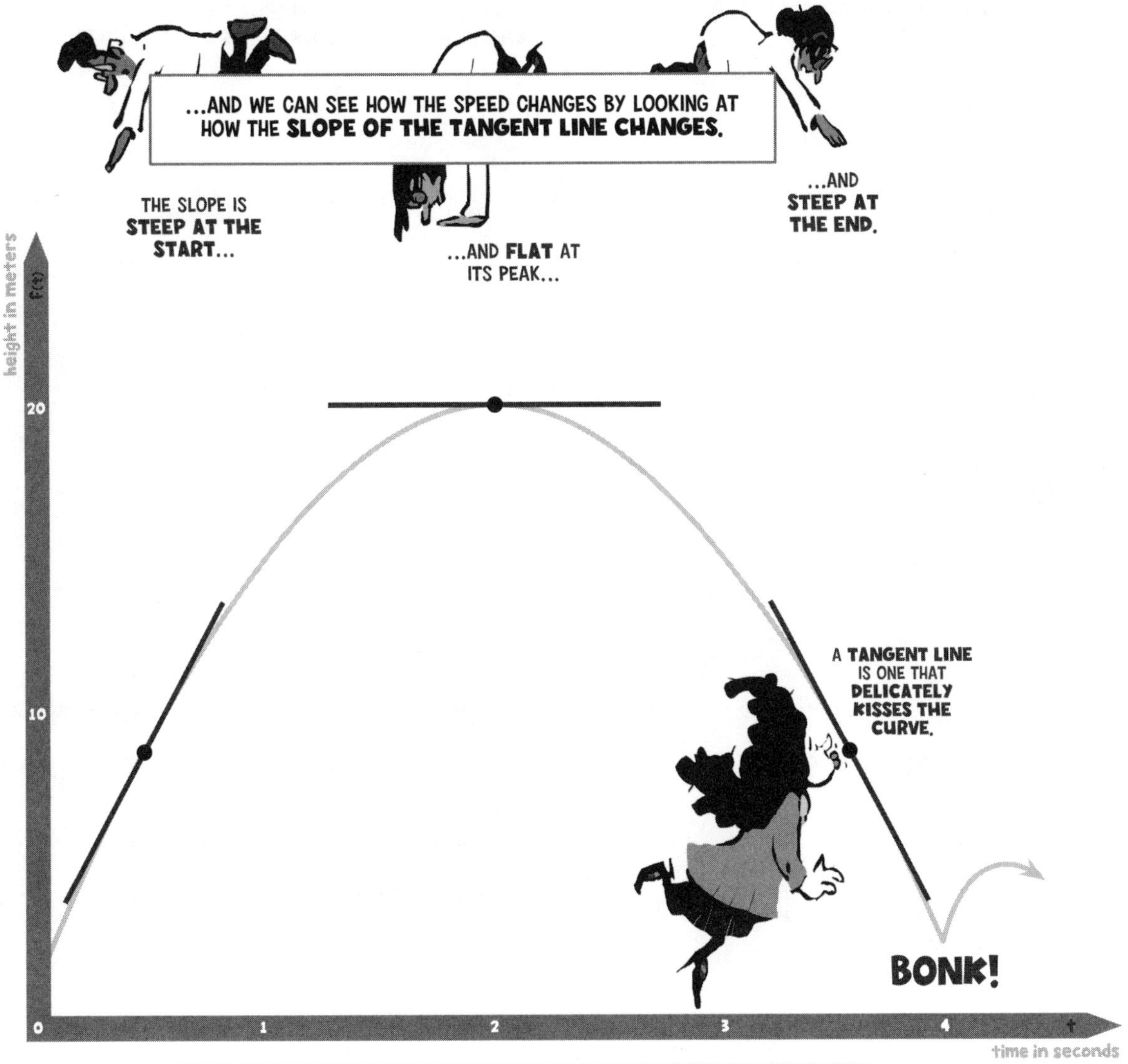
...AND WE CAN SEE HOW THE SPEED CHANGES BY LOOKING AT HOW THE **SLOPE OF THE TANGENT LINE CHANGES.**
THE SLOPE IS **STEEP AT THE START...**
...AND **FLAT** AT ITS PEAK...
...AND **STEEP AT THE END.**
A **TANGENT LINE** IS ONE THAT **DELICATELY KISSES THE CURVE.**
BONK!
height in meters
f(t)
20
10
0
1
2
3
4
t
time in seconds

BUT HOW DO WE **CALCULATE THE SLOPE** OF A TANGENT LINE?

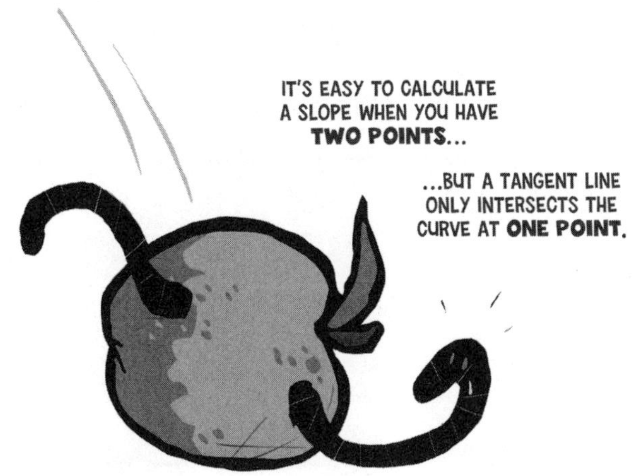
IT'S EASY TO CALCULATE A SLOPE WHEN YOU HAVE **TWO POINTS...**
...BUT A TANGENT LINE ONLY INTERSECTS THE CURVE AT **ONE POINT.**

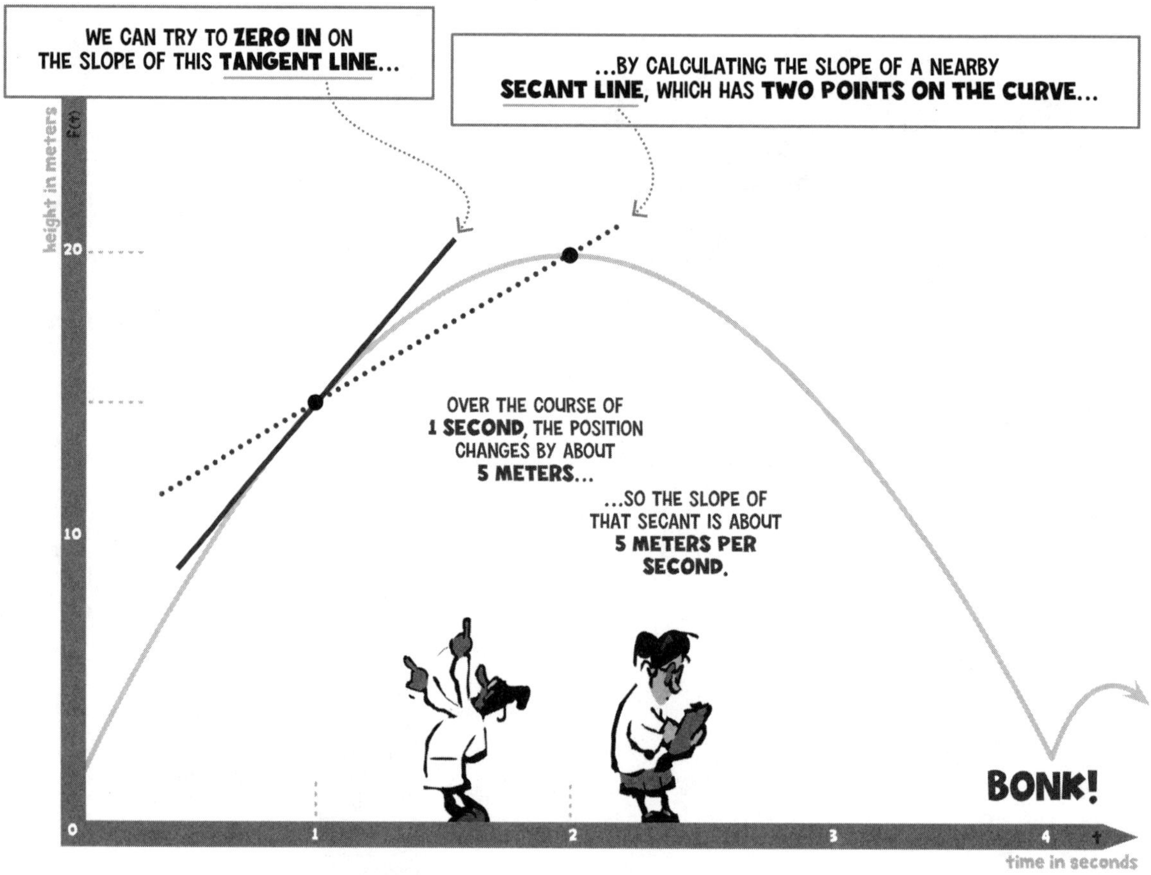
WE CAN TRY TO **ZERO IN** ON THE SLOPE OF THIS **TANGENT LINE**...
...BY CALCULATING THE SLOPE OF A NEARBY **SECANT LINE**, WHICH HAS **TWO POINTS ON THE CURVE**...
height in meters
f(t)
20
10
0
OVER THE COURSE OF **1 SECOND**, THE POSITION CHANGES BY ABOUT **5 METERS**...
...SO THE SLOPE OF THAT SECANT IS ABOUT **5 METERS PER SECOND.**
BONK!
1
2
3
4
t
time in seconds

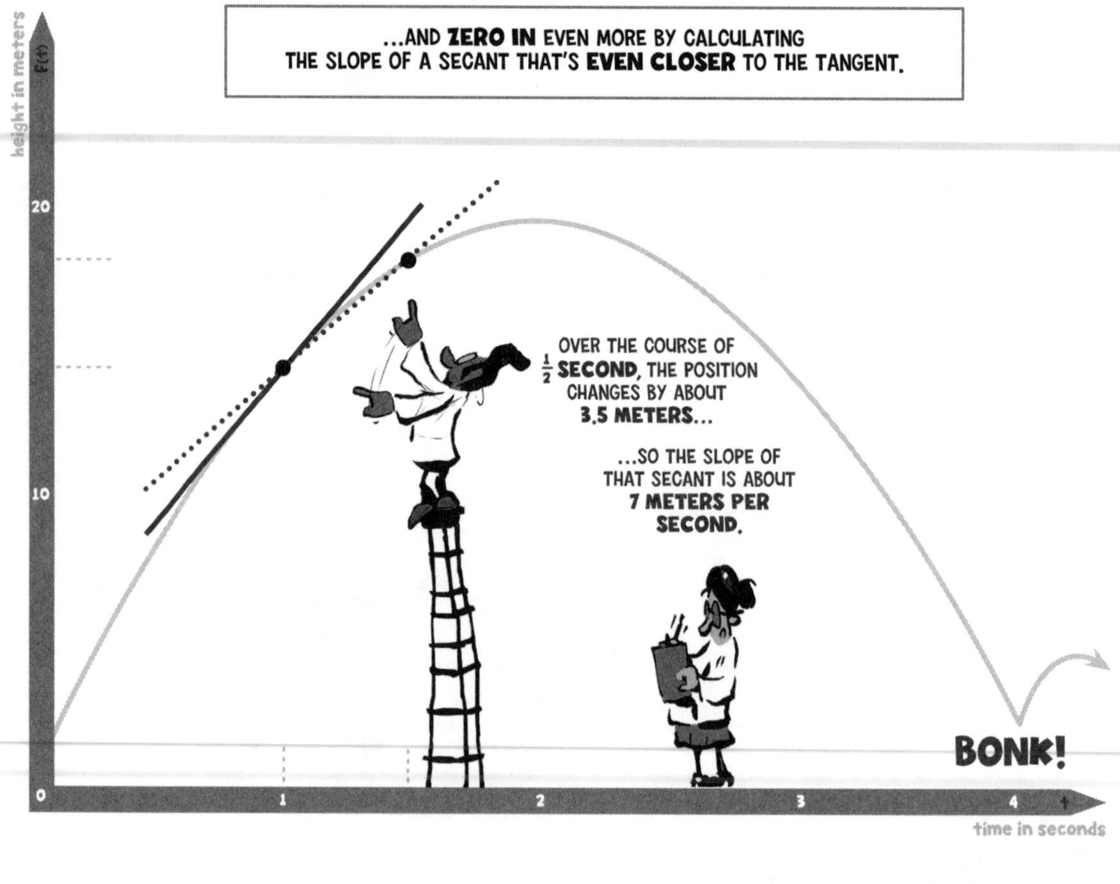
...AND **ZERO IN** EVEN MORE BY CALCULATING THE SLOPE OF A SECANT THAT'S **EVEN CLOSER** TO THE TANGENT.
height in meters
f(t)
20
10
0
OVER THE COURSE OF $\frac{1}{2}$ **SECOND**, THE POSITION CHANGES BY ABOUT **3.5 METERS**...
...SO THE SLOPE OF THAT SECANT IS ABOUT **7 METERS PER SECOND.**
BONK!
1
2
3
4
t
time in seconds

THE MORE WE **ZERO IN**, THE **MORE PRECISE** OUR ESTIMATE WILL BE...

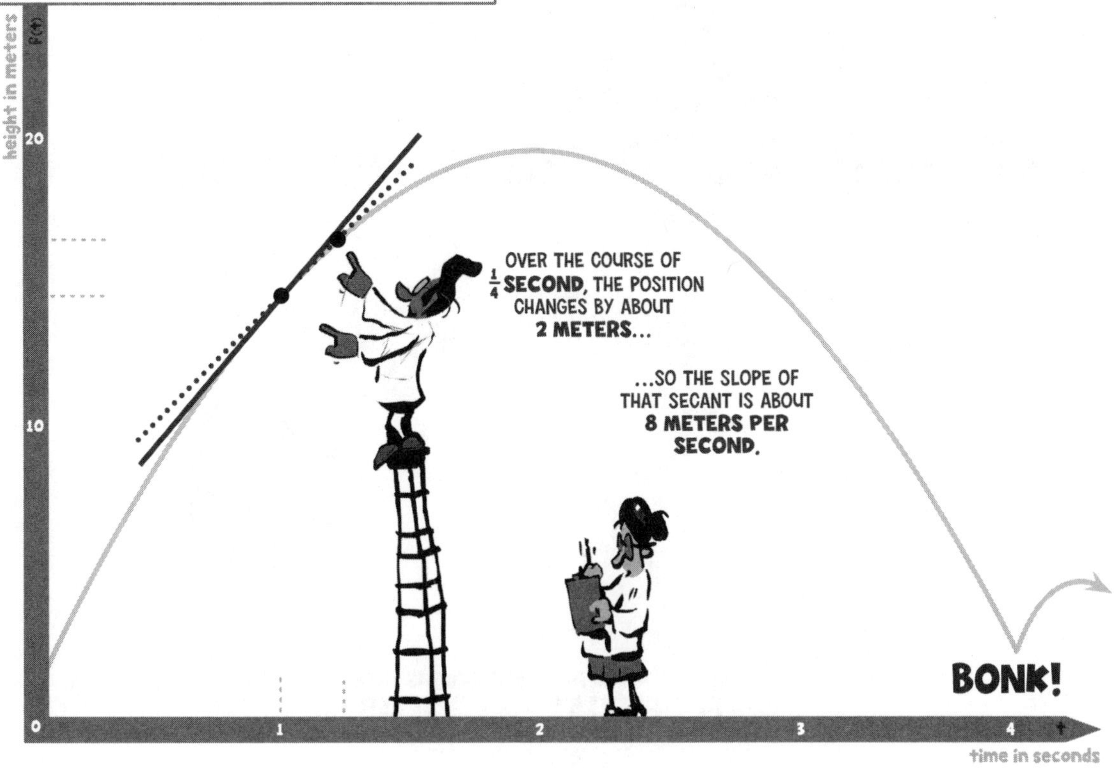

height in meters
f(t)
20
10
OVER THE COURSE OF $\frac{1}{4}$ **SECOND**, THE POSITION CHANGES BY ABOUT **2 METERS**...
...SO THE SLOPE OF THAT SECANT IS ABOUT **8 METERS PER SECOND.**
BONK!
0
1
2
3
4
t
time in seconds

...BUT WE'LL NEVER BE ABLE TO CALCULATE THE SLOPE OF THE TANGENT WITH **ABSOLUTE PRECISION.**

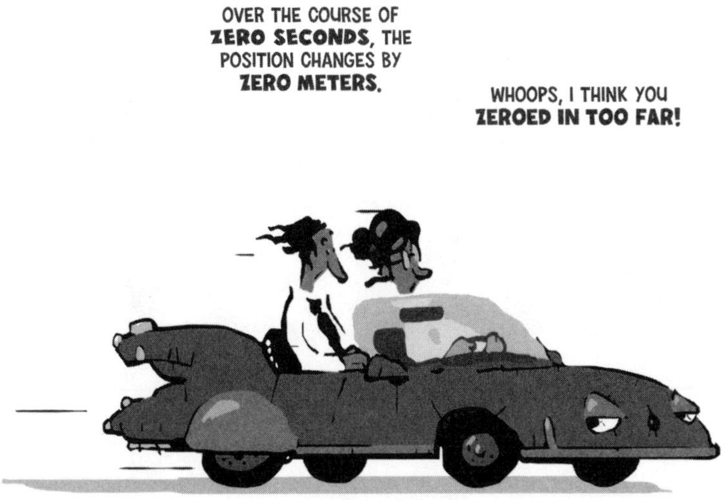

OVER THE COURSE OF **ZERO SECONDS**, THE POSITION CHANGES BY **ZERO METERS.**
WHOOPS, I THINK YOU **ZEROED IN TOO FAR!**

IT'S JUST LIKE WHAT WE SAW ON **PAGE 19!**

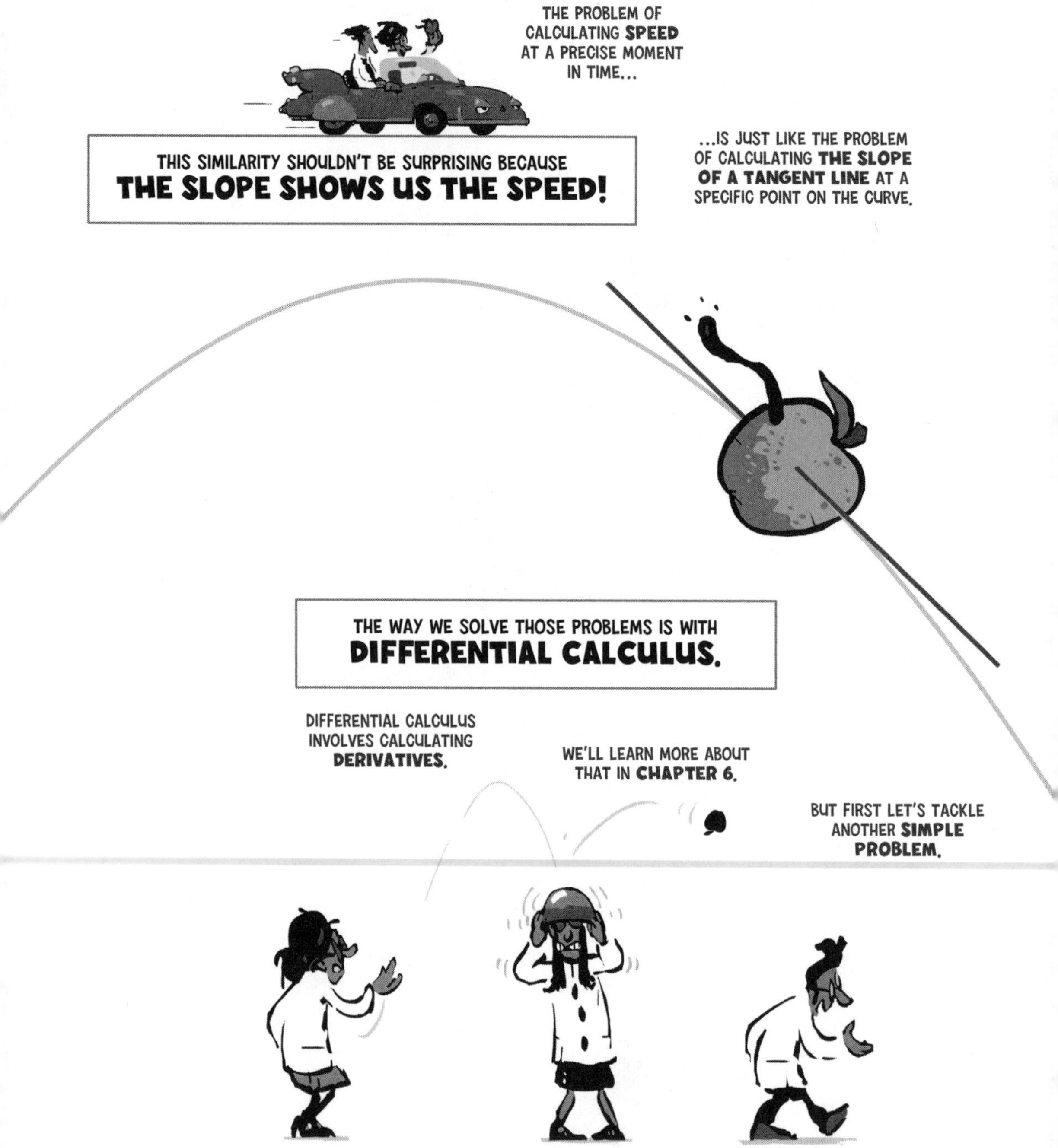
THE PROBLEM OF CALCULATING SPEED AT A PRECISE MOMENT IN TIME...
...IS JUST LIKE THE PROBLEM OF CALCULATING THE SLOPE OF A TANGENT LINE AT A SPECIFIC POINT ON THE CURVE.
THIS SIMILARITY SHOULDN'T BE SURPRISING BECAUSE THE SLOPE SHOWS US THE SPEED!
THE WAY WE SOLVE THOSE PROBLEMS IS WITH DIFFERENTIAL CALCULUS.
DIFFERENTIAL CALCULUS INVOLVES CALCULATING DERIVATIVES.
WE'LL LEARN MORE ABOUT THAT IN CHAPTER 6.
BUT FIRST LET'S TACKLE ANOTHER SIMPLE PROBLEM.

CHAPTER 3
AREA

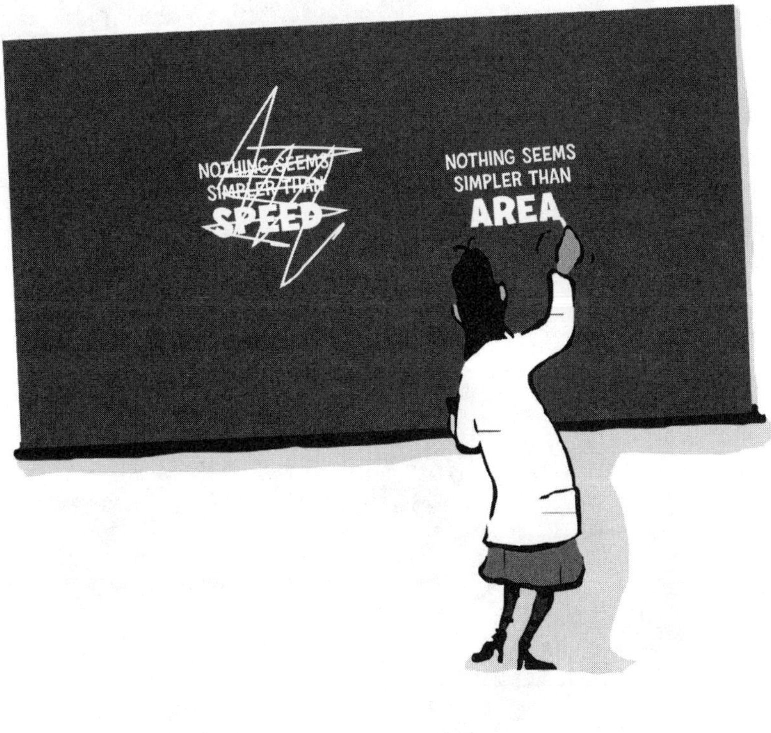
NOTHING SEEMS
SIMPLER THAN
SPEED
NOTHING SEEMS
SIMPLER THAN
AREA

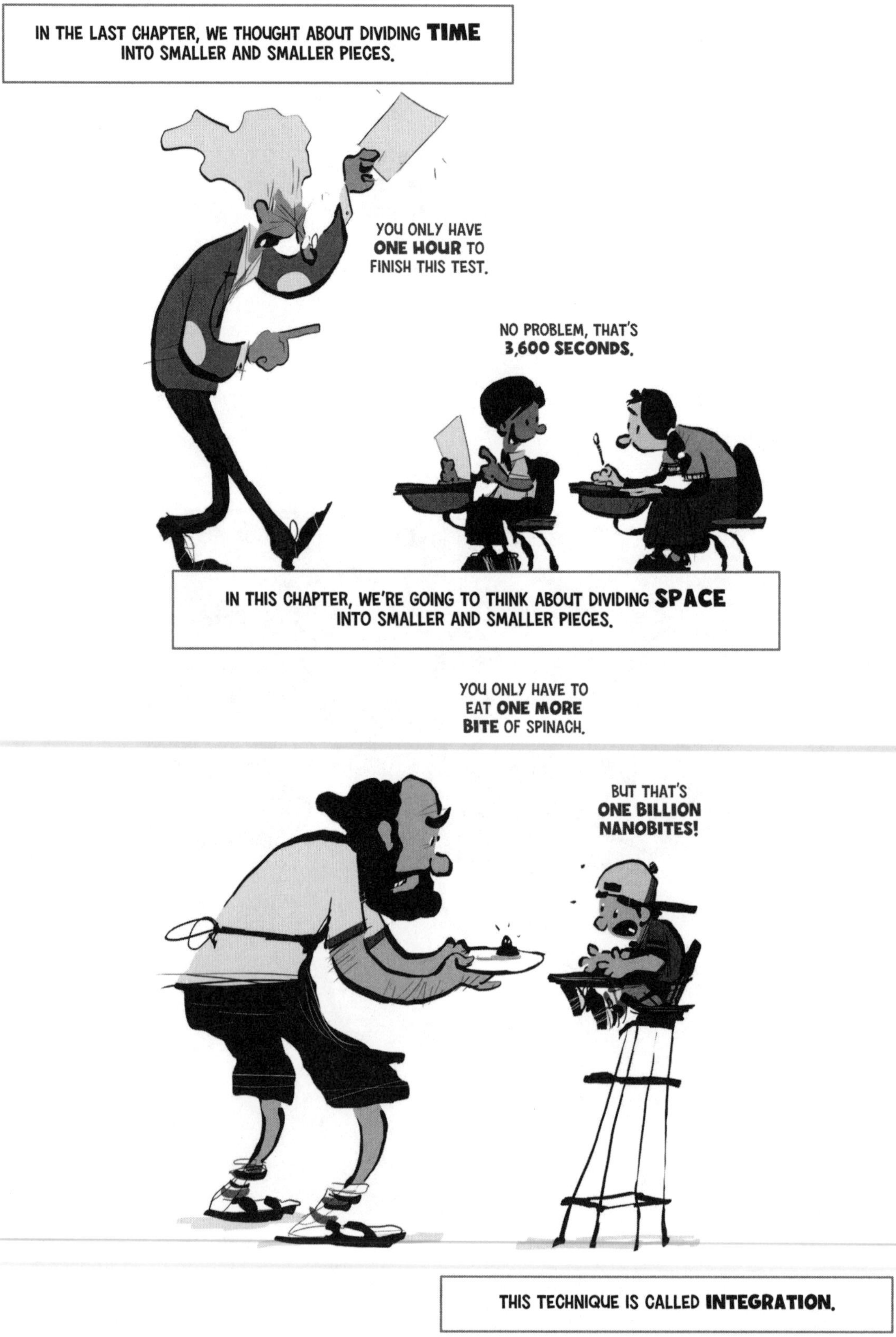
IN THE LAST CHAPTER, WE THOUGHT ABOUT DIVIDING **TIME** INTO SMALLER AND SMALLER PIECES.
YOU ONLY HAVE **ONE HOUR** TO FINISH THIS TEST.
NO PROBLEM, THAT'S **3,600 SECONDS.**
IN THIS CHAPTER, WE'RE GOING TO THINK ABOUT DIVIDING **SPACE** INTO SMALLER AND SMALLER PIECES.
YOU ONLY HAVE TO EAT **ONE MORE BITE** OF SPINACH.
BUT THAT'S **ONE BILLION NANOBITES!**
THIS TECHNIQUE IS CALLED **INTEGRATION.**

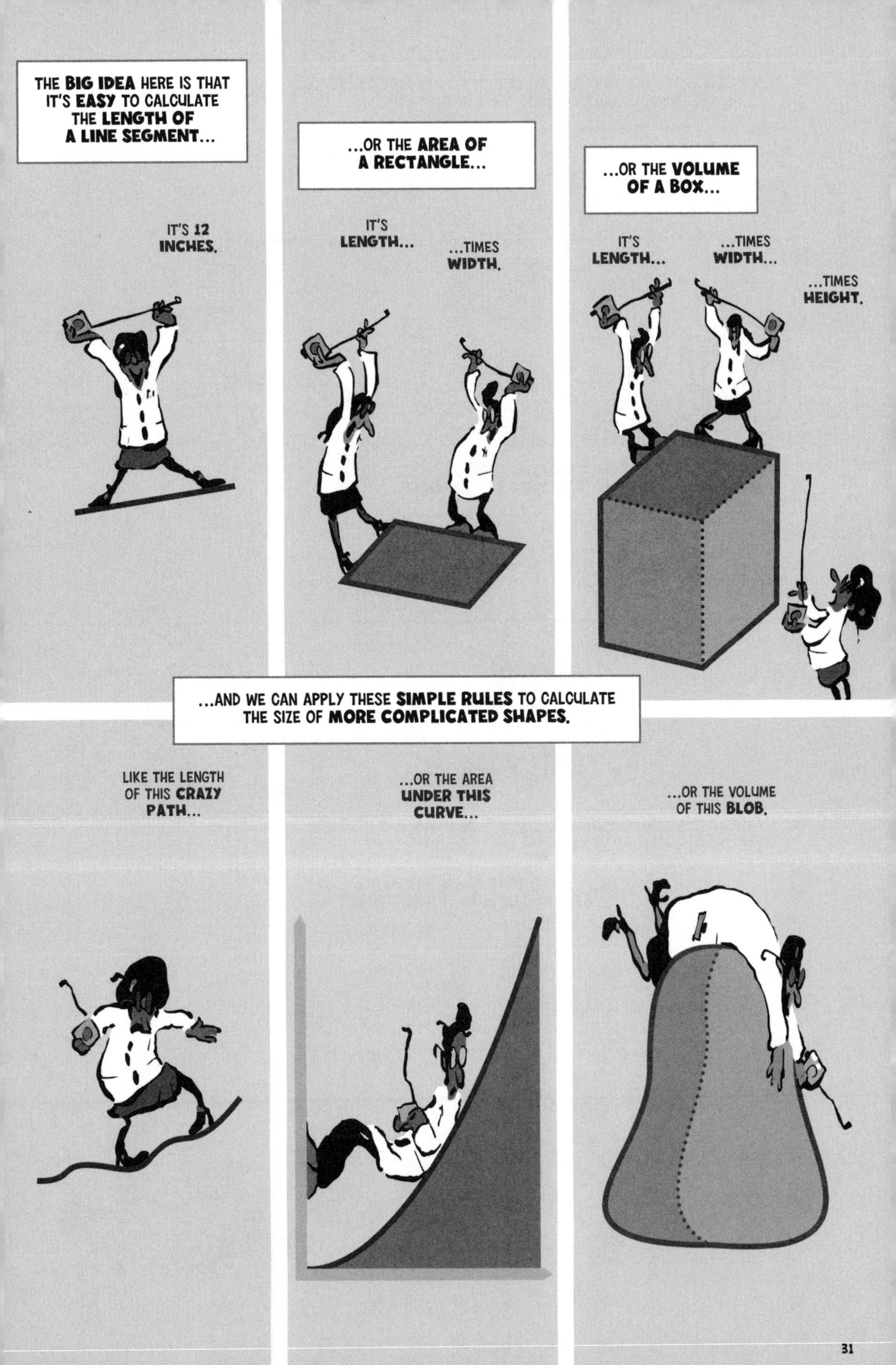

THE BIG IDEA HERE IS THAT IT'S EASY TO CALCULATE THE LENGTH OF A LINE SEGMENT...
IT'S 12 INCHES.
...OR THE AREA OF A RECTANGLE...
IT'S LENGTH...
...TIMES WIDTH.
...OR THE VOLUME OF A BOX...
IT'S LENGTH...
...TIMES WIDTH...
...TIMES HEIGHT.
...AND WE CAN APPLY THESE SIMPLE RULES TO CALCULATE THE SIZE OF MORE COMPLICATED SHAPES.
LIKE THE LENGTH OF THIS CRAZY PATH...
...OR THE AREA UNDER THIS CURVE...
...OR THE VOLUME OF THIS BLOB.

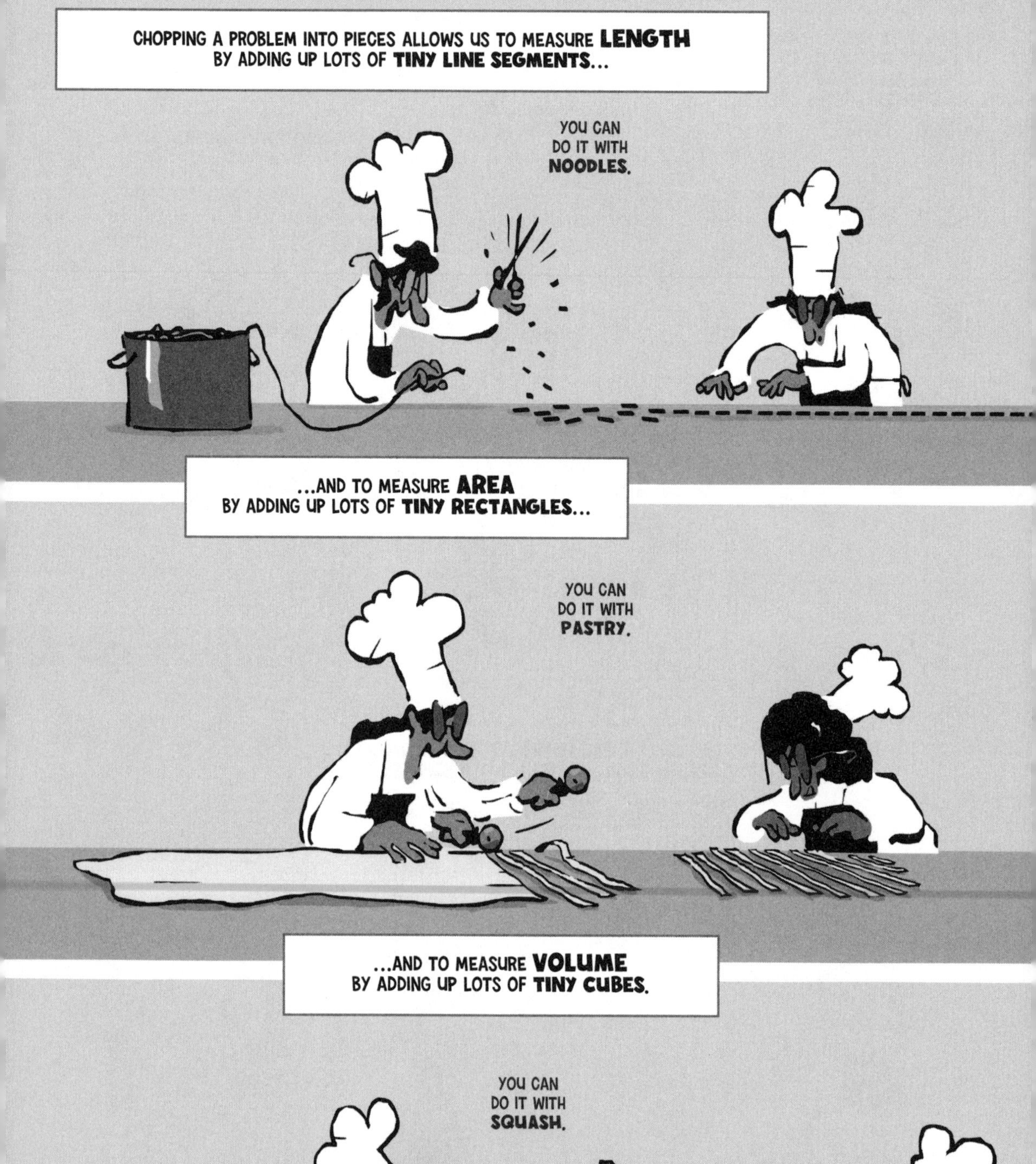
CHOPPING A PROBLEM INTO PIECES ALLOWS US TO MEASURE LENGTH BY ADDING UP LOTS OF TINY LINE SEGMENTS...
YOU CAN DO IT WITH NOODLES.
...AND TO MEASURE AREA BY ADDING UP LOTS OF TINY RECTANGLES...
YOU CAN DO IT WITH PASTRY.
...AND TO MEASURE VOLUME BY ADDING UP LOTS OF TINY CUBES.
YOU CAN DO IT WITH SQUASH.

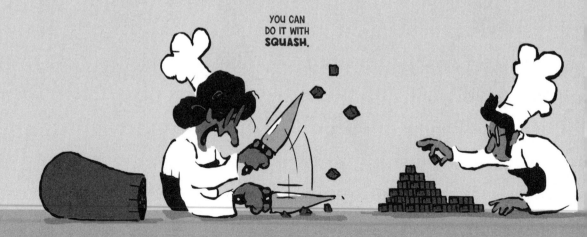

THE BASIC IDEA WAS NICELY EXPRESSED BY THE ITALIAN MATHEMATICIAN **BONAVENTURA CAVALIERI** (1598–1647).

"[**TWO-DIMENSIONAL OBJECTS**] SHOULD BE CONCEIVED BY US IN THE SAME MANNER AS **CLOTHS ARE MADE UP OF PARALLEL THREADS...**

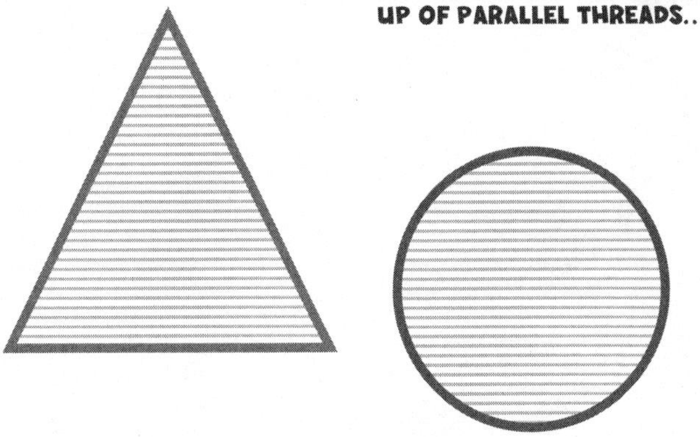

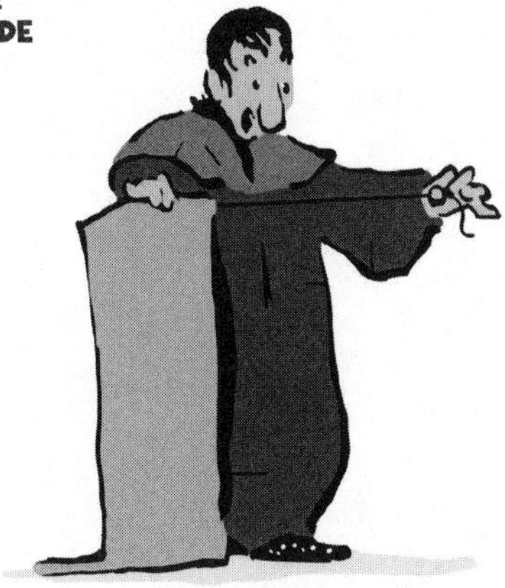

...AND [**THREE-DIMENSIONAL OBJECTS**] ARE IN FACT LIKE **BOOKS, WHICH ARE COMPOSED OF PARALLEL PAGES.**"

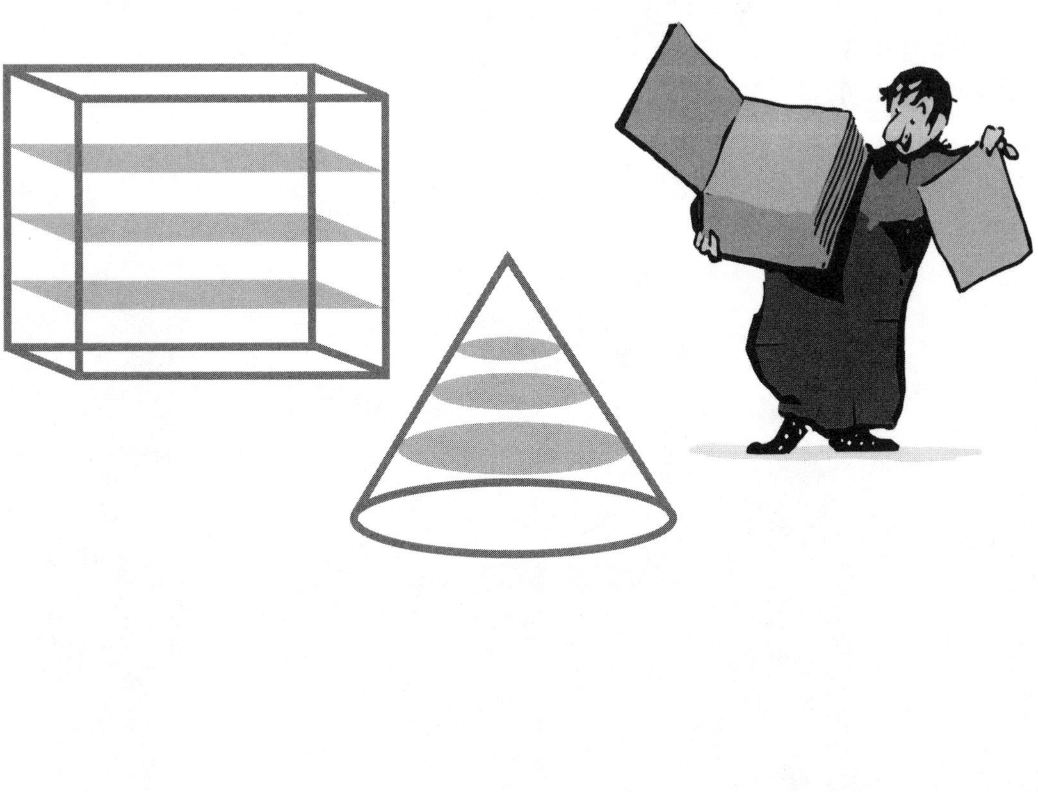

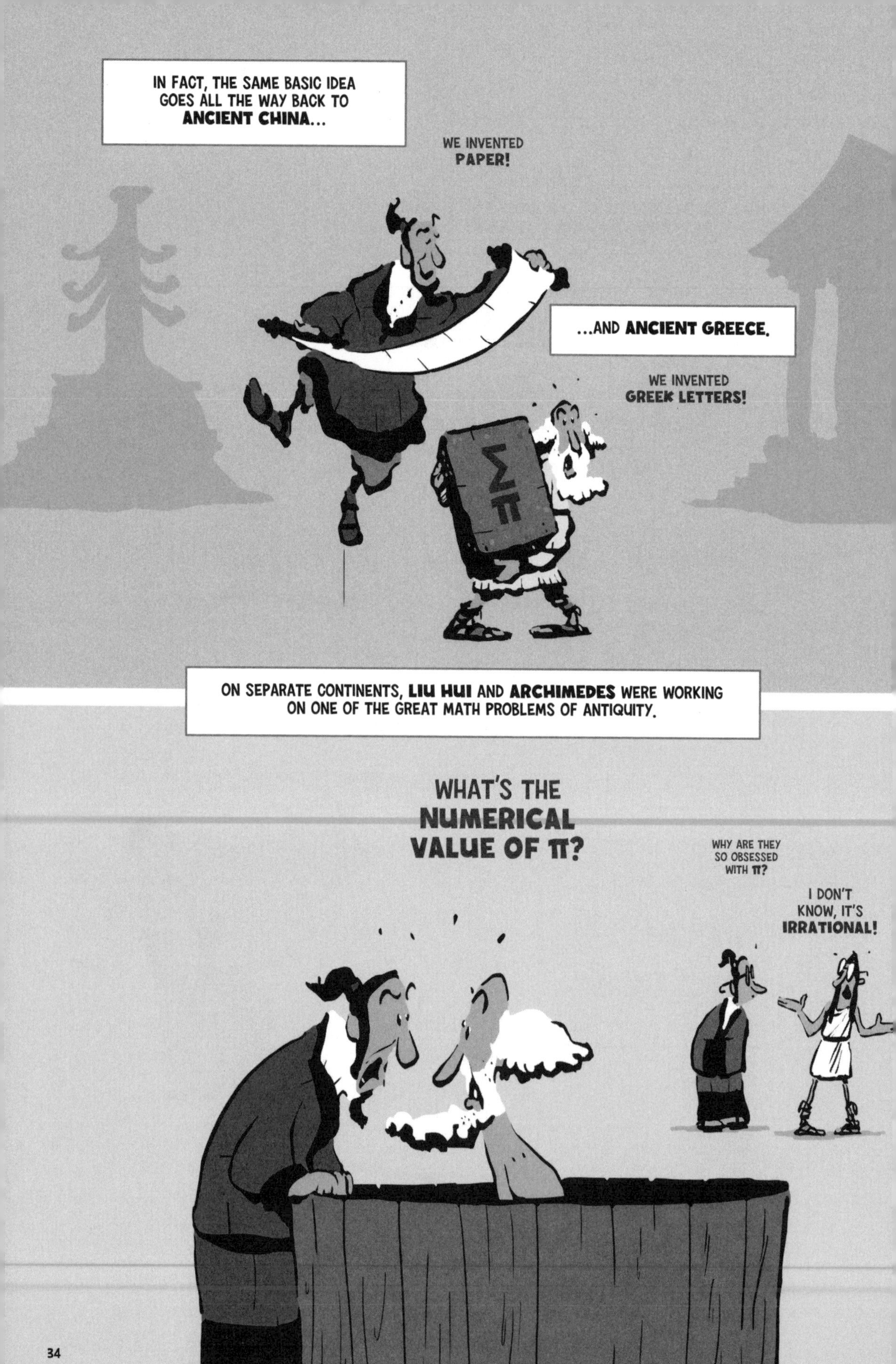
IN FACT, THE SAME BASIC IDEA GOES ALL THE WAY BACK TO **ANCIENT CHINA**...
WE INVENTED **PAPER!**
...AND **ANCIENT GREECE.**
WE INVENTED **GREEK LETTERS!**
Σ π
ON SEPARATE CONTINENTS, **LIU HUI** AND **ARCHIMEDES** WERE WORKING ON ONE OF THE GREAT MATH PROBLEMS OF ANTIQUITY.
WHAT'S THE **NUMERICAL VALUE OF π?**
WHY ARE THEY SO OBSESSED WITH **π?**
I DON'T KNOW, IT'S **IRRATIONAL!**

THEIR BASIC APPROACH WAS TO TAKE A **CIRCLE OF RADIUS 1**...

...AND **PUT A SQUARE INSIDE IT.**

THE AREA OF ANY CIRCLE = $\pi \cdot r^2$...

...SO THE AREA OF THIS CIRCLE = π...

$r=1$

...AND THIS SQUARE'S AREA IS SMALLER THAN π.

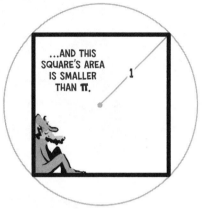

1

THEN THEY CALCULATED THE **AREA OF THE SQUARE**...

WE CAN CHOP A SQUARE INTO **FOUR IDENTICAL TRIANGLES**...

...AND THE FORMULA

$$\frac{base \cdot height}{2}$$

GIVES US THE AREA OF A TRIANGLE.

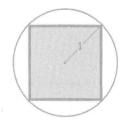

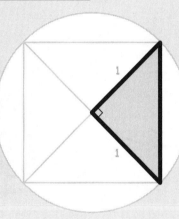

1

1

SO THE AREA OF EACH TRIANGLE = $\frac{1}{2}$...

$\frac{1 \cdot 1}{2} = \frac{1}{2}$

...SO THE AREA OF THE SQUARE = **2.**

$4 \cdot (\frac{1}{2}) = 2$

...AND CONCLUDED THAT $\pi > 2$.

$\pi > 2$?

THIS IS WHY YOU'RE **GENIUSES??**

1

NO, WE CAN **ALSO** SHOW...

...THAT $\pi < 4$!

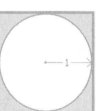

1

THEIR **REAL GENIUS** WAS IN THE WAY THEY **CONTINUOUSLY IMPROVED THIS KIND OF ESTIMATE.**

CHOP CHOP
CHOP CHOP
CHOP CHOP
CHOP CHOP
CHOP CHOP

THIS IS **NOT** CONTINUOUSLY IMPROVING MY **BLISTERS!**

THEIR IDEA ALLOWS US TO MOVE FROM A **SQUARE**...

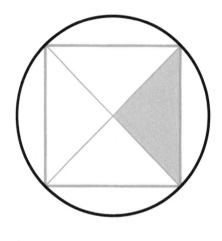

THERE ARE **4 PIECES,** EACH WITH AREA $\frac{1}{2}$.

SO $\pi > 2$.

...TO AN **OCTAGON**...

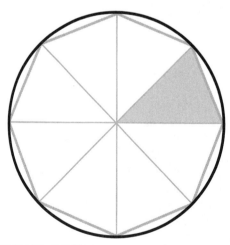

THERE ARE **8 PIECES,** EACH WITH AREA 0.35355

SO $\pi > 2.828427$

...TO A **HEXADECAGON**...

THERE ARE **16** PIECES,
EACH WITH AREA
0.1913417

SO **π>3.0614674**

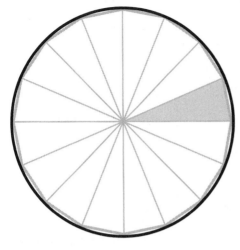

...AND **SO ON AND SO ON**...

32 SIDES:
π>3.1214453

64 SIDES:
π>3.1365488

128 SIDES:
π>3.1403331

...AS **FAR AS WE WANT TO GO!**

DIVIDING A
CIRCLE INTO **TINY
SLIVERS**...

...IS SIMILAR TO MY
IDEA OF **THREADS
AND PAGES**...

...OR CHOPPING
UP A **PIZZA!**

IT'S OKAY TO **SKIP** THESE PAGES IF YOU'RE NOT A GENIUS.

THERE ARE **TWO OBVIOUS QUESTIONS**. THE FIRST IS
HOW DID THEY DO IT?

TO SEE HOW, LET'S GO FROM A **SQUARE**...

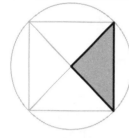

...TO AN **OCTAGON**.

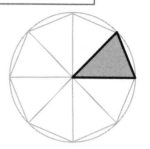

NOTICE THAT EACH PIECE OF THE OCTAGON CAN BE CUT INTO **TWO RIGHT TRIANGLES**...

...AND WE **ALREADY KNOW THE AREA OF THE INSIDE ONE**.

IT'S **HALF THE AREA** OF ONE OF THE **SQUARE'S PIECES!**

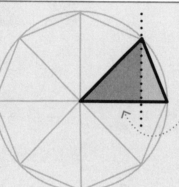

TO GET THE AREA OF THE **OUTSIDE TRIANGLE**...

...WE NEED TO MULTIPLY $\frac{1}{2} \cdot$ **HEIGHT** $\cdot$ **BASE**.

BUT HOW DO WE GET **THOSE VALUES?**

THAT'S WHY WE'RE **GENIUSES.**

height

?

base

TO DO THIS, WE CALCULATE THE **HEIGHT**...

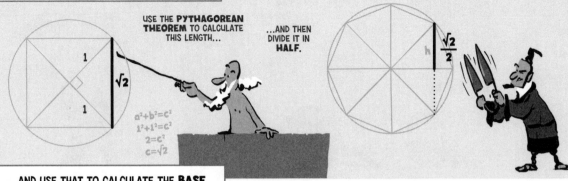

USE THE **PYTHAGOREAN THEOREM** TO CALCULATE THIS LENGTH...

1

√2

1

$a^2+b^2=c^2$
$1^2+1^2=c^2$
$2=c^2$
$c=\sqrt{2}$

...AND THEN DIVIDE IT IN **HALF**.

h $\frac{\sqrt{2}}{2}$

...AND USE THAT TO CALCULATE THE **BASE**.

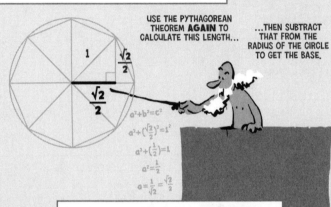

USE THE PYTHAGOREAN THEOREM **AGAIN** TO CALCULATE THIS LENGTH...

1    $\frac{\sqrt{2}}{2}$

$\frac{\sqrt{2}}{2}$

$a^2+b^2=c^2$
$a^2+(\frac{\sqrt{2}}{2})^2=1^2$
$a^2+(\frac{1}{2})=1$
$a^2=\frac{1}{2}$
$a=\frac{1}{\sqrt{2}}=\frac{\sqrt{2}}{2}$

...THEN SUBTRACT THAT FROM THE RADIUS OF THE CIRCLE TO GET THE BASE.

b

$1-\frac{\sqrt{2}}{2}$

NOW WE CAN **ADD UP** ALL THE AREAS...

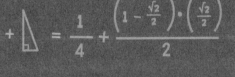

$$\triangle + \triangle = \frac{1}{4} + \frac{\left(1-\frac{\sqrt{2}}{2}\right)\cdot\left(\frac{\sqrt{2}}{2}\right)}{2}$$

$= 0.3535533$

...TO IMPROVE OUR **ESTIMATE OF π**.

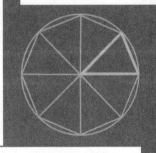

π > 8 • 0.3535533

SO **π > 2.8284271**

AND WE CAN REPEAT THIS **OVER AND OVER AGAIN**.

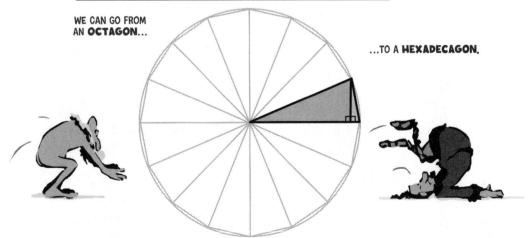

WE CAN GO FROM AN **OCTAGON**...

...TO A **HEXADECAGON**.

39

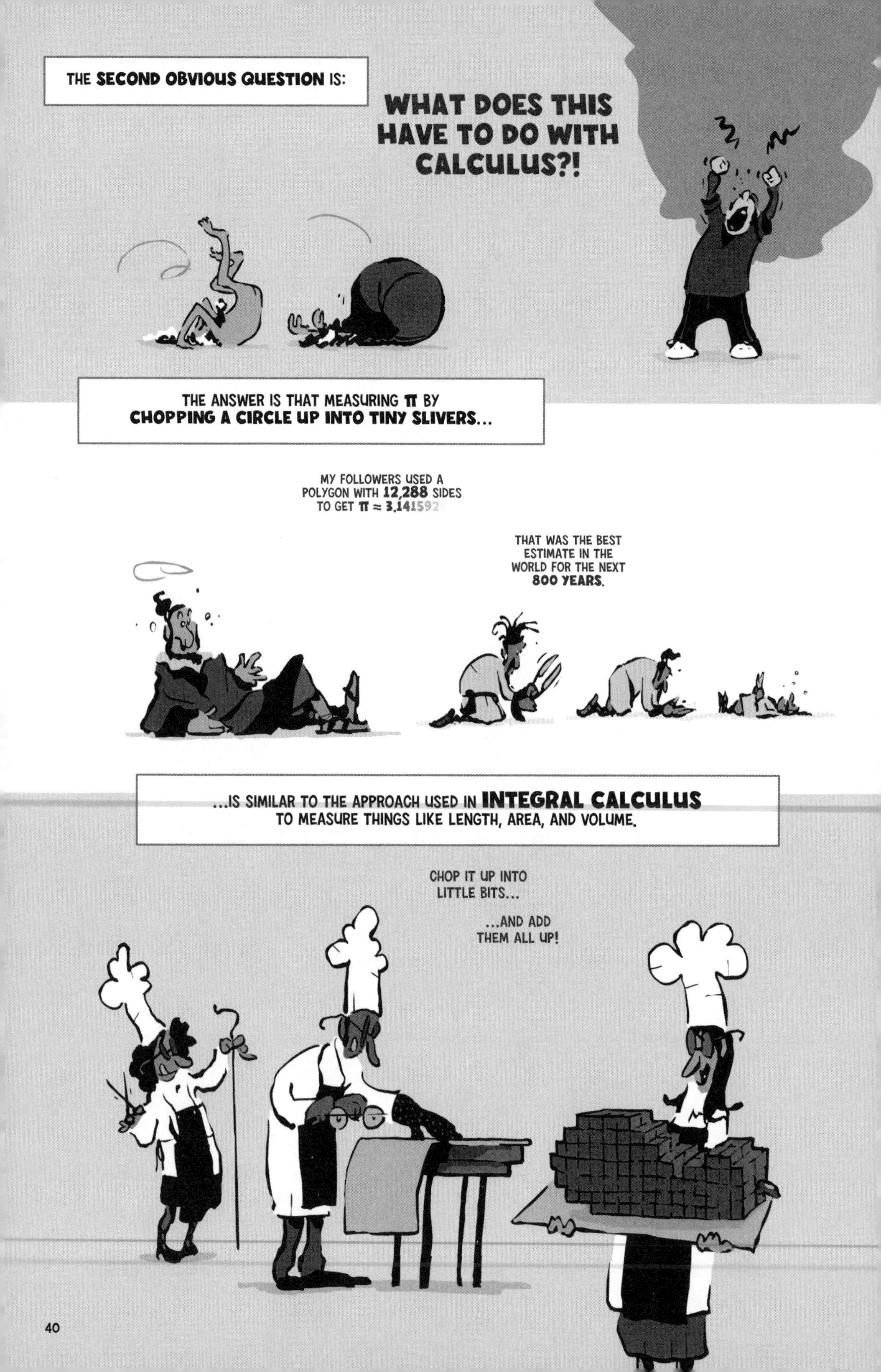

THE SECOND OBVIOUS QUESTION IS:
WHAT DOES THIS HAVE TO DO WITH CALCULUS?!
THE ANSWER IS THAT MEASURING π BY CHOPPING A CIRCLE UP INTO TINY SLIVERS...
MY FOLLOWERS USED A POLYGON WITH 12,288 SIDES TO GET π ≈ 3.141592
THAT WAS THE BEST ESTIMATE IN THE WORLD FOR THE NEXT 800 YEARS.
...IS SIMILAR TO THE APPROACH USED IN INTEGRAL CALCULUS TO MEASURE THINGS LIKE LENGTH, AREA, AND VOLUME.
CHOP IT UP INTO LITTLE BITS...
...AND ADD THEM ALL UP!

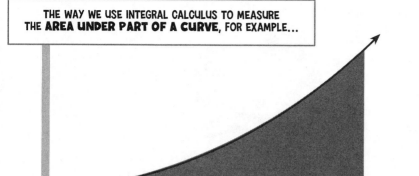

THE WAY WE USE INTEGRAL CALCULUS TO MEASURE THE **AREA UNDER PART OF A CURVE**, FOR EXAMPLE...

...IS TO GET AN **INITIAL ESTIMATE** BY ADDING UP THE AREAS OF **RECTANGLES LIKE THESE**...

...AND THEN GET AN **EVEN BETTER ESTIMATE** BY ADDING UP THE AREAS OF **EVEN SMALLER RECTANGLES**...

...AND **ON** AND **ON** AND **ON!**

ARE WE THERE YET?

NO, BUT WE ARE GETTING CLOSER, IN A **LIMITED SORT OF WAY**.

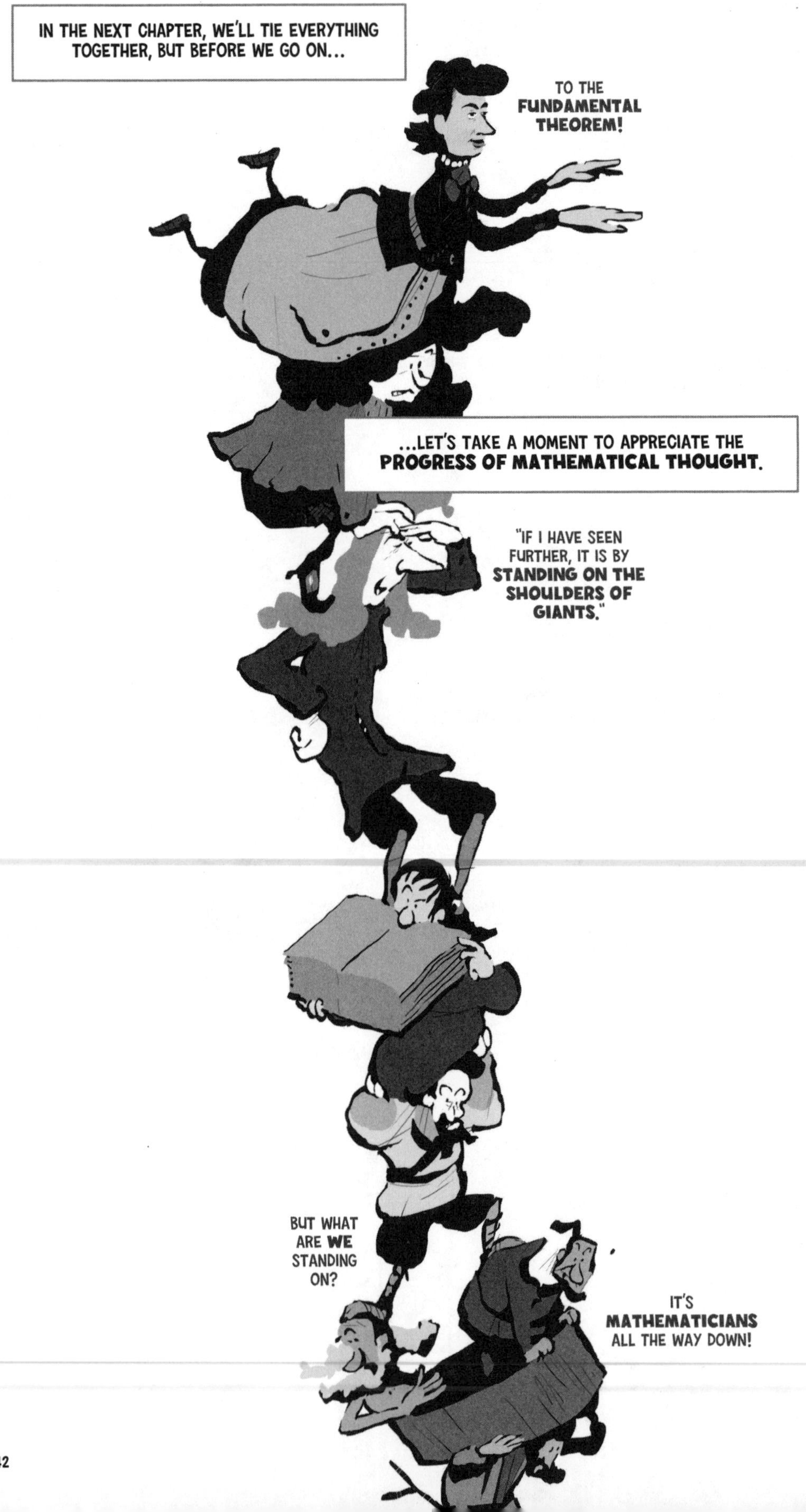
IN THE NEXT CHAPTER, WE'LL TIE EVERYTHING TOGETHER, BUT BEFORE WE GO ON...
TO THE **FUNDAMENTAL THEOREM!**
...LET'S TAKE A MOMENT TO APPRECIATE THE **PROGRESS OF MATHEMATICAL THOUGHT**.
"IF I HAVE SEEN FURTHER, IT IS BY **STANDING ON THE SHOULDERS OF GIANTS.**"
BUT WHAT ARE **WE** STANDING ON?
IT'S **MATHEMATICIANS** ALL THE WAY DOWN!

CHAPTER 4

THE FUNDAMENTAL THEOREM OF CALCULUS

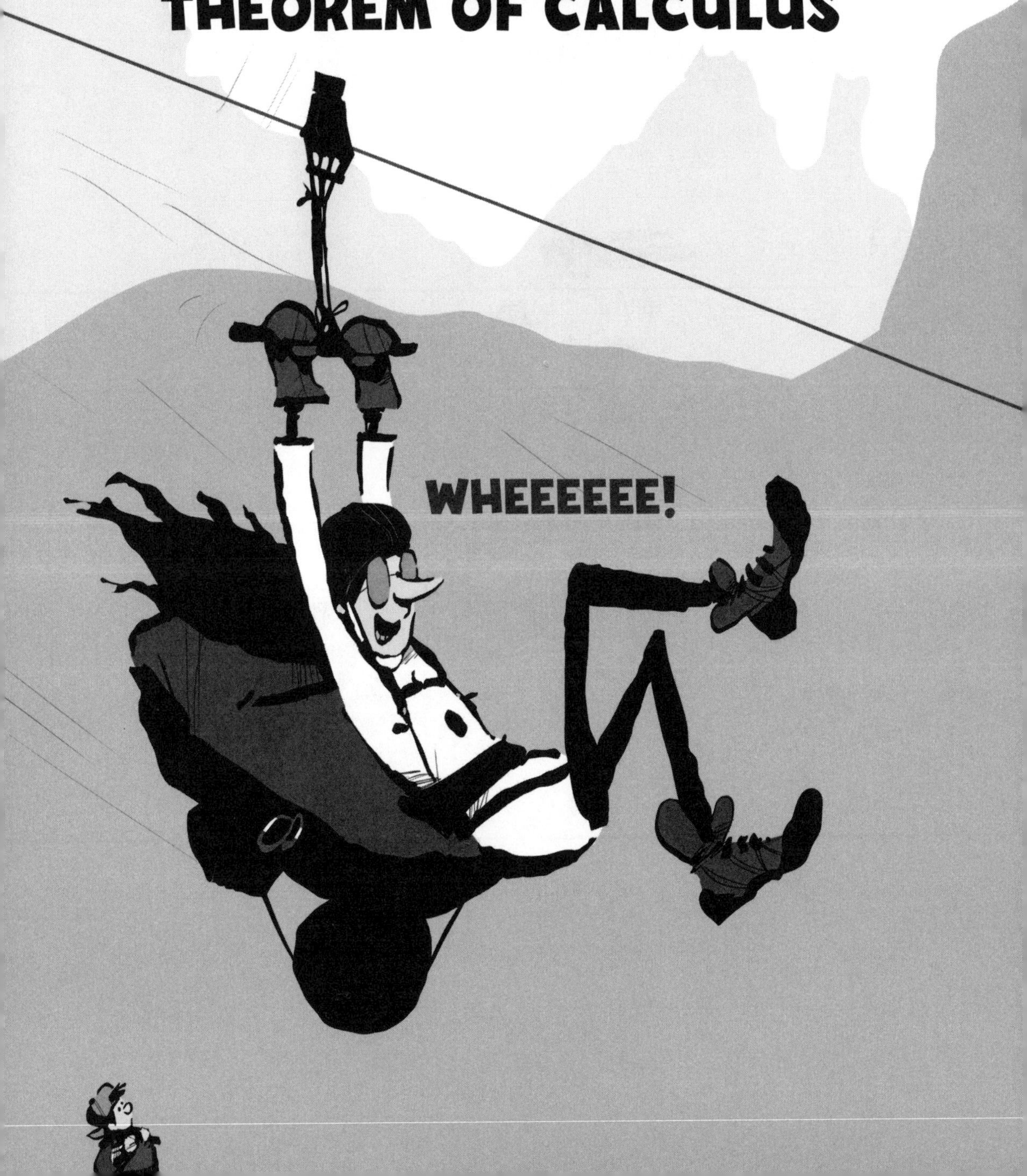
WHEEEEEE!

SO FAR, WE'VE LEARNED THAT
DERIVATIVES
INVOLVE TAKING SOMETHING
REALLY SMALL...

...AND **DIVIDING** IT BY SOMETHING
ELSE **REALLY SMALL**.

SPEED IS
DISTANCE DIVIDED
BY TIME...

...SO LET'S TAKE THIS
TINY DISTANCE YOU
JUST TRAVELED...

...AND DIVIDE IT BY THIS
TINY AMOUNT OF TIME
THAT JUST ELAPSED.

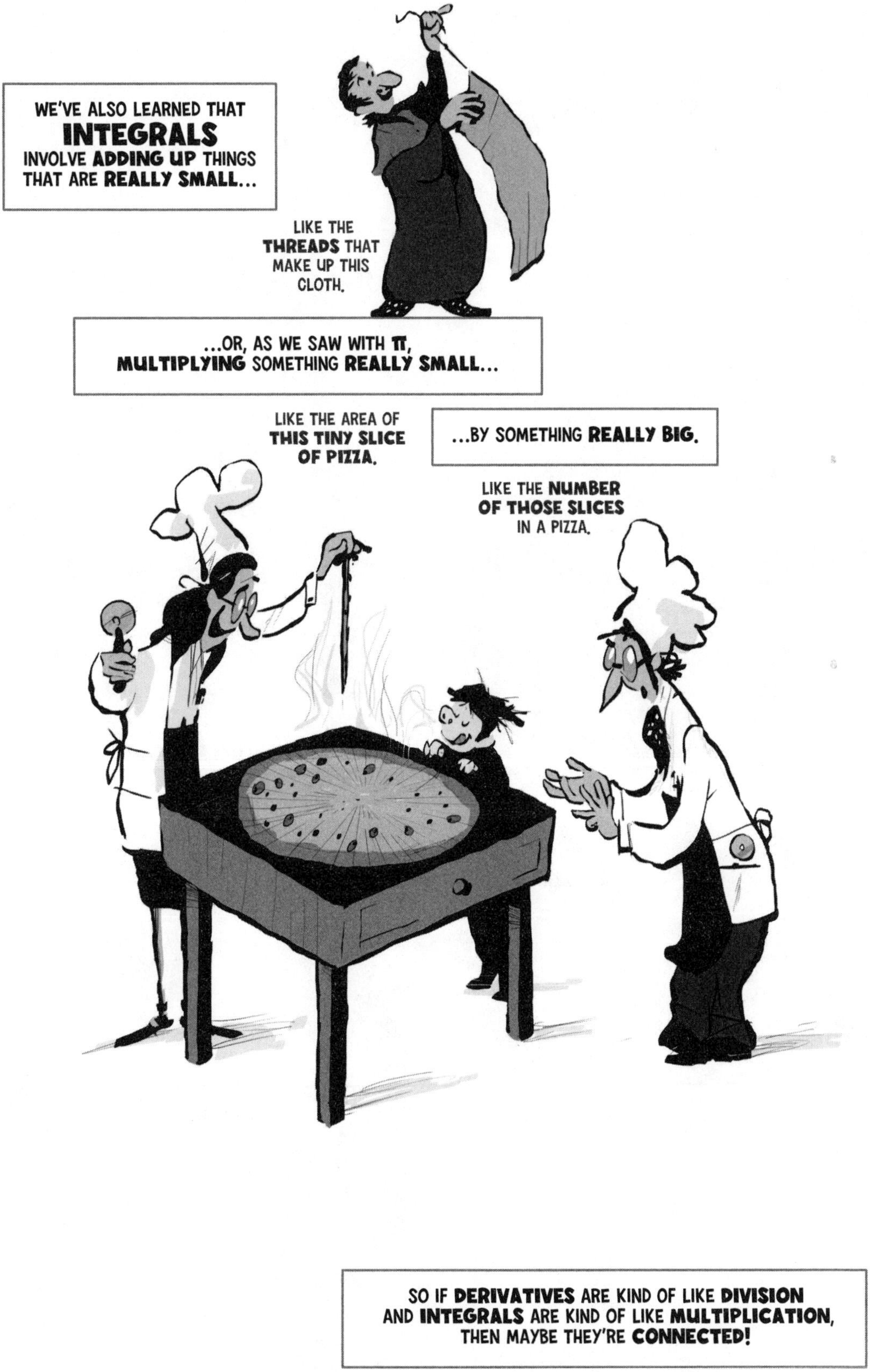

WE'VE ALSO LEARNED THAT INTEGRALS INVOLVE ADDING UP THINGS THAT ARE REALLY SMALL...
LIKE THE THREADS THAT MAKE UP THIS CLOTH.
...OR, AS WE SAW WITH π, MULTIPLYING SOMETHING REALLY SMALL...
LIKE THE AREA OF THIS TINY SLICE OF PIZZA.
...BY SOMETHING REALLY BIG.
LIKE THE NUMBER OF THOSE SLICES IN A PIZZA.
SO IF DERIVATIVES ARE KIND OF LIKE DIVISION AND INTEGRALS ARE KIND OF LIKE MULTIPLICATION, THEN MAYBE THEY'RE CONNECTED!

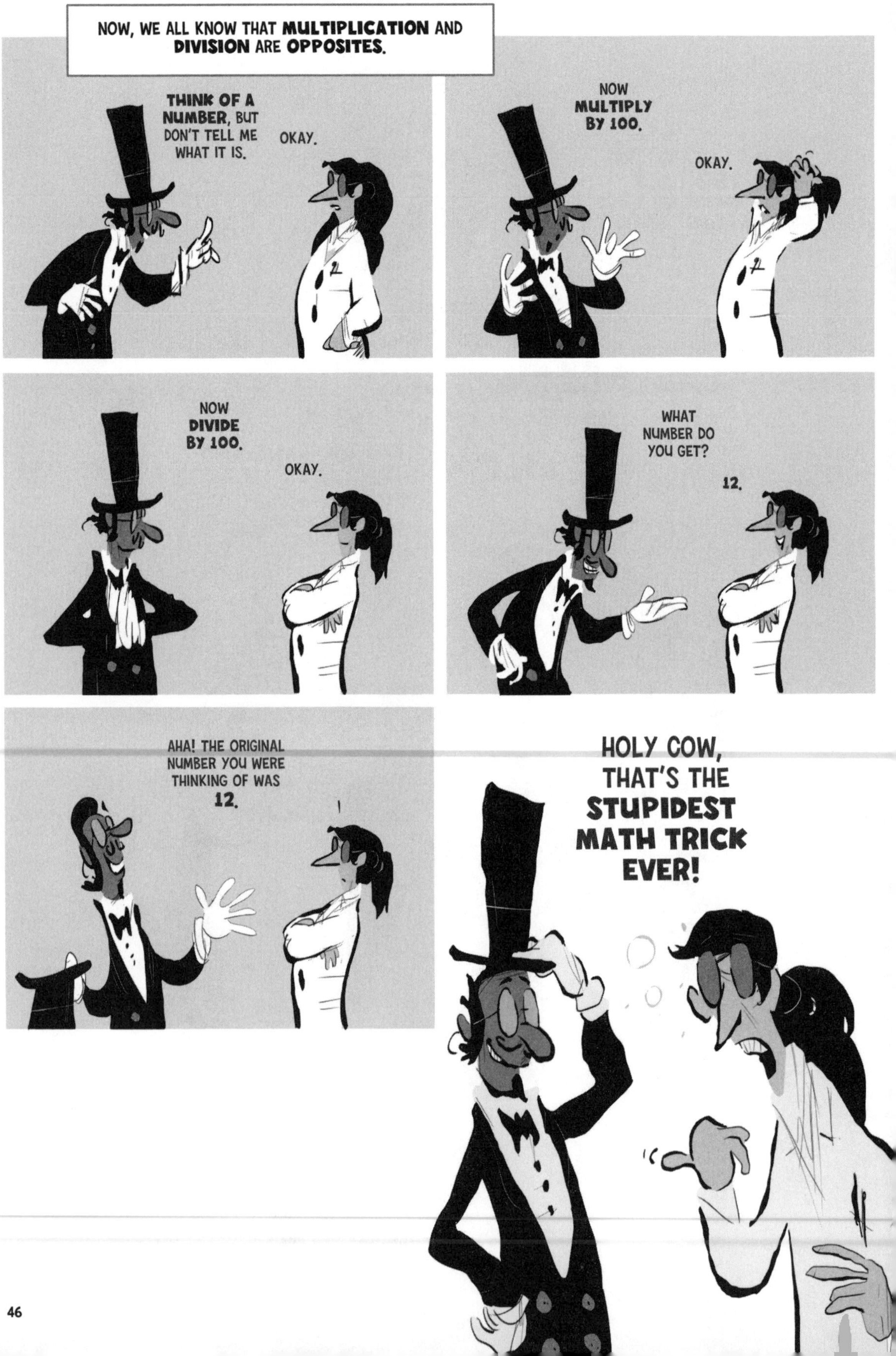
NOW, WE ALL KNOW THAT MULTIPLICATION AND DIVISION ARE OPPOSITES.
THINK OF A NUMBER, BUT DON'T TELL ME WHAT IT IS.
OKAY.
NOW MULTIPLY BY 100.
OKAY.
NOW DIVIDE BY 100.
OKAY.
WHAT NUMBER DO YOU GET?
12.
AHA! THE ORIGINAL NUMBER YOU WERE THINKING OF WAS 12.
!
HOLY COW, THAT'S THE STUPIDEST MATH TRICK EVER!

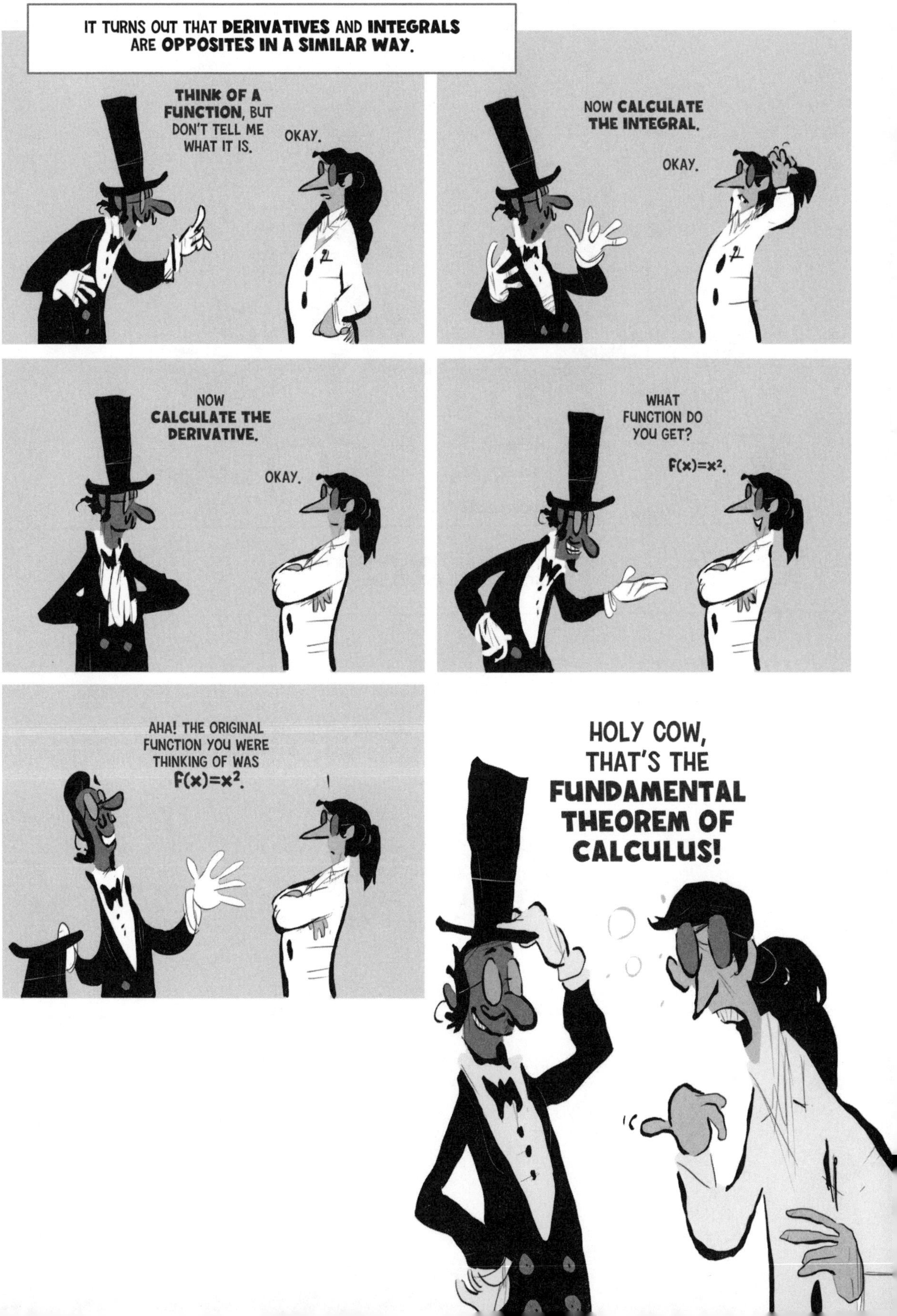
IT TURNS OUT THAT DERIVATIVES AND INTEGRALS ARE OPPOSITES IN A SIMILAR WAY.
THINK OF A FUNCTION, BUT DON'T TELL ME WHAT IT IS.
OKAY.
NOW CALCULATE THE INTEGRAL.
OKAY.
NOW CALCULATE THE DERIVATIVE.
OKAY.
WHAT FUNCTION DO YOU GET?
F(x)=x².
AHA! THE ORIGINAL FUNCTION YOU WERE THINKING OF WAS F(x)=x².
!
HOLY COW, THAT'S THE FUNDAMENTAL THEOREM OF CALCULUS!

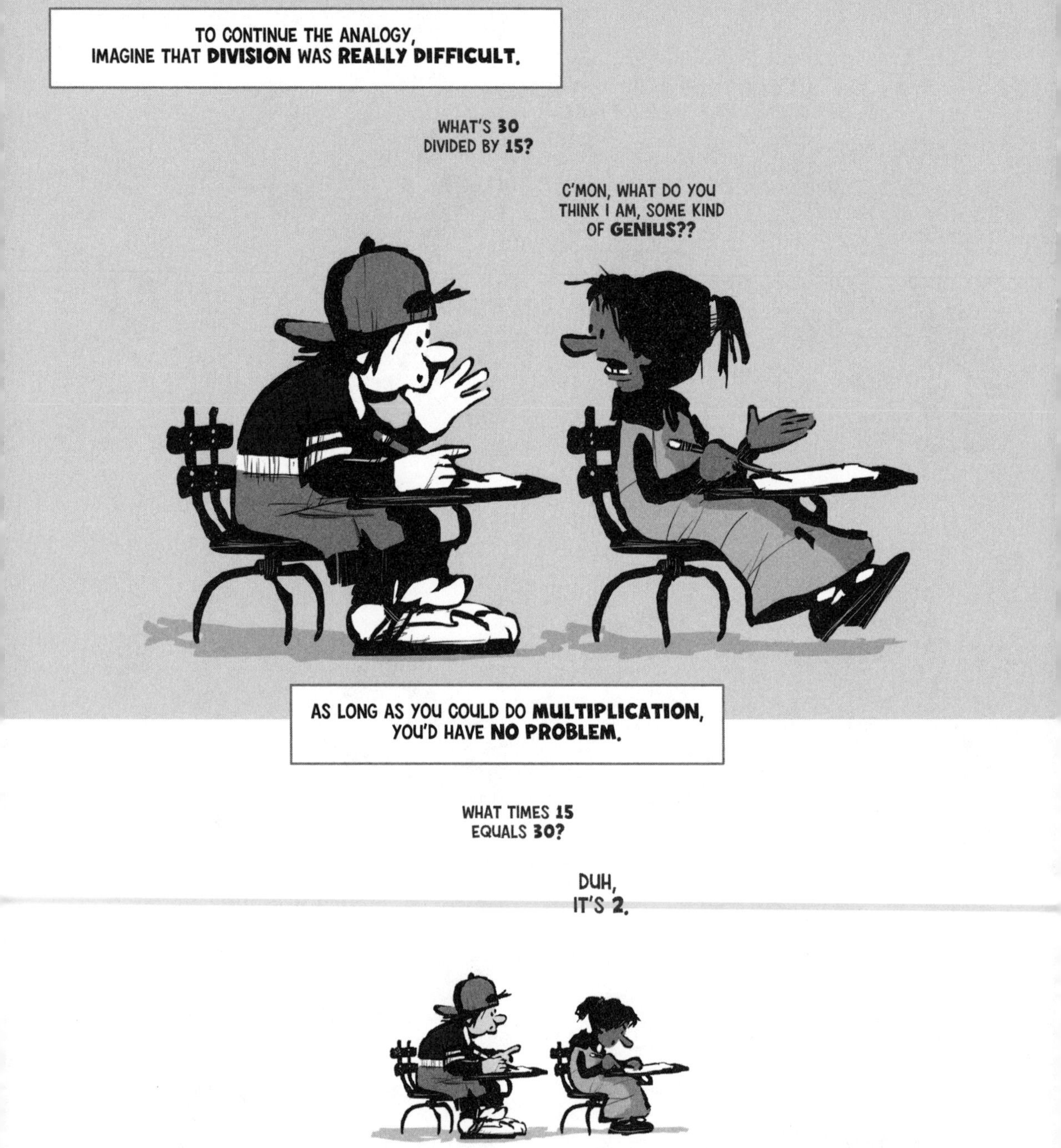
TO CONTINUE THE ANALOGY, IMAGINE THAT DIVISION WAS REALLY DIFFICULT.
WHAT'S 30 DIVIDED BY 15?
C'MON, WHAT DO YOU THINK I AM, SOME KIND OF GENIUS??
AS LONG AS YOU COULD DO MULTIPLICATION, YOU'D HAVE NO PROBLEM.
WHAT TIMES 15 EQUALS 30?
DUH, IT'S 2.

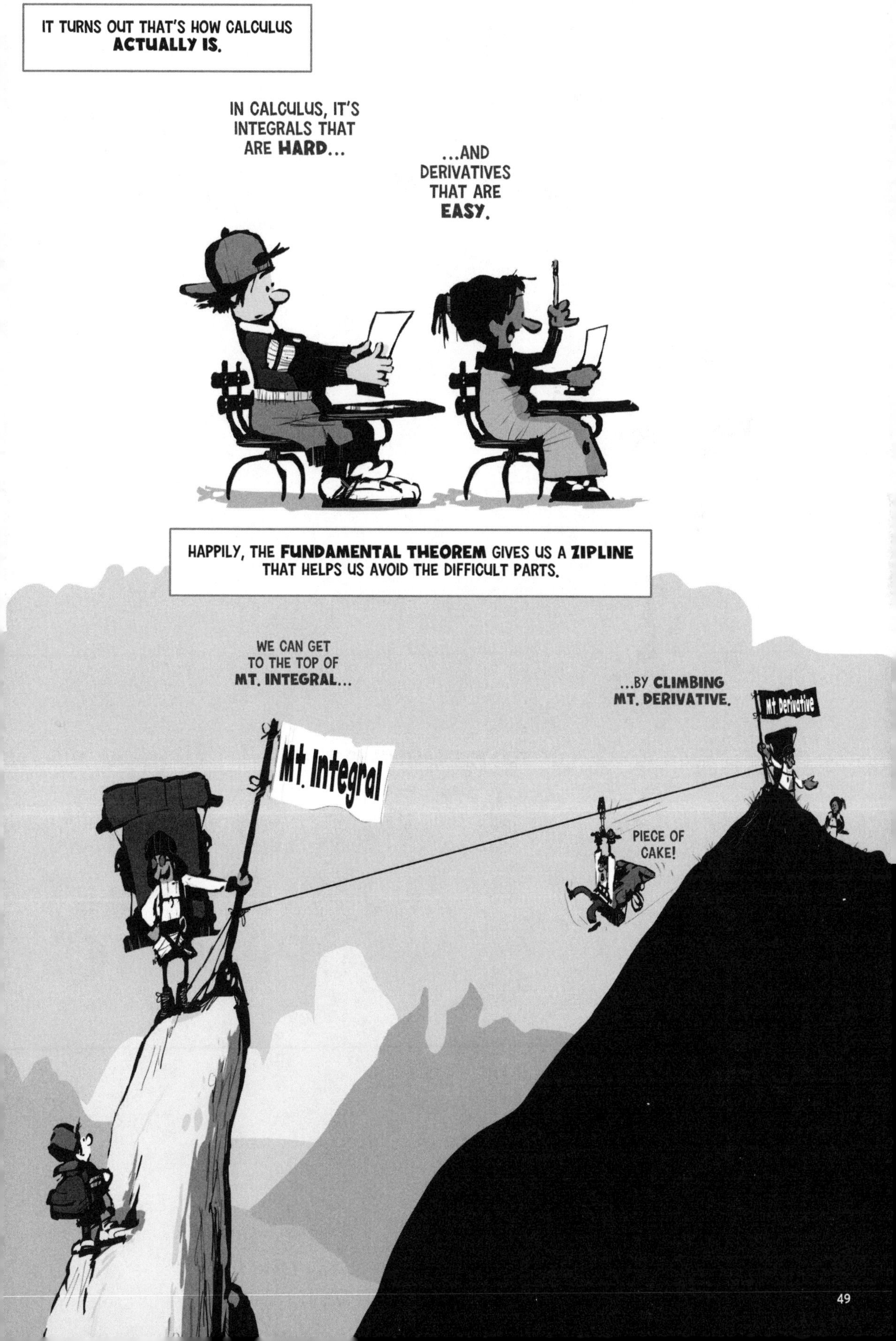
IT TURNS OUT THAT'S HOW CALCULUS **ACTUALLY IS.**
IN CALCULUS, IT'S INTEGRALS THAT ARE **HARD**...
...AND DERIVATIVES THAT ARE **EASY.**
HAPPILY, THE **FUNDAMENTAL THEOREM** GIVES US A **ZIPLINE** THAT HELPS US AVOID THE DIFFICULT PARTS.
WE CAN GET TO THE TOP OF **MT. INTEGRAL**...
...BY **CLIMBING MT. DERIVATIVE.**
Mt. Integral
Mt Derivative
PIECE OF CAKE!

THE **MATH** OF
THE FUNDAMENTAL THEOREM
LOOKS **DAUNTING**...

AND TO MAKE
MATTERS WORSE, THE
FUNDAMENTAL THEOREM
HAS **TWO PARTS.**

The **Fundamental Theorem of Calculus (Part One)** states that if f(x) is a continuous function, then:

$$\frac{d}{dx}\left[\int_a^x f(t)\,dt\right] = f(x)$$

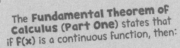

The **Fundamental Theorem of Calculus (Part Two)** states that if F(x) is an anti-derivative of a continuous function f(x), then:

$$\int_a^b f(x)\,dx = F(b) - F(a)$$

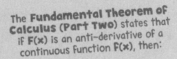

...BUT THE **MEANING** IS SIMPLE.

DERIVATIVES AND **INTEGRALS** ARE **OPPOSITES.**

THERE'S EVEN AN EASY WAY TO UNDERSTAND THE **INTUITION** BEHIND THE FUNDAMENTAL THEOREM.

LET'S FOCUS ON THE **FIRST** PART.

The **Fundamental Theorem of Calculus (Part One)** states that if f(x) is a continuous function, then:

$$\frac{d}{dx}\left[\int_a^x f(t)\,dt\right] = f(x)$$

TO GET THAT INTUITION, LET'S CONSIDER A **FUNCTION**.

SEE THE GLOSSARY FOR A **FORMAL DEFINITION** OF A FUNCTION...

...BUT FOR THE INTUITION JUST THINK OF A **SQUIGGLY LINE** THAT STAYS ABOVE THE **x-AXIS**...

...LIKE **THIS ONE!**

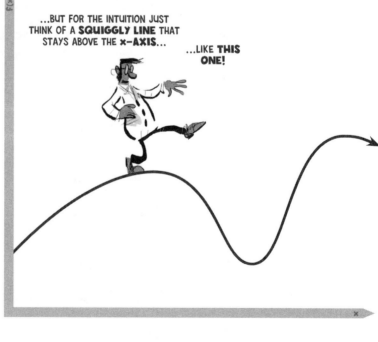

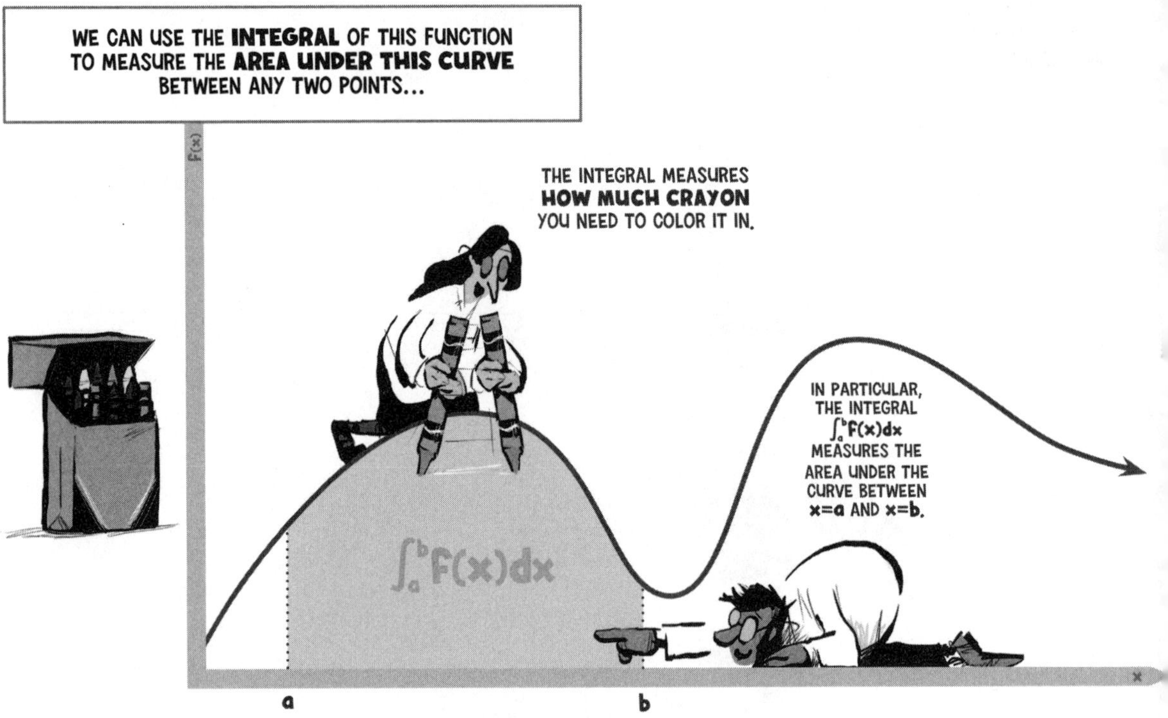
WE CAN USE THE **INTEGRAL** OF THIS FUNCTION TO MEASURE THE **AREA UNDER THIS CURVE** BETWEEN ANY TWO POINTS...
THE INTEGRAL MEASURES **HOW MUCH CRAYON** YOU NEED TO COLOR IT IN.
IN PARTICULAR, THE INTEGRAL $\int_a^b F(x)dx$ MEASURES THE AREA UNDER THE CURVE BETWEEN **x=a** AND **x=b**.
F(x)
$\int_a^b F(x)dx$
a
b
x

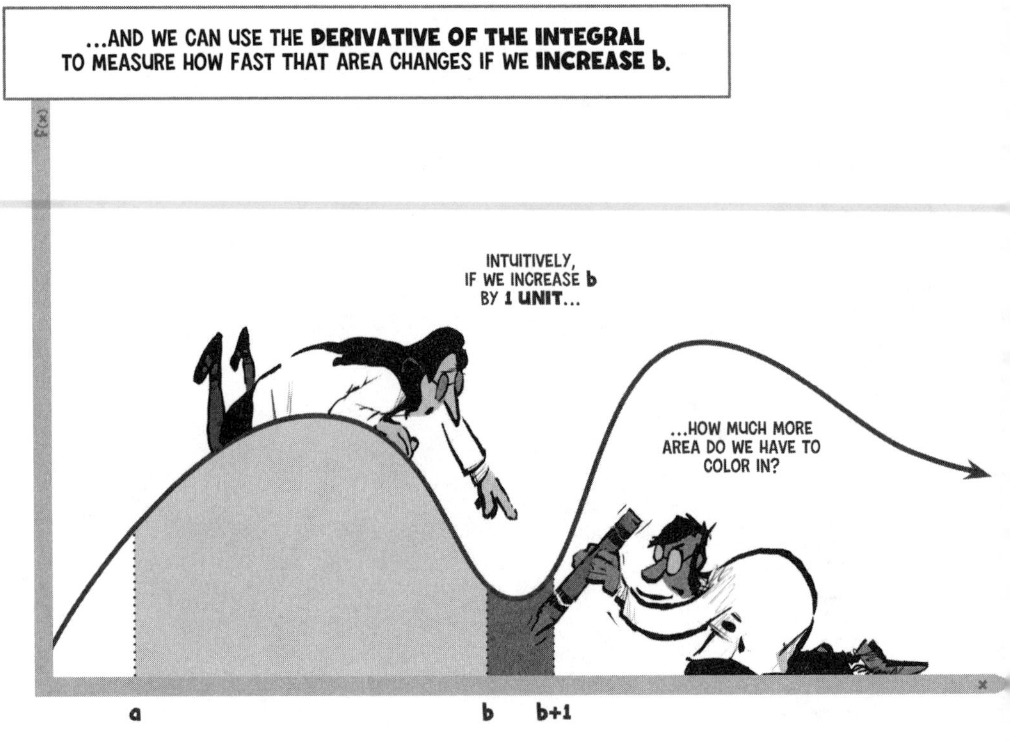
...AND WE CAN USE THE **DERIVATIVE OF THE INTEGRAL** TO MEASURE HOW FAST THAT AREA CHANGES IF WE **INCREASE b**.
INTUITIVELY, IF WE INCREASE **b** BY **1 UNIT**...
...HOW MUCH MORE AREA DO WE HAVE TO COLOR IN?
F(x)
a
b
b+1
x

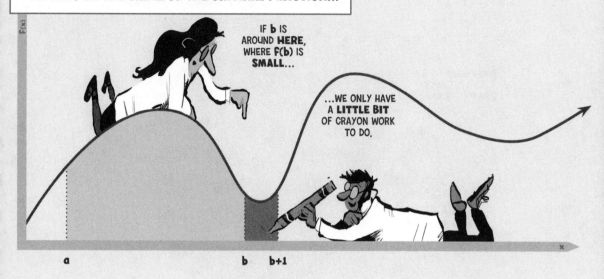

IT MAKES SENSE THAT THE RATE OF CHANGE OF THE AREA **DEPENDS ON THE VALUE OF THE ORIGINAL FUNCTION...**
F(x)
IF **b** IS AROUND **HERE**, WHERE **F(b)** IS **SMALL...**
...WE ONLY HAVE A **LITTLE BIT** OF CRAYON WORK TO DO.
a
b
b+1
x
F(x)
IF **b** IS AROUND **HERE**, WHERE **F(b)** IS **LARGE...**
...WE HAVE A **LOT MORE** CRAYON WORK TO DO.
a
b
b+1
x
...AND THAT'S EXACTLY WHAT THE **FIRST PART OF THE FUNDAMENTAL THEOREM SAYS!**
IF YOU TAKE A FUNCTION **F(x)**...
...AND CALCULATE AN **INTEGRAL**...
...AND THEN CALCULATE A **DERIVATIVE**...
...YOU GET BACK **F(x)**!

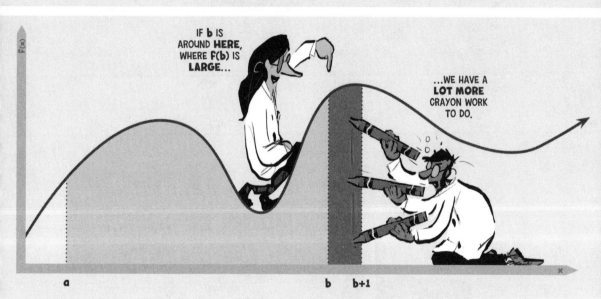

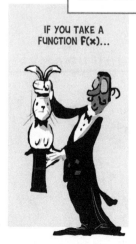

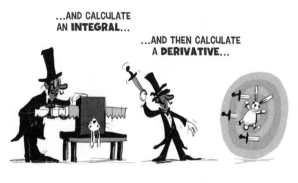

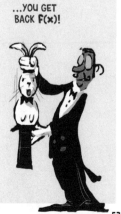

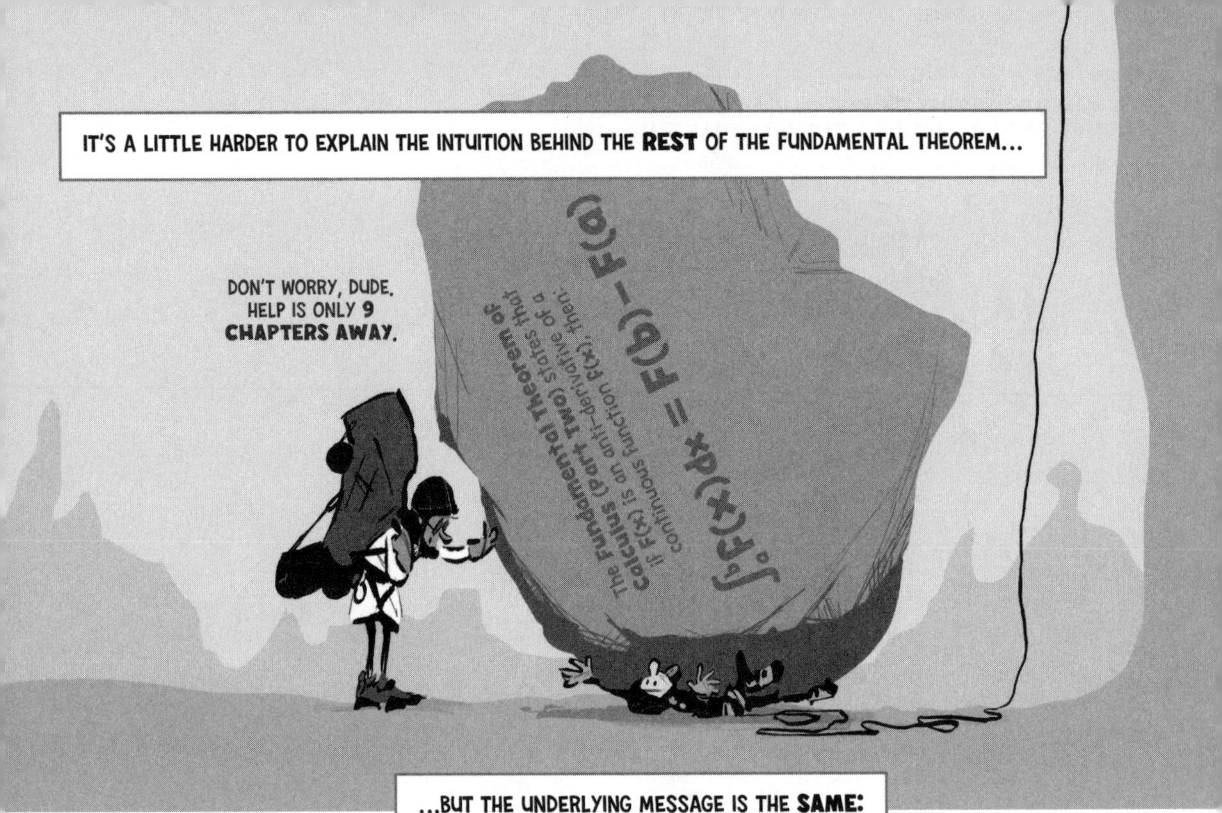
IT'S A LITTLE HARDER TO EXPLAIN THE INTUITION BEHIND THE **REST** OF THE FUNDAMENTAL THEOREM...
DON'T WORRY, DUDE. HELP IS ONLY **9 CHAPTERS AWAY.**
The **Fundamental Theorem of Calculus (Part Two)** states that if F(x) is an anti-derivative of a continuous function f(x), then:
$\int_a^b f(x)dx = F(b) - F(a)$

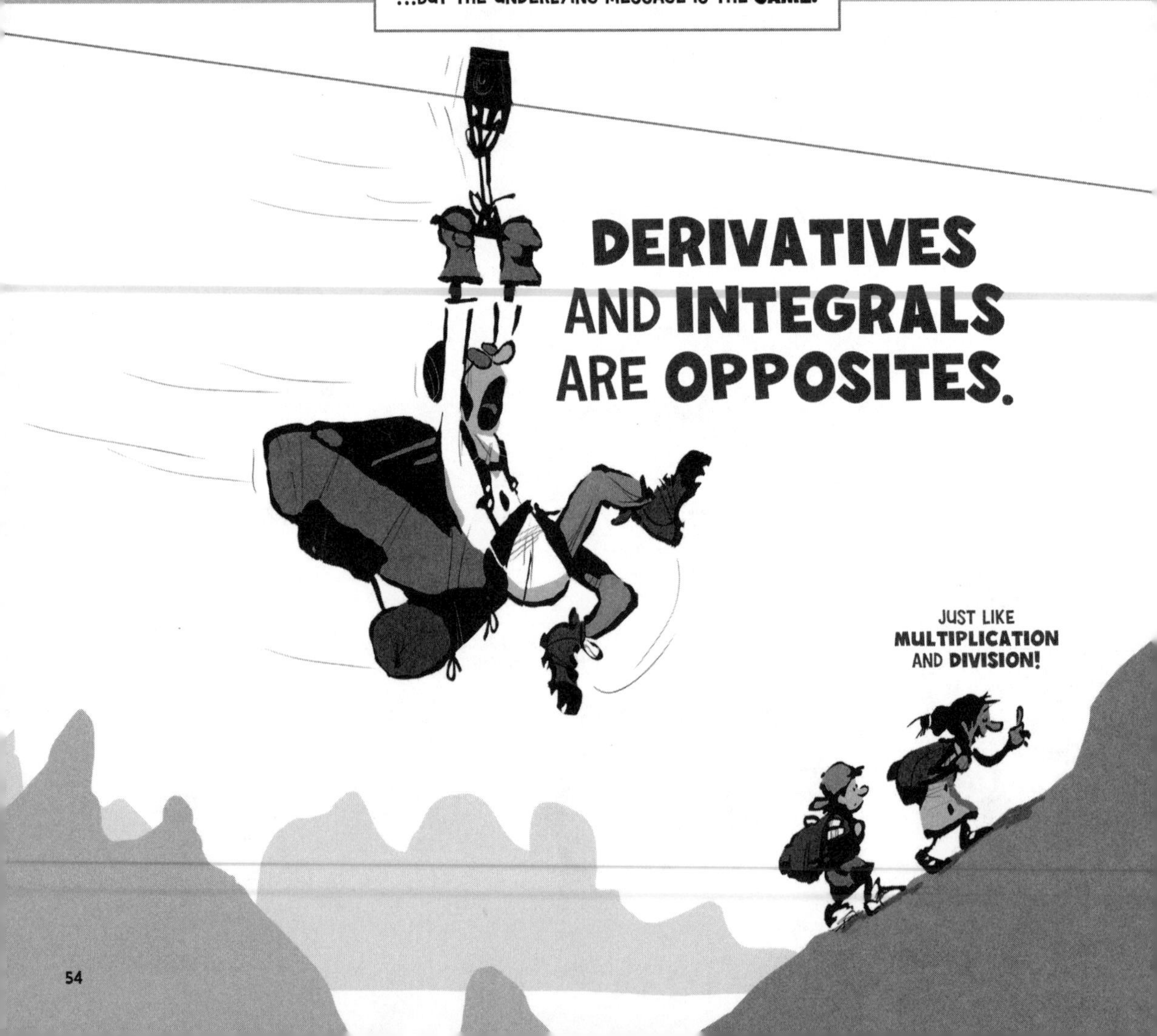
...BUT THE UNDERLYING MESSAGE IS THE **SAME:**
DERIVATIVES AND **INTEGRALS** ARE **OPPOSITES.**
JUST LIKE **MULTIPLICATION** AND **DIVISION!**

CHAPTER 5
LIMITS

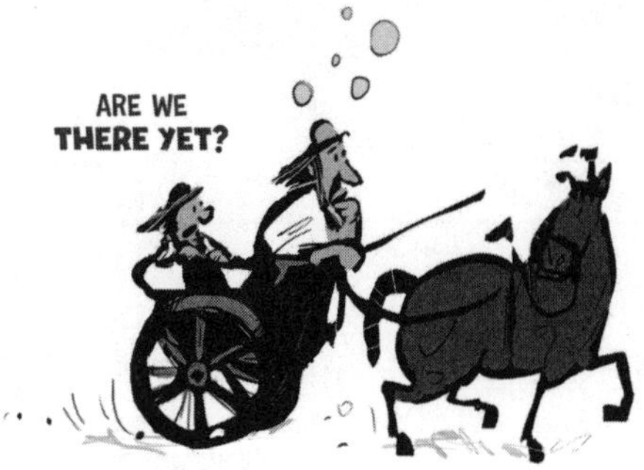
ARE WE
THERE YET?

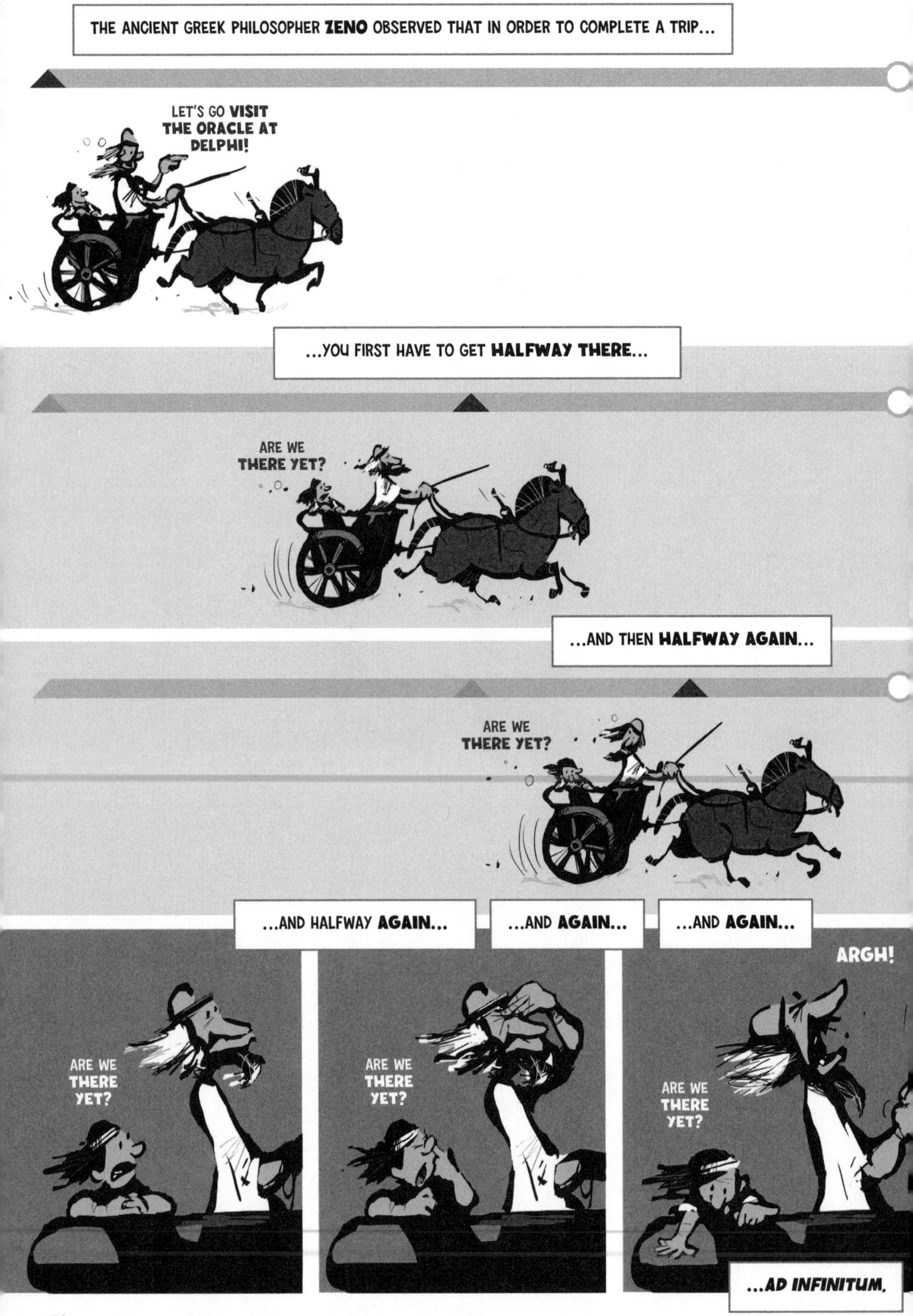
THE ANCIENT GREEK PHILOSOPHER **ZENO** OBSERVED THAT IN ORDER TO COMPLETE A TRIP...
LET'S GO **VISIT THE ORACLE AT DELPHI!**
...YOU FIRST HAVE TO GET **HALFWAY THERE...**
ARE WE **THERE YET?**
...AND THEN **HALFWAY AGAIN...**
ARE WE **THERE YET?**
...AND HALFWAY **AGAIN...**
ARE WE **THERE YET?**
...AND **AGAIN...**
ARE WE **THERE YET?**
...AND **AGAIN...**
ARGH!
ARE WE **THERE YET?**
...AD INFINITUM.

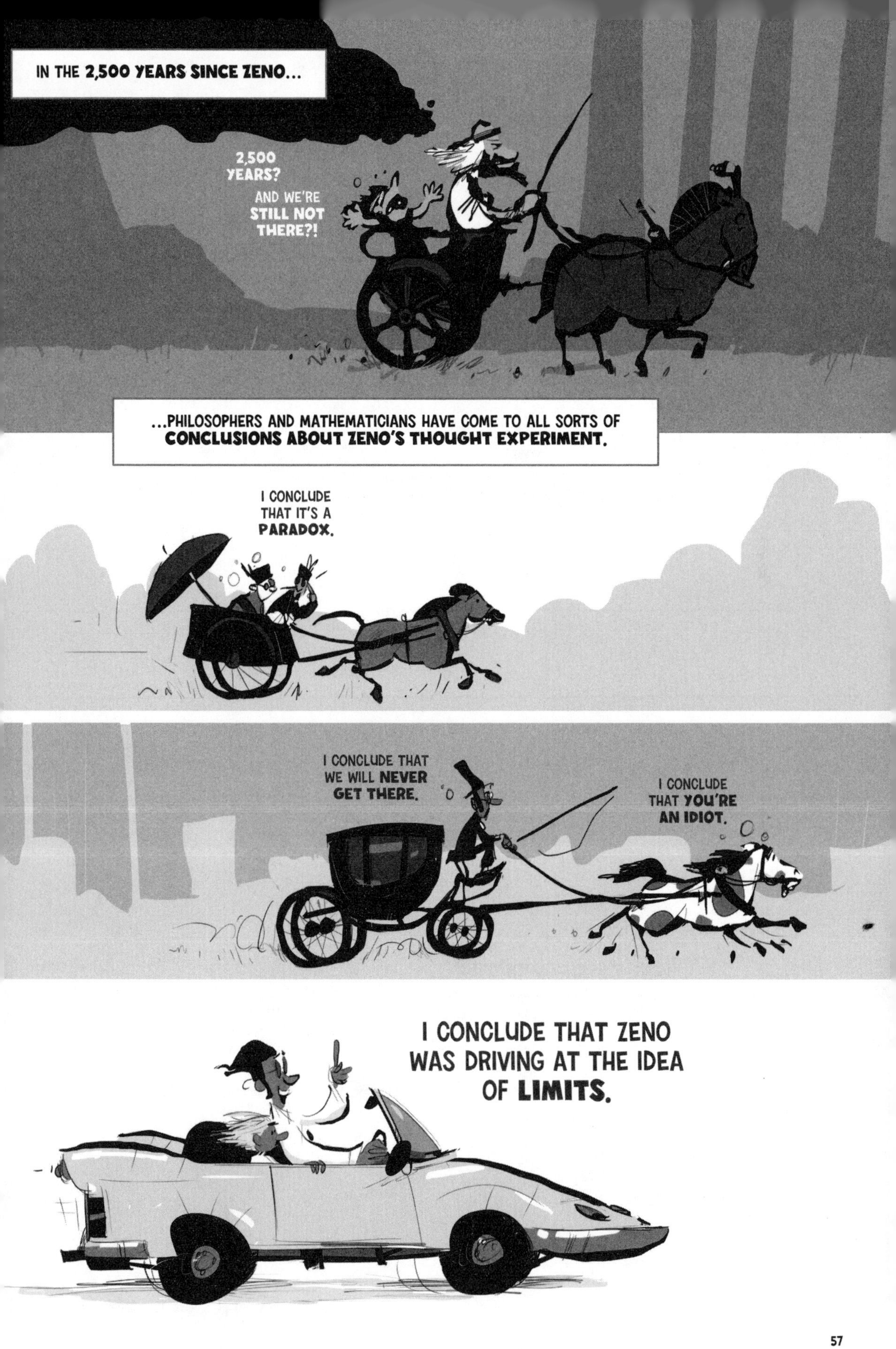
IN THE 2,500 YEARS SINCE ZENO...
2,500 YEARS?
AND WE'RE STILL NOT THERE?!
...PHILOSOPHERS AND MATHEMATICIANS HAVE COME TO ALL SORTS OF CONCLUSIONS ABOUT ZENO'S THOUGHT EXPERIMENT.
I CONCLUDE THAT IT'S A PARADOX.
I CONCLUDE THAT WE WILL NEVER GET THERE.
I CONCLUDE THAT YOU'RE AN IDIOT.
I CONCLUDE THAT ZENO WAS DRIVING AT THE IDEA OF LIMITS.

THE IDEA OF **LIMITS** LIES AT THE HEART OF CALCULUS.

AND PLENTY OF **OTHER MATHEMATICAL TOPICS TOO!**

IN THE FOLLOWING CHAPTERS, WE'RE GOING TO USE LIMITS TO DEFINE **SPEED**...

WHAT'S THE **LIMIT** AS t→0 OF DISTANCE DIVIDED BY TIME?

...AND TO **MEASURE THE SLOPE OF TANGENT LINES**...

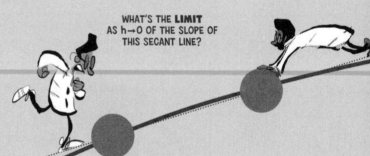

WHAT'S THE **LIMIT** AS h→0 OF THE SLOPE OF THIS SECANT LINE?

...AND TO **CALCULATE AREA.**

WHAT'S THE **LIMIT** OF THE SUM OF THESE RECTANGULAR AREAS...

...AS THE WIDTH OF EACH RECTANGLE GOES TO ZERO?

BUT FIRST LET'S GET THE **INTUITION.**

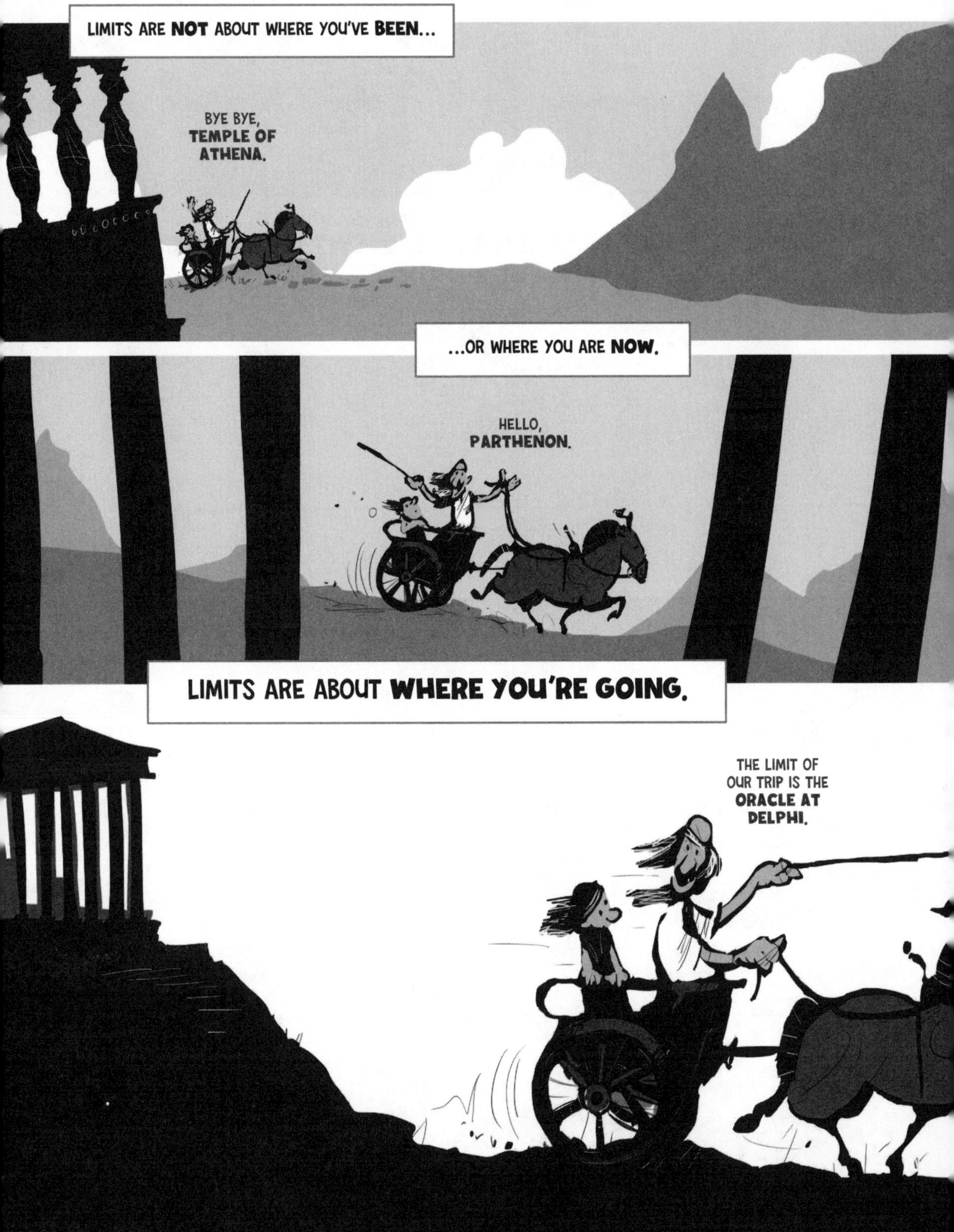

LIMITS ARE **NOT** ABOUT WHERE YOU'VE **BEEN...**
BYE BYE, **TEMPLE OF ATHENA.**
...OR WHERE YOU ARE **NOW.**
HELLO, **PARTHENON.**
LIMITS ARE ABOUT **WHERE YOU'RE GOING.**
THE LIMIT OF OUR TRIP IS THE **ORACLE AT DELPHI.**

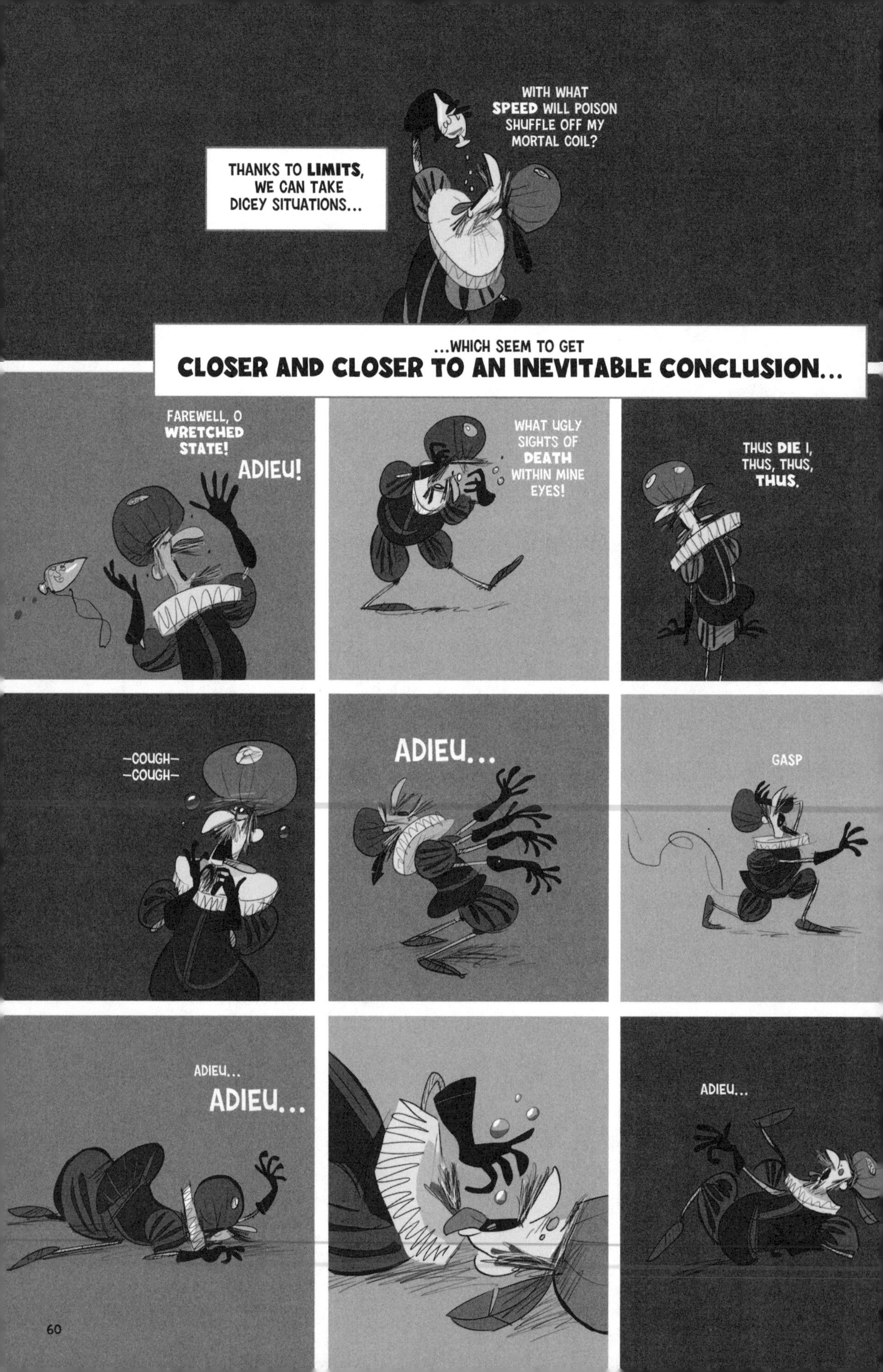
WITH WHAT SPEED WILL POISON SHUFFLE OFF MY MORTAL COIL?
THANKS TO LIMITS, WE CAN TAKE DICEY SITUATIONS...
...WHICH SEEM TO GET CLOSER AND CLOSER TO AN INEVITABLE CONCLUSION...
FAREWELL, O WRETCHED STATE!
ADIEU!
WHAT UGLY SIGHTS OF DEATH WITHIN MINE EYES!
THUS DIE I, THUS, THUS, THUS.
-COUGH-
-COUGH-
ADIEU...
GASP
ADIEU...
ADIEU...
ADIEU...

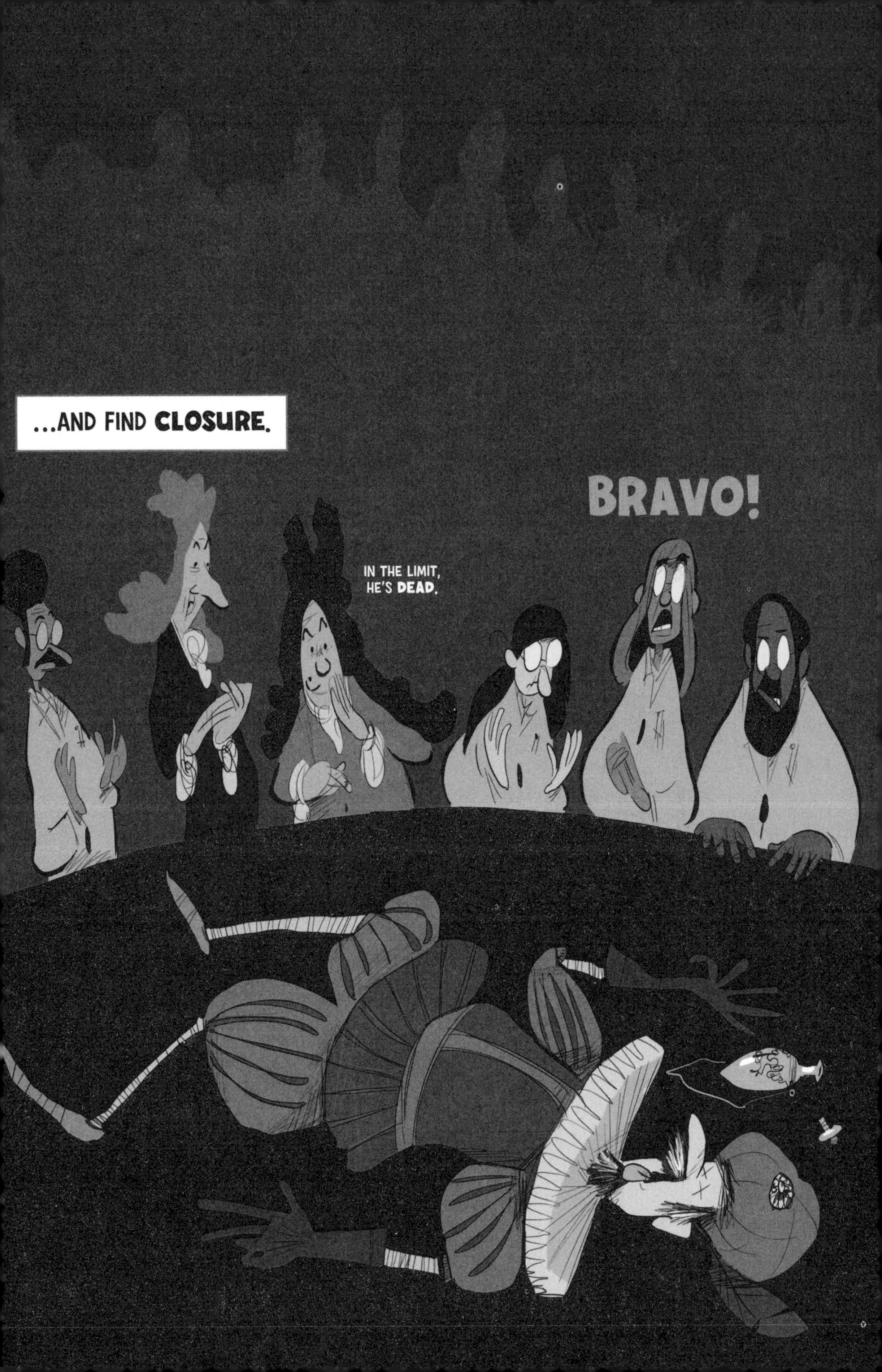
...AND FIND **CLOSURE.**
BRAVO!
IN THE LIMIT, HE'S **DEAD.**

FOR AN **EASY EXAMPLE**, CONSIDER THIS INFINITE SEQUENCE: $\frac{1}{2}, \frac{3}{4}, \frac{7}{8}, \frac{15}{16}, \frac{31}{32}, \frac{63}{64}, \frac{127}{128} \ldots$

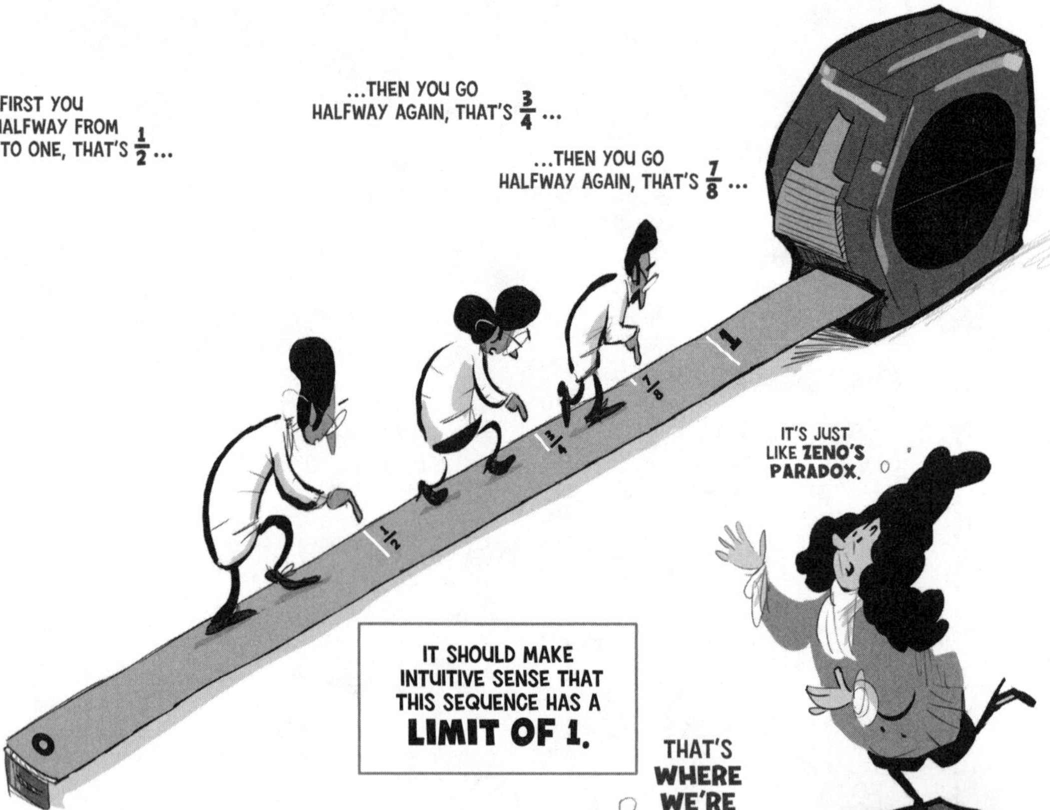

FIRST YOU GO HALFWAY FROM ZERO TO ONE, THAT'S 1/2...
...THEN YOU GO HALFWAY AGAIN, THAT'S 3/4...
...THEN YOU GO HALFWAY AGAIN, THAT'S 7/8...
1/2
3/4
7/8
1
IT'S JUST LIKE ZENO'S PARADOX.

IT SHOULD MAKE INTUITIVE SENSE THAT THIS SEQUENCE HAS A **LIMIT OF 1.**

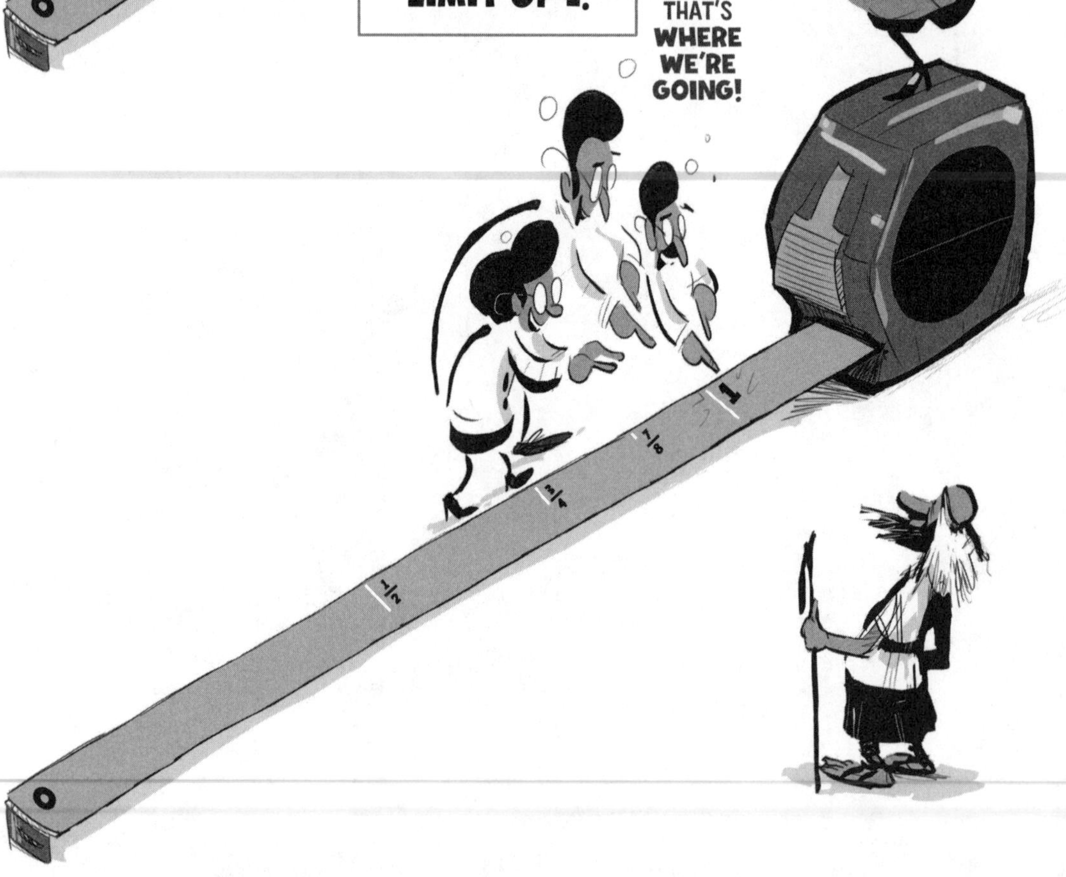

THAT'S WHERE WE'RE GOING!
1/2
3/4
7/8
1

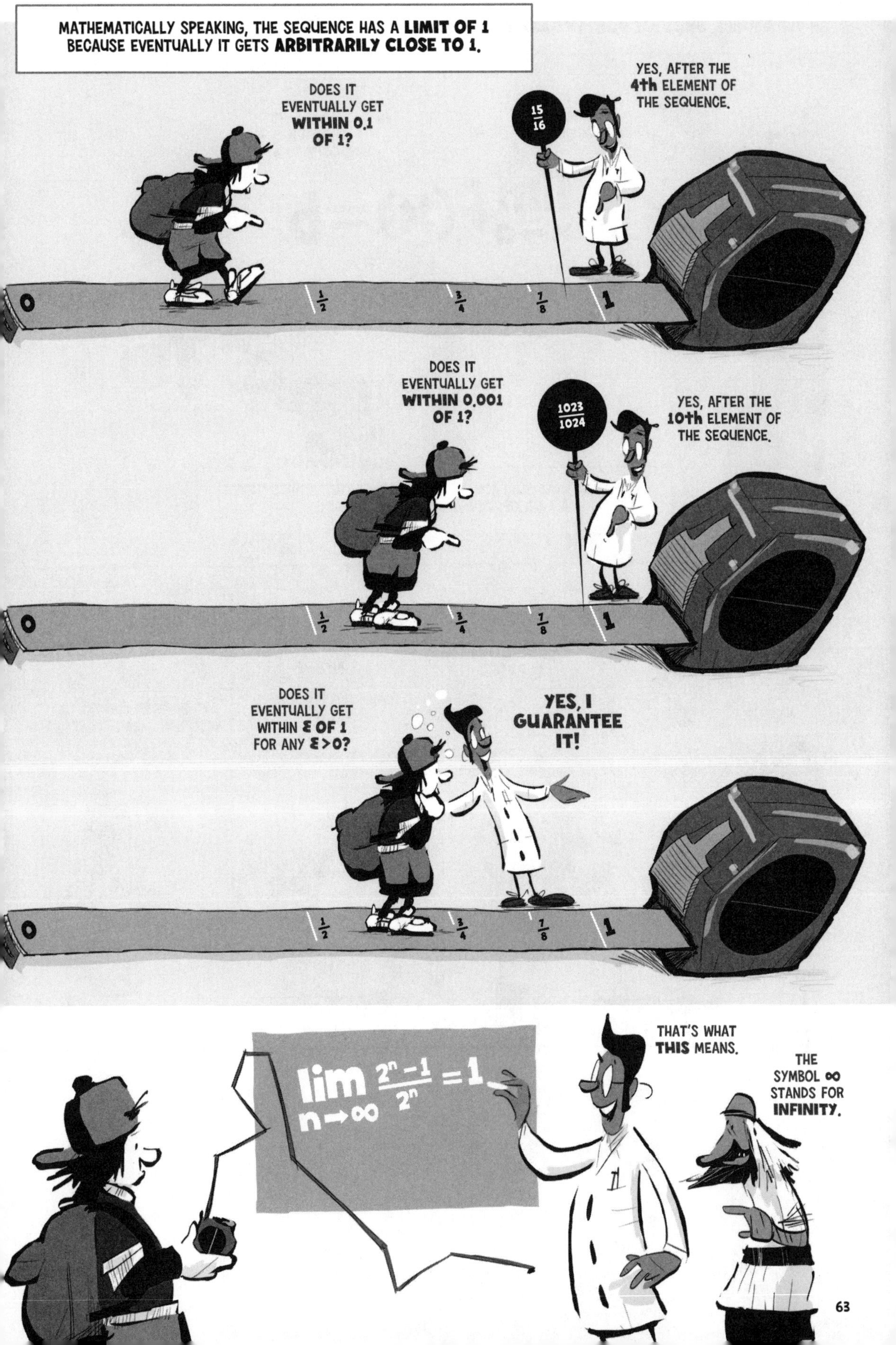
MATHEMATICALLY SPEAKING, THE SEQUENCE HAS A **LIMIT OF 1** BECAUSE EVENTUALLY IT GETS **ARBITRARILY CLOSE TO 1.**
DOES IT EVENTUALLY GET **WITHIN 0.1 OF 1?**
15/16
YES, AFTER THE **4th** ELEMENT OF THE SEQUENCE.
0
1/2
3/4
7/8
1
DOES IT EVENTUALLY GET **WITHIN 0.001 OF 1?**
1023/1024
YES, AFTER THE **10th** ELEMENT OF THE SEQUENCE.
0
1/2
3/4
7/8
1
DOES IT EVENTUALLY GET WITHIN **ε OF 1** FOR ANY **ε > 0?**
YES, I GUARANTEE IT!
0
1/2
3/4
7/8
1
$\lim_{n\to\infty} \frac{2^n-1}{2^n} = 1$
THAT'S WHAT **THIS** MEANS.
THE SYMBOL ∞ STANDS FOR **INFINITY.**

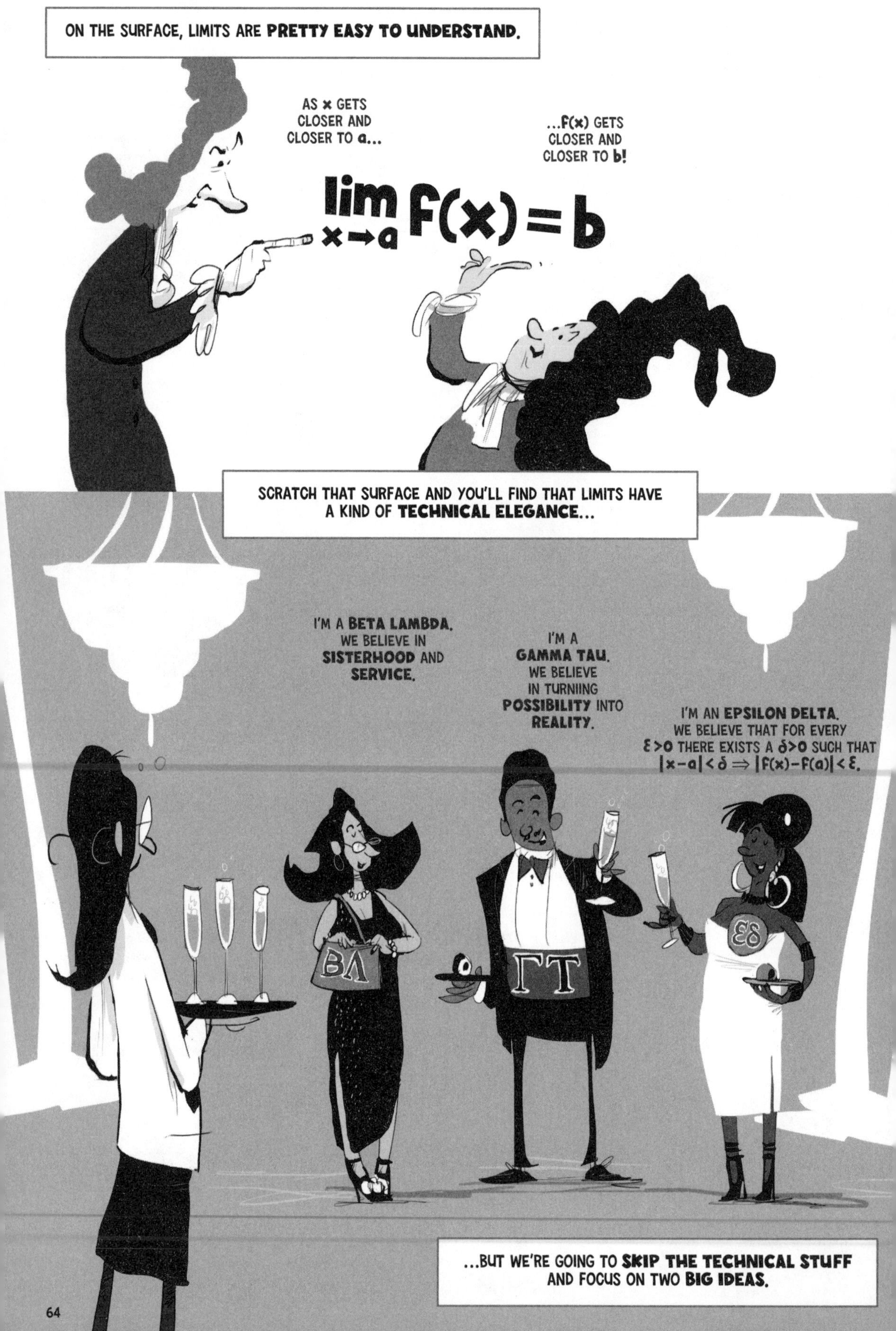
ON THE SURFACE, LIMITS ARE PRETTY EASY TO UNDERSTAND.
AS x GETS CLOSER AND CLOSER TO a...
...f(x) GETS CLOSER AND CLOSER TO b!
$\lim_{x \to a} f(x) = b$
SCRATCH THAT SURFACE AND YOU'LL FIND THAT LIMITS HAVE A KIND OF TECHNICAL ELEGANCE...
I'M A BETA LAMBDA. WE BELIEVE IN SISTERHOOD AND SERVICE.
I'M A GAMMA TAU. WE BELIEVE IN TURNIING POSSIBILITY INTO REALITY.
I'M AN EPSILON DELTA. WE BELIEVE THAT FOR EVERY $\varepsilon > 0$ THERE EXISTS A $\delta > 0$ SUCH THAT $|x-a| < \delta \Rightarrow |f(x)-f(a)| < \varepsilon$.
ΒΛ
ΓΤ
εδ
...BUT WE'RE GOING TO SKIP THE TECHNICAL STUFF AND FOCUS ON TWO BIG IDEAS.

ONE BIG IDEA IS THAT **LIMITS FOLLOW SIMPLE RULES**, JUST LIKE MULTIPLICATION AND OTHER **MATHEMATICAL TOOLS**.

MULTIPLICATION **ALWAYS** WORKS THIS WAY, **NO MATTER WHAT!**

$$c \cdot (a+b) = c \cdot a + c \cdot b$$

FOR EXAMPLE, LIMITS **PASS THROUGH PLUS AND MINUS SIGNS**...

THE **LIMIT OF A SUM**...

...IS THE **SUM OF THE LIMITS**.

$$\lim_{x \to a} [f(x) + g(x)] = \lim_{x \to a} f(x) + \lim_{x \to a} g(x)$$

...AND THEY **PASS THROUGH CONSTANT MULTIPLES**.

MULTIPLYING BY A CONSTANT **INSIDE** THE LIMIT...

...IS THE SAME AS MULTIPLYING BY THAT CONSTANT **OUTSIDE** THE LIMIT.

$$\lim_{x \to a} [c \cdot f(x)] = c \cdot \lim_{x \to a} f(x)$$

WE'LL BE USING **BOTH OF THESE RULES** LATER.

ANOTHER BIG IDEA IS THAT LIMITS MAKE IT POSSIBLE TO DESCRIBE SIMPLE-SOUNDING CONCEPTS...
CONTINUOUS MEANS THAT YOU DON'T HAVE TO LIFT YOUR PENCIL OFF THE PAPER.
THIS FUNCTION IS CONTINUOUS AT POINT a...
...AND EVERYWHERE ELSE.
F(x)
x
a
THIS FUNCTION IS NOT CONTINUOUS AT POINT a...
...BUT IT IS CONTINUOUS EVERYWHERE ELSE.
F(x)
x
a
DIFFERENTIABLE MEANS CONTINUOUS AND SMOOTH.
THIS FUNCTION IS DIFFERENTIABLE EVERYWHERE.
F(x)
x
THIS FUNCTION IS NOT DIFFERENTIABLE AT POINTS a OR b.
F(x)
x
a
b

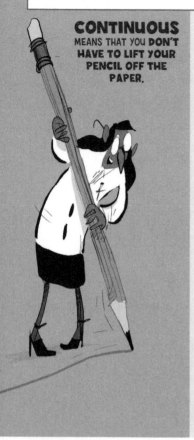

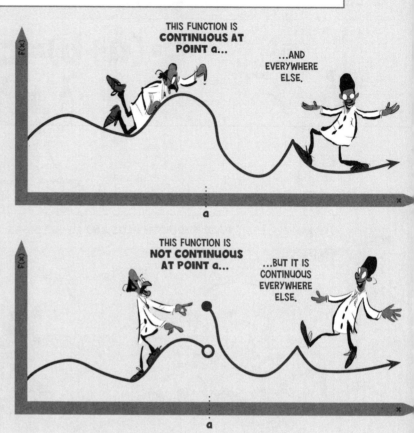

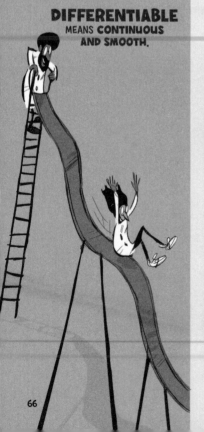

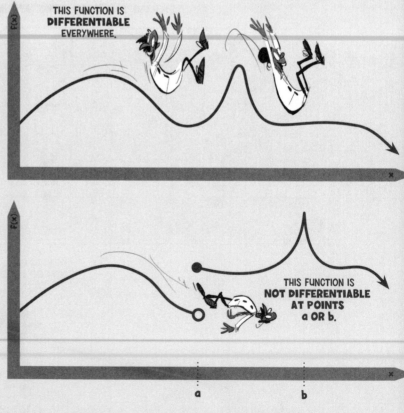

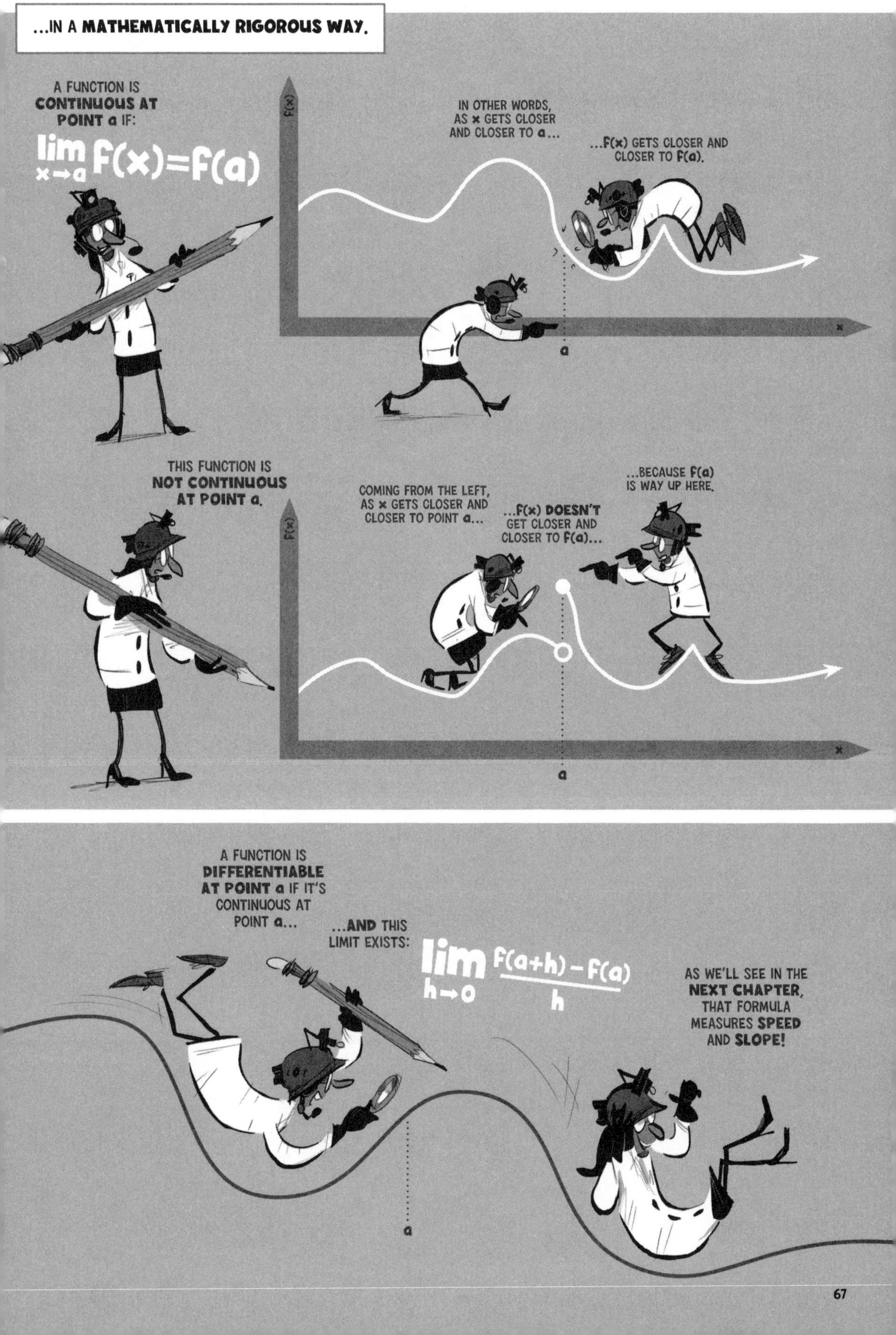
...IN A **MATHEMATICALLY RIGOROUS WAY.**
A FUNCTION IS **CONTINUOUS AT POINT a** IF:
$$\lim_{x \to a} f(x) = f(a)$$
f(x)
IN OTHER WORDS, AS x GETS CLOSER AND CLOSER TO a...
...f(x) GETS CLOSER AND CLOSER TO f(a).
x
a
THIS FUNCTION IS **NOT CONTINUOUS AT POINT a.**
f(x)
COMING FROM THE LEFT, AS x GETS CLOSER AND CLOSER TO POINT a...
...f(x) **DOESN'T** GET CLOSER AND CLOSER TO f(a)...
...BECAUSE f(a) IS WAY UP HERE.
x
a
A FUNCTION IS **DIFFERENTIABLE AT POINT a** IF IT'S CONTINUOUS AT POINT a...
...**AND** THIS LIMIT EXISTS:
$$\lim_{h \to 0} \frac{f(a+h) - f(a)}{h}$$
AS WE'LL SEE IN THE **NEXT CHAPTER,** THAT FORMULA MEASURES **SPEED** AND **SLOPE!**
a

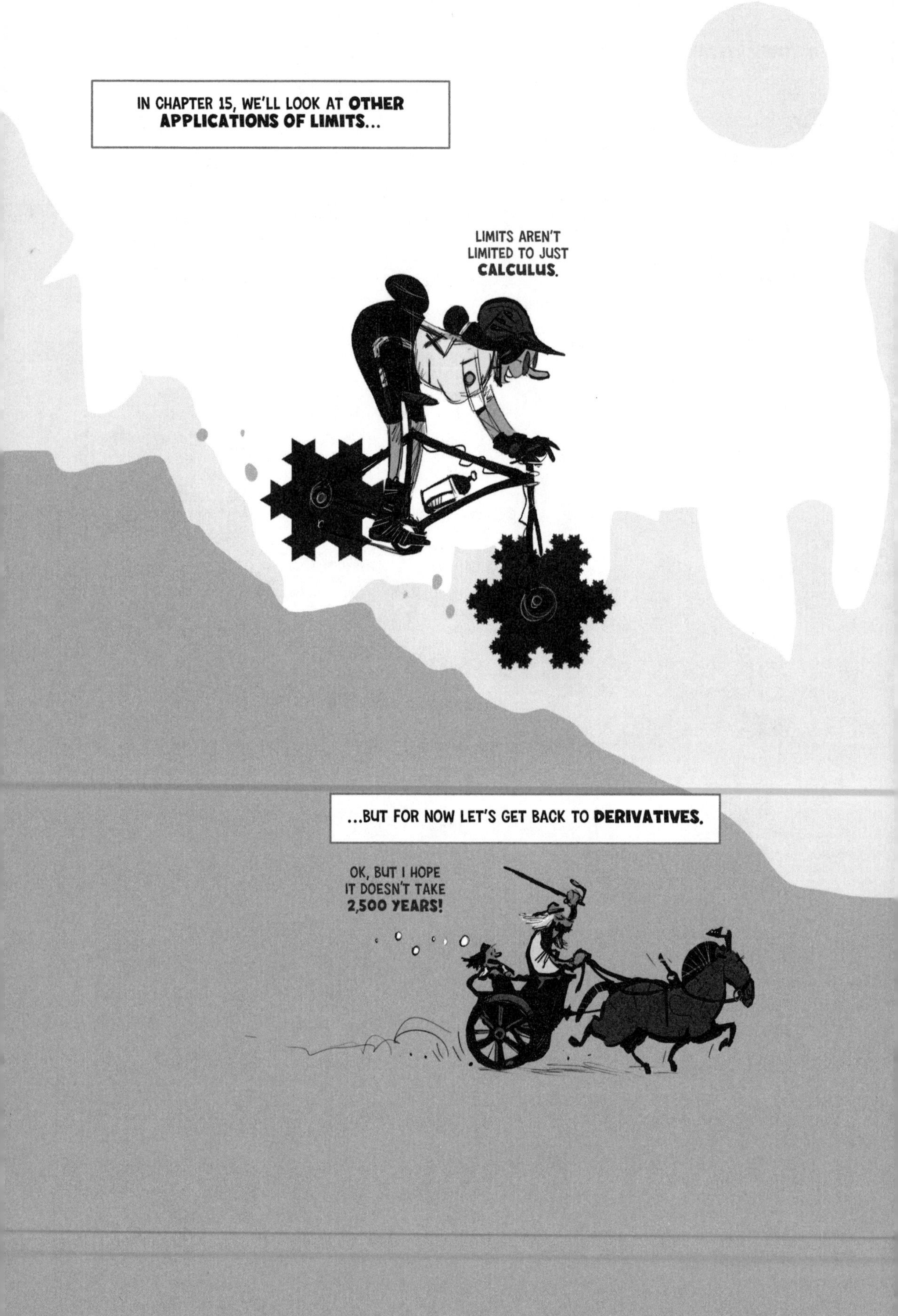
IN CHAPTER 15, WE'LL LOOK AT **OTHER APPLICATIONS OF LIMITS**...
LIMITS AREN'T LIMITED TO JUST **CALCULUS**.
...BUT FOR NOW LET'S GET BACK TO **DERIVATIVES**.
OK, BUT I HOPE IT DOESN'T TAKE **2,500 YEARS!**

PART TWO
DERIVATIVES

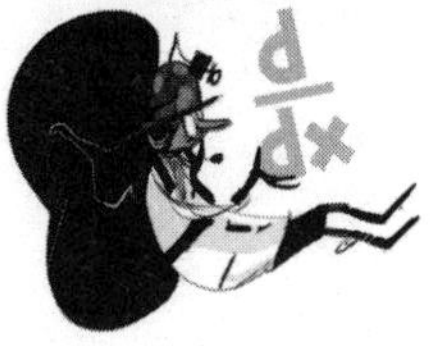
d/dx

CHAPTER 6
LIMITS AND DERIVATIVES

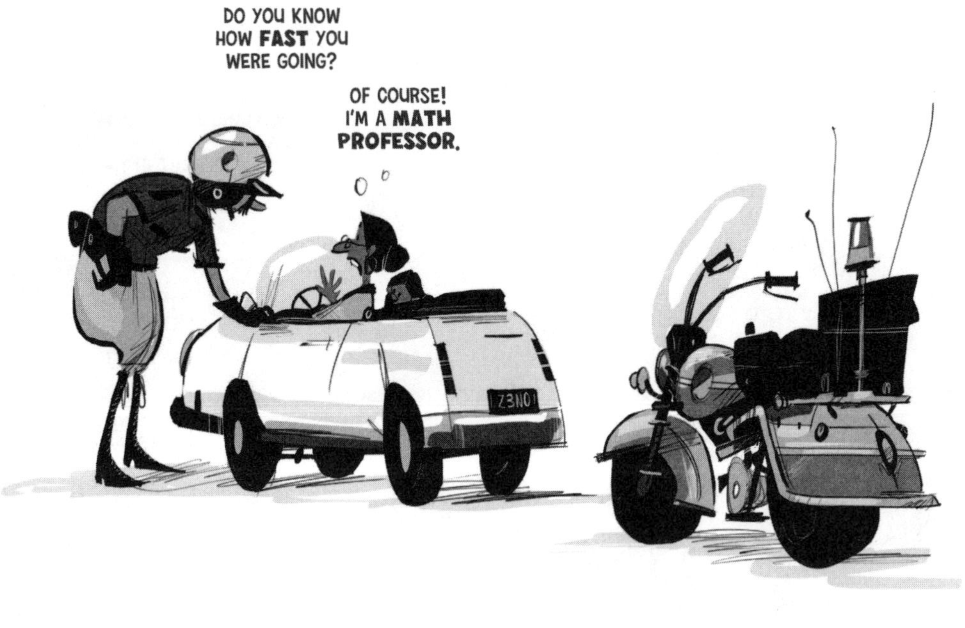
DO YOU KNOW HOW **FAST** YOU WERE GOING?
OF COURSE! I'M A **MATH PROFESSOR**.
Z3N0

IN CHAPTER 2, WE ENCOUNTERED THE PROBLEM OF **DEFINING SPEED...**

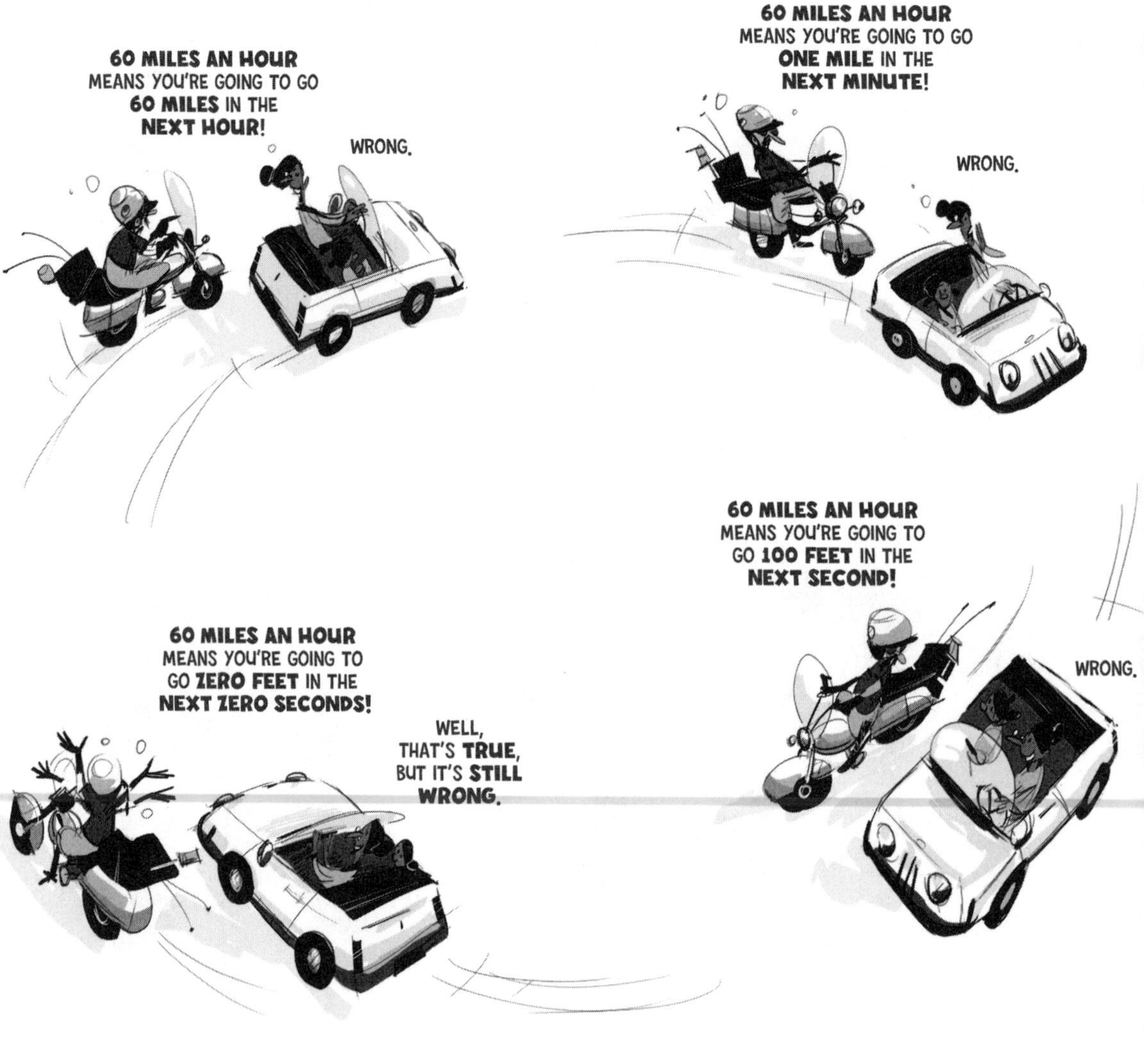
60 MILES AN HOUR MEANS YOU'RE GOING TO GO 60 MILES IN THE NEXT HOUR!
WRONG.
60 MILES AN HOUR MEANS YOU'RE GOING TO GO ONE MILE IN THE NEXT MINUTE!
WRONG.
60 MILES AN HOUR MEANS YOU'RE GOING TO GO 100 FEET IN THE NEXT SECOND!
WRONG.
60 MILES AN HOUR MEANS YOU'RE GOING TO GO ZERO FEET IN THE NEXT ZERO SECONDS!
WELL, THAT'S TRUE, BUT IT'S STILL WRONG.

...OR, AS A PHYSICIST WOULD PUT IT, THE PROBLEM OF DEFINING **VELOCITY**.

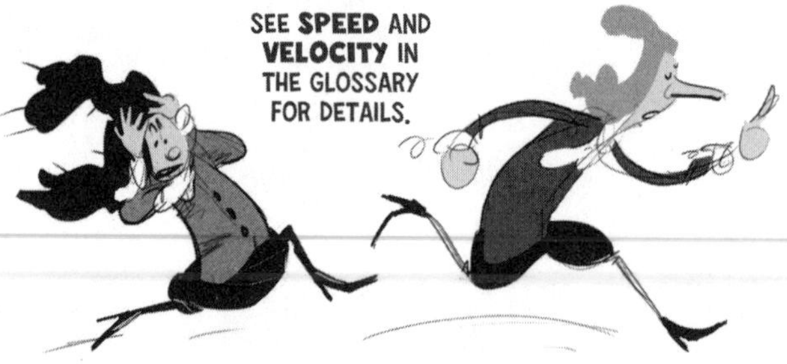
SEE SPEED AND VELOCITY IN THE GLOSSARY FOR DETAILS.

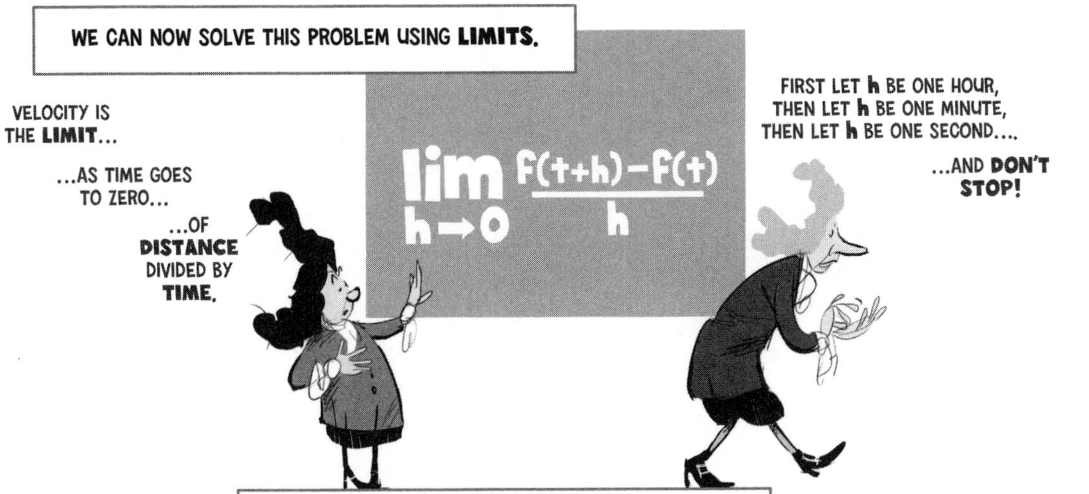
WE CAN NOW SOLVE THIS PROBLEM USING **LIMITS.**
VELOCITY IS THE **LIMIT**...
...AS TIME GOES TO ZERO...
...OF **DISTANCE** DIVIDED BY **TIME.**
$\lim_{h \to 0} \frac{f(t+h)-f(t)}{h}$
FIRST LET **h** BE ONE HOUR, THEN LET **h** BE ONE MINUTE, THEN LET **h** BE ONE SECOND....
...AND **DON'T STOP!**

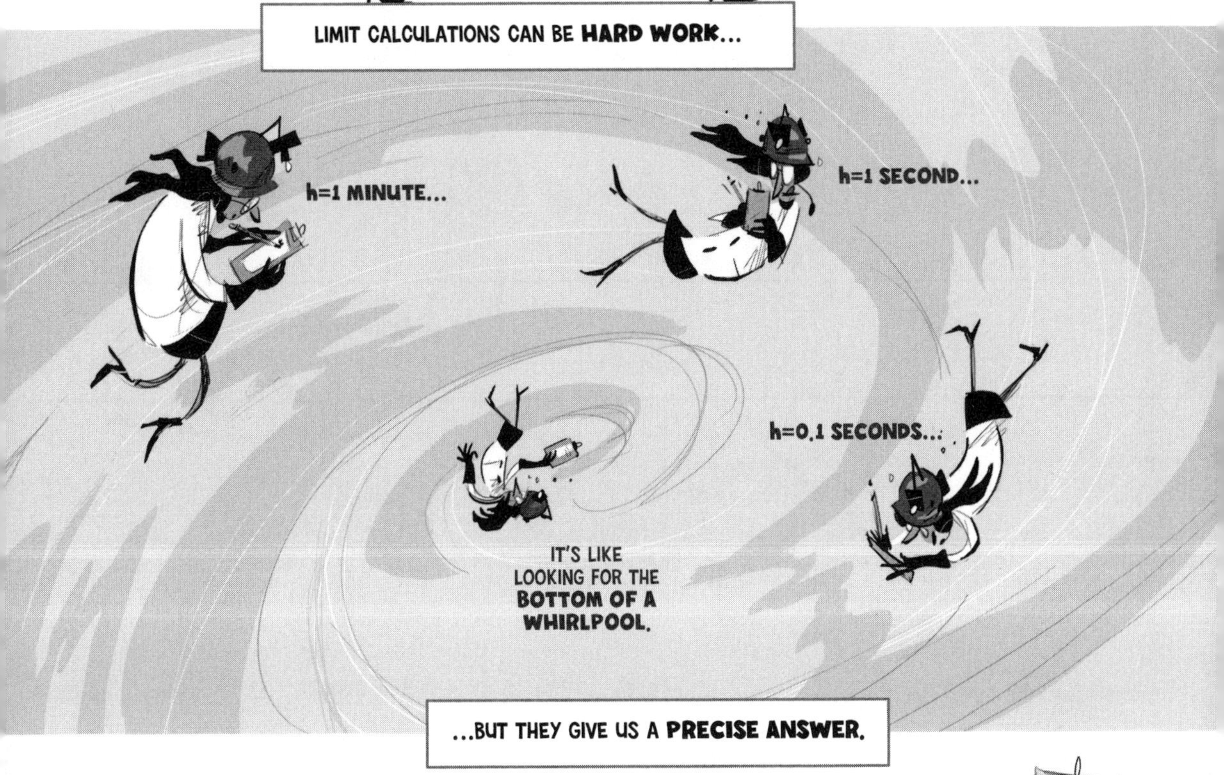
LIMIT CALCULATIONS CAN BE **HARD WORK**...
h=1 MINUTE...
h=1 SECOND...
h=0.1 SECONDS...
IT'S LIKE LOOKING FOR THE **BOTTOM OF A WHIRLPOOL.**

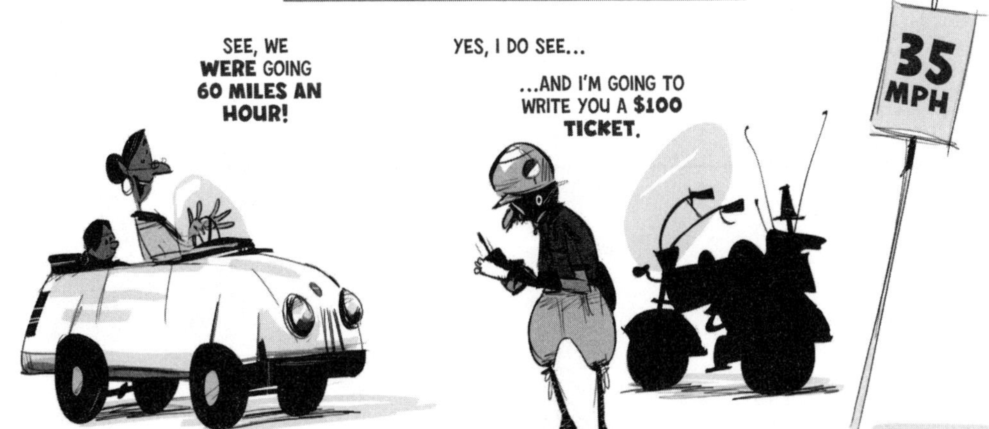
...BUT THEY GIVE US A **PRECISE ANSWER.**
SEE, WE **WERE** GOING **60 MILES AN HOUR!**
YES, I DO SEE...
...AND I'M GOING TO WRITE YOU A **$100 TICKET.**
35 MPH

IN CHAPTER 2, WE **ALSO** ENCOUNTERED THE PROBLEM OF **CALCULATING THE SLOPE OF A TANGENT LINE.**

f(x)
WHAT'S THE SLOPE **HERE?**
OR **HERE?**
OR **HERE?**
x

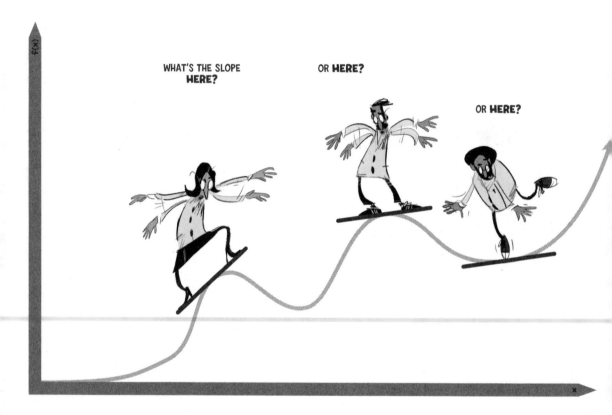

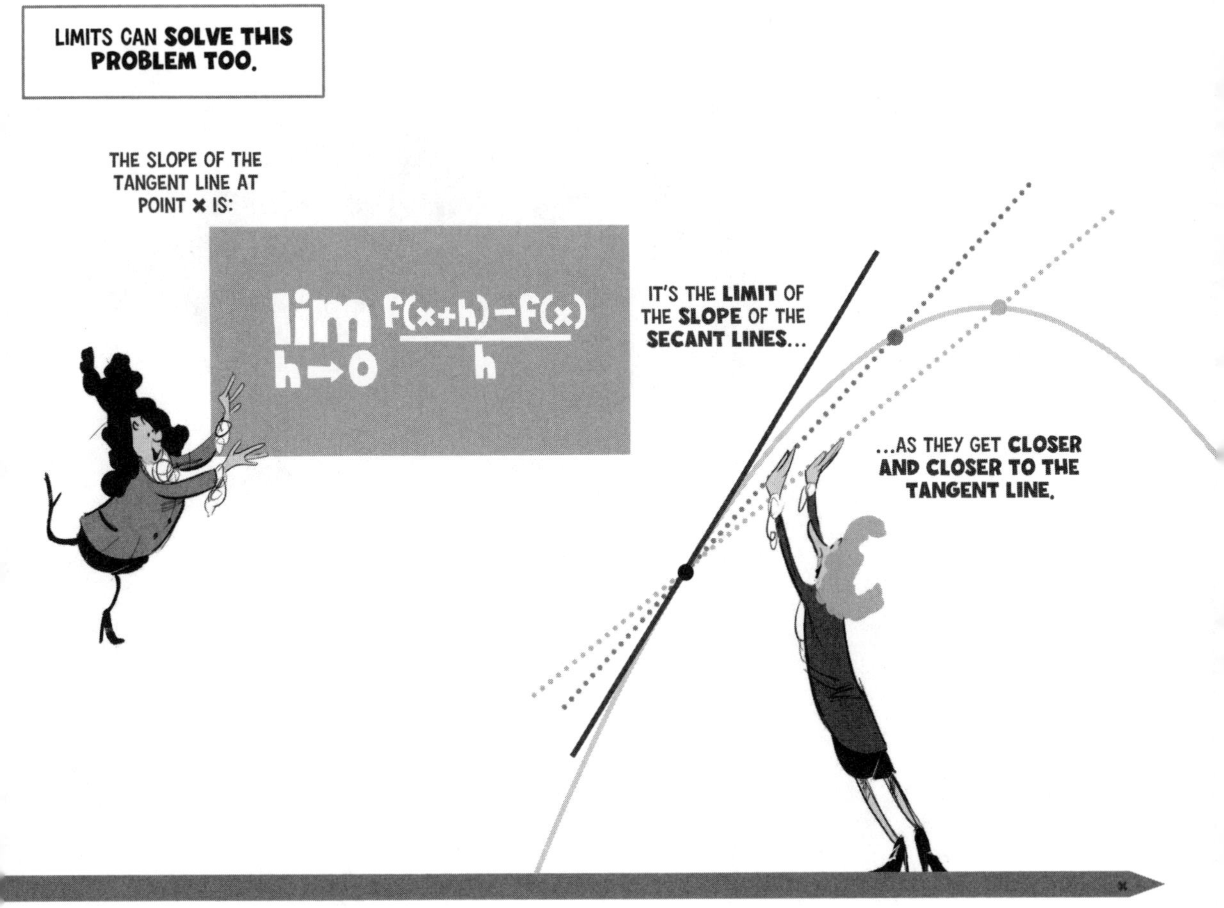
LIMITS CAN **SOLVE THIS PROBLEM TOO.**
THE SLOPE OF THE TANGENT LINE AT POINT x IS:
$\lim_{h \to 0} \frac{f(x+h)-f(x)}{h}$
IT'S THE **LIMIT** OF THE **SLOPE** OF THE **SECANT LINES...**
...AS THEY GET **CLOSER AND CLOSER TO THE TANGENT LINE.**

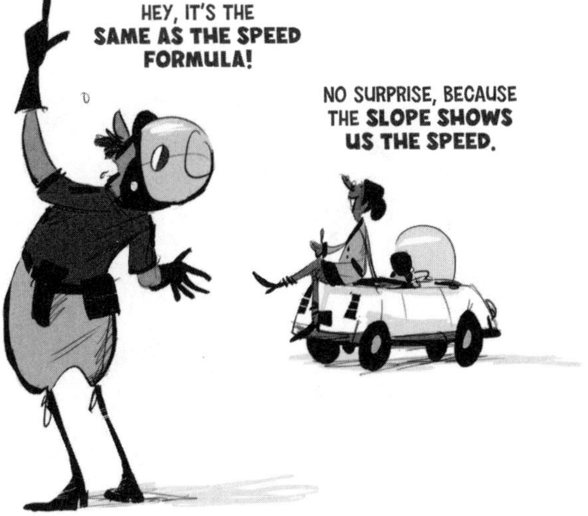
HEY, IT'S THE **SAME AS THE SPEED FORMULA!**
NO SURPRISE, BECAUSE THE **SLOPE SHOWS US THE SPEED.**

THIS **BASIC FORMULA** IS CALLED THE **DERIVATIVE.**

THE **DERIVATIVE** OF F(x) IS:

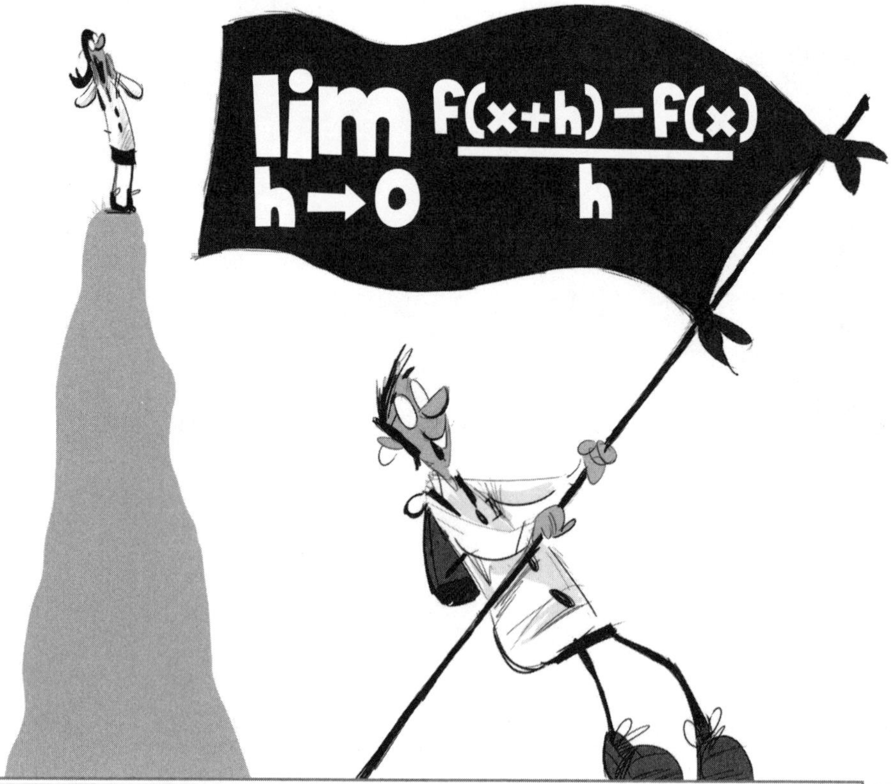

$$\lim_{h \to 0} \frac{f(x+h) - f(x)}{h}$$

IT'S THE **MOST IMPORTANT IDEA** IN THIS BOOK.

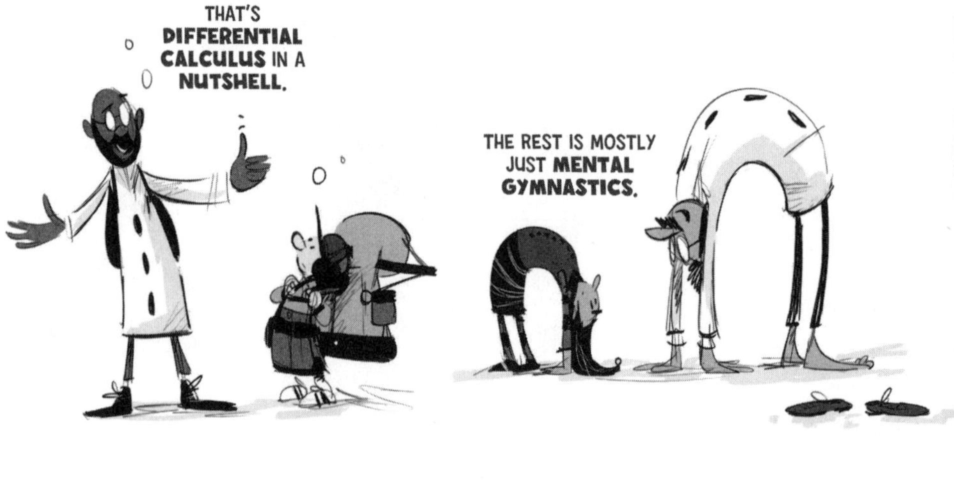

THAT'S **DIFFERENTIAL CALCULUS** IN A **NUTSHELL.**
THE REST IS MOSTLY JUST **MENTAL GYMNASTICS.**

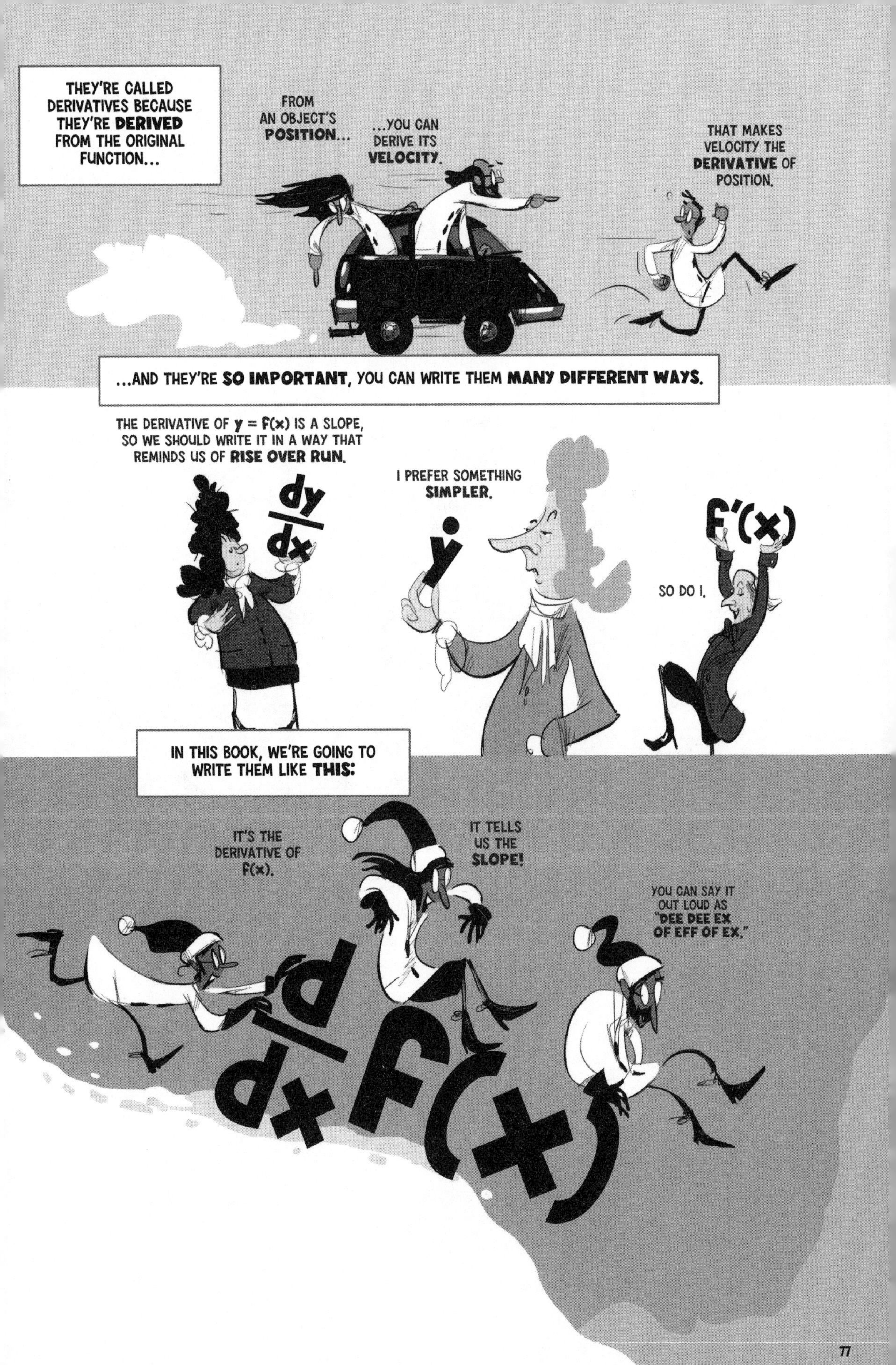
THEY'RE CALLED DERIVATIVES BECAUSE THEY'RE DERIVED FROM THE ORIGINAL FUNCTION...
FROM AN OBJECT'S POSITION...
...YOU CAN DERIVE ITS VELOCITY.
THAT MAKES VELOCITY THE DERIVATIVE OF POSITION.
...AND THEY'RE SO IMPORTANT, YOU CAN WRITE THEM MANY DIFFERENT WAYS.
THE DERIVATIVE OF y = f(x) IS A SLOPE, SO WE SHOULD WRITE IT IN A WAY THAT REMINDS US OF RISE OVER RUN.
dy/dx
I PREFER SOMETHING SIMPLER.
ẏ
f′(x)
SO DO I.
IN THIS BOOK, WE'RE GOING TO WRITE THEM LIKE THIS:
IT'S THE DERIVATIVE OF f(x).
IT TELLS US THE SLOPE!
YOU CAN SAY IT OUT LOUD AS "DEE DEE EX OF EFF OF EX."
d/dx f(x)

FOR A SIMPLE EXAMPLE, LET'S CALCULATE THE DERIVATIVE OF $f(x) = c$.

WE CALL THIS A **CONSTANT FUNCTION**, SINCE ITS VALUE DOESN'T CHANGE.

WE CAN USE THE **BASIC FORMULA**...

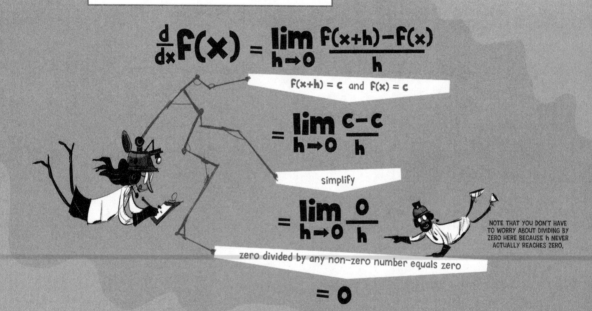

$$\frac{d}{dx} f(x) = \lim_{h \to 0} \frac{f(x+h) - f(x)}{h}$$

$f(x+h) = c$ and $f(x) = c$

$$= \lim_{h \to 0} \frac{c - c}{h}$$

simplify

$$= \lim_{h \to 0} \frac{0}{h}$$

NOTE THAT YOU DON'T HAVE TO WORRY ABOUT DIVIDING BY ZERO HERE BECAUSE h NEVER ACTUALLY REACHES ZERO.

zero divided by any non-zero number equals zero

$$= 0$$

...TO SEE THAT THE DERIVATIVE EQUALS **ZERO** FOR EVERY **x**.

THE **SLOPE** IS **ZERO**.

OUR **POSITION ISN'T CHANGING**, SO OUR VELOCITY IS **ZERO MILES PER HOUR**.

NEXT, LET'S CALCULATE THE DERIVATIVE OF $f(x) = x$.

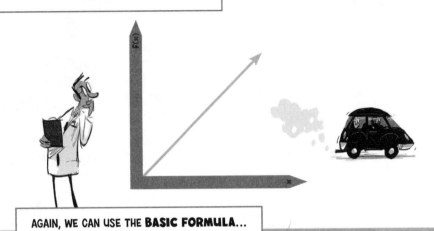

AGAIN, WE CAN USE THE **BASIC FORMULA**...

$$\frac{d}{dx}f(x) = \lim_{h \to 0} \frac{f(x+h) - f(x)}{h}$$

$f(x+h) = x+h$ and $f(x) = x$

$$= \lim_{h \to 0} \frac{(x+h) - x}{h}$$

simplify

$$= \lim_{h \to 0} \frac{h}{h}$$

any non-zero number divided by itself equals one

$$= 1$$

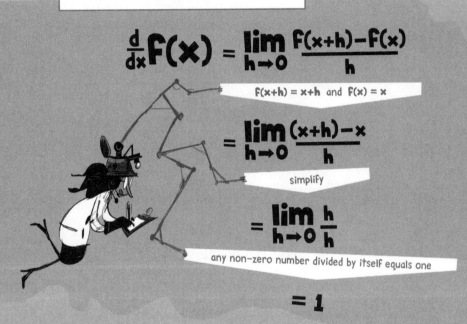

...TO SEE THAT THE DERIVATIVE EQUALS **ONE** FOR EVERY x.

THE **SLOPE** IS **ONE**!

WE'RE GOING **ONE MILE PER HOUR**.

WHEE!!

THERE'S ALSO SOME NICE **INTUITION** BEHIND THE FORMULA FOR DERIVATIVES.

$$\frac{d}{dx}f(x) = \lim_{h \to 0} \frac{f(x+h) - f(x)}{h}$$

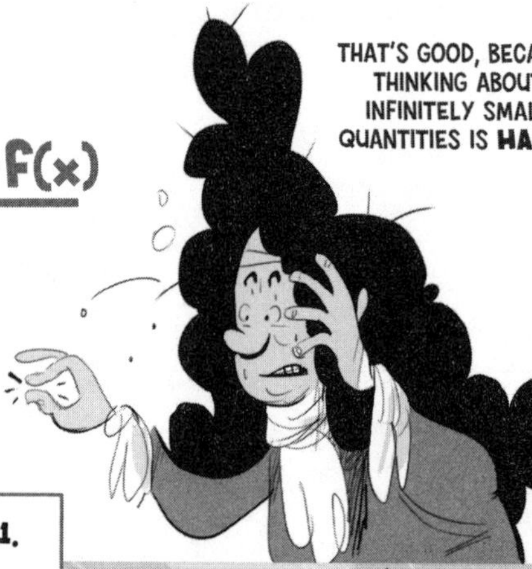
THAT'S GOOD, BECAUSE THINKING ABOUT INFINITELY SMALL QUANTITIES IS **HARD!**

AS WE'RE MOVING TOWARD **h = 0**, LET'S STOP AT **h = 1**.

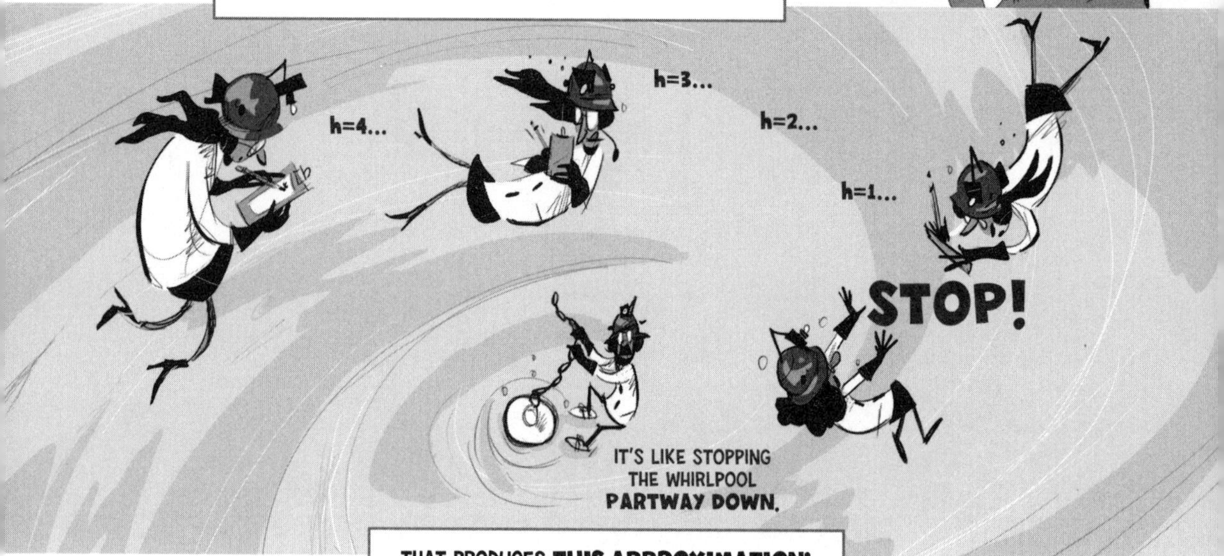
h=4...
h=3...
h=2...
h=1...
STOP!
IT'S LIKE STOPPING THE WHIRLPOOL **PARTWAY DOWN.**

THAT PRODUCES **THIS APPROXIMATION:**

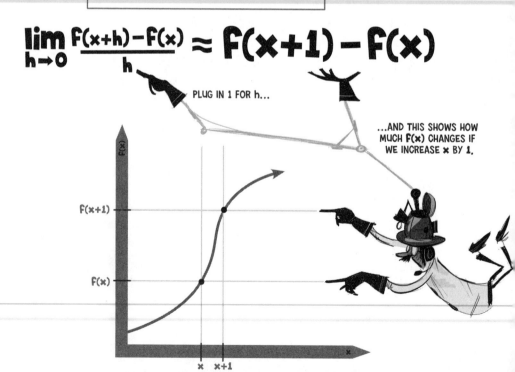

$$\lim_{h \to 0} \frac{f(x+h) - f(x)}{h} \approx f(x+1) - f(x)$$

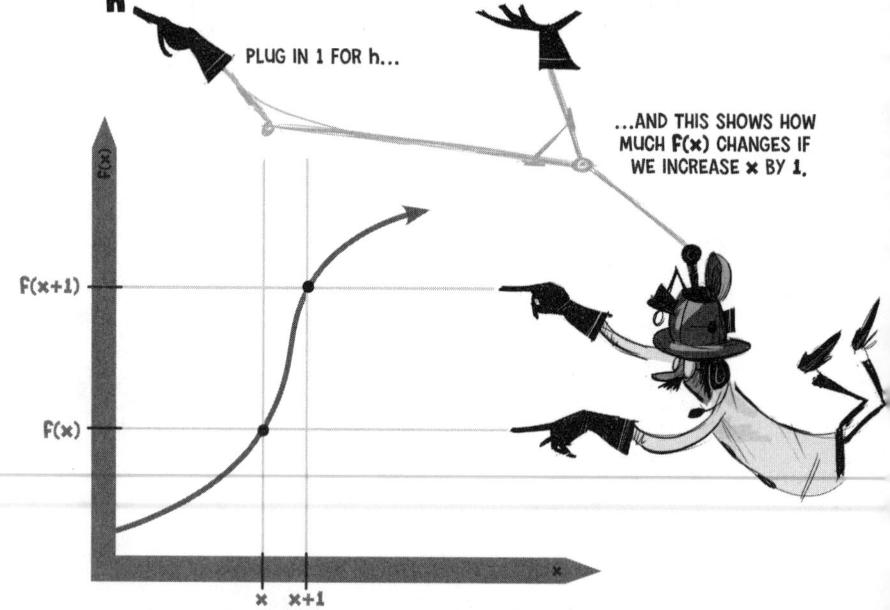
PLUG IN 1 FOR h...
...AND THIS SHOWS HOW MUCH f(x) CHANGES IF WE INCREASE x BY 1.
f(x)
f(x+1)
f(x)
x
x
x+1

IN **PHYSICS**, FOR EXAMPLE, IF **f(t)** IS THE HEIGHT AFTER **t** SECONDS OF AN APPLE THROWN STRAIGHT UPWARD...

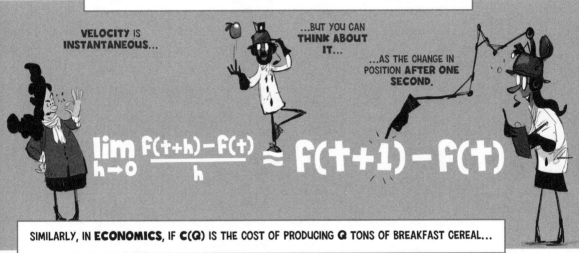

...WE CAN USE **h=1** TO GET AN **INTUITIVE SENSE** OF THE APPLE'S VELOCITY.

VELOCITY IS **INSTANTANEOUS**...

...BUT YOU CAN **THINK ABOUT IT**...

...AS THE CHANGE IN POSITION **AFTER ONE SECOND**.

$$\lim_{h \to 0} \frac{f(t+h) - f(t)}{h} \approx f(t+1) - f(t)$$

SIMILARLY, IN **ECONOMICS**, IF **C(Q)** IS THE COST OF PRODUCING **Q** TONS OF BREAKFAST CEREAL...

Klog's
Integrity
O's!

...WE CAN USE **h=1** TO GET AN **INTUITIVE SENSE** OF HOW MUCH IT WILL COST TO **PRODUCE A LITTLE BIT MORE CEREAL**.

AS WE'LL SEE IN CHAPTER 10, **MARGINAL COST** IS MEASURED WITH **INFINITESIMALS**...

...BUT YOU CAN **THINK ABOUT IT**...

...AS THE COST OF PRODUCING **ONE MORE TON OF CEREAL**.

$$\lim_{h \to 0} \frac{C(Q+h) - C(Q)}{h} \approx C(Q+1) - C(Q)$$

APPROXIMATIONS AND **INTUITION** ARE ALL WELL AND GOOD...

THEY HELP US **THINK** ABOUT CALCULUS.

...BUT AT THE END OF THE DAY, WE NEED TO DO SOME **ACTUAL CALCULATIONS.**

WE ALREADY CALCULATED THE DERIVATIVE OF $f(x) = c$...

...AND $f(x) = x$.

SO OUR NEXT TASK IS TO WRESTLE WITH SOME DERIVATIVES THAT ARE **MORE COMPLICATED.**

LIKE **ACTUALLY CALCULATING** THE VELOCITY OF AN APPLE BEING THROWN STRAIGHT UP INTO THE AIR...

...WHILE BEING PULLED DOWN BY **GRAVITY.**

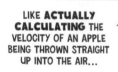

IF THIS FUNCTION DESCRIBES ITS POSITION...

$$f(t) = 2 + 19.6t - 4.9t^2$$

...WHAT FUNCTION DESCRIBES ITS **VELOCITY?**

$$\frac{d}{dt}f(t) = ?$$

CHAPTER 7
THE CALCULUS TOOLKIT

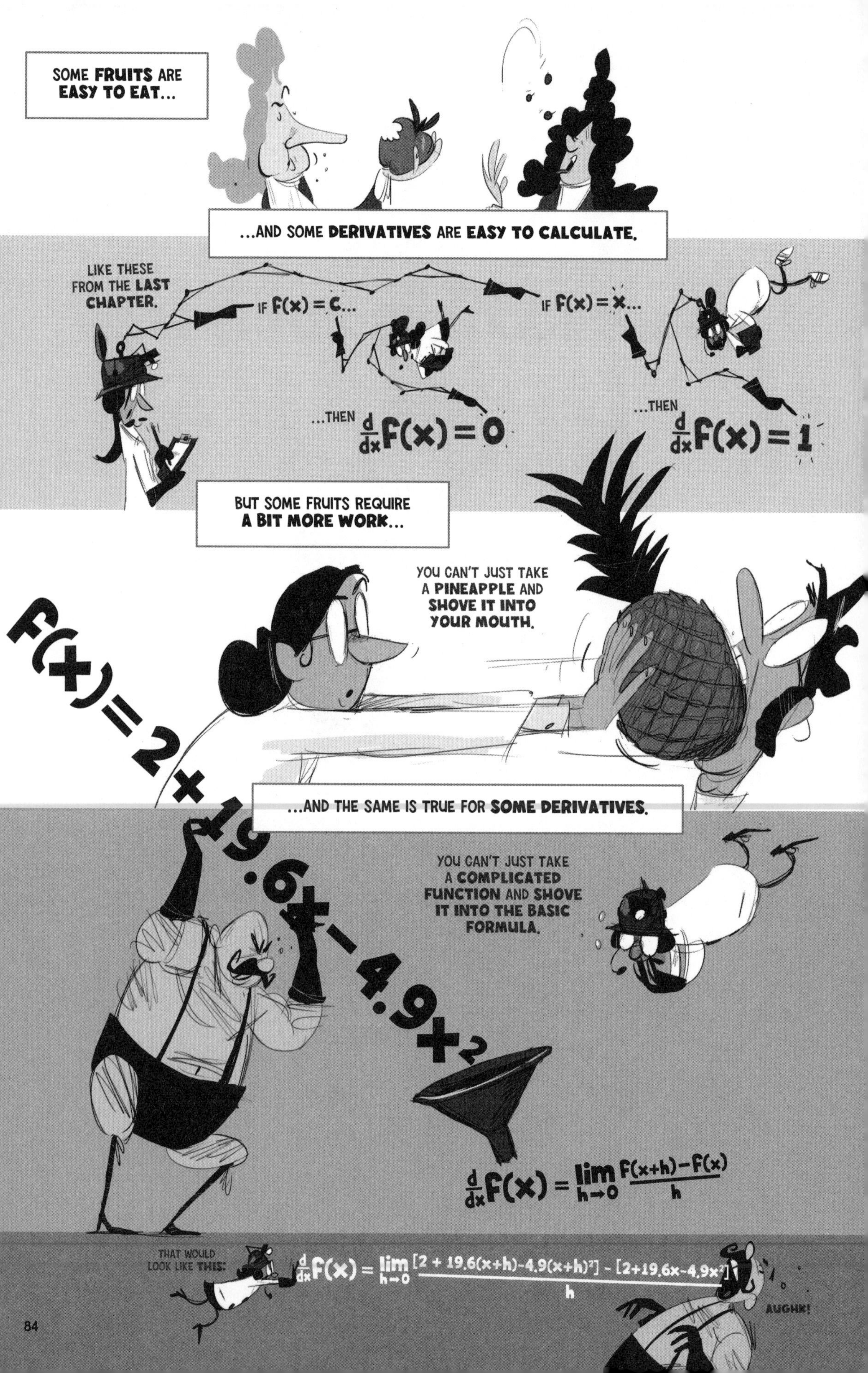

SOME **FRUITS** ARE **EASY TO EAT...**
...AND SOME **DERIVATIVES** ARE **EASY TO CALCULATE.**
LIKE THESE FROM THE **LAST CHAPTER.**
IF $f(x) = c$...
...THEN $\frac{d}{dx}f(x) = 0$
IF $f(x) = x$...
...THEN $\frac{d}{dx}f(x) = 1$
BUT SOME FRUITS REQUIRE **A BIT MORE WORK...**
YOU CAN'T JUST TAKE A **PINEAPPLE** AND **SHOVE IT INTO YOUR MOUTH.**
$f(x) = 2 + 19.6x - 4.9x^2$
...AND THE SAME IS TRUE FOR **SOME DERIVATIVES.**
YOU CAN'T JUST TAKE A **COMPLICATED FUNCTION** AND **SHOVE IT INTO THE BASIC FORMULA.**
$\frac{d}{dx}f(x) = \lim_{h \to 0} \frac{f(x+h) - f(x)}{h}$
THAT WOULD LOOK LIKE **THIS:**
$\frac{d}{dx}f(x) = \lim_{h \to 0} \frac{[2 + 19.6(x+h) - 4.9(x+h)^2] - [2 + 19.6x - 4.9x^2]}{h}$
AUGHK!

WHAT YOU NEED ARE THE RIGHT **TOOLS.**

IN THIS CHAPTER, WE'RE GOING TO LEARN TO USE TOOLS LIKE THE **SUM RULE**...

$$\frac{d}{dx}\left[\;\rule{0.6cm}{0cm}+\rule{0.9cm}{0cm}\;\right] = \frac{d}{dx}\rule{0.6cm}{0cm} + \frac{d}{dx}\rule{0.9cm}{0cm}$$

...AND THE **CONSTANT MULTIPLE RULE**...

$$\frac{d}{dx}\left[c \cdot \rule{0.8cm}{0cm}\right] = c \cdot \frac{d}{dx}\rule{0.8cm}{0cm}$$

...TO TURN **COMPLICATED DERIVATIVES** INTO **FRUIT SALAD.**

$$F(t) = 2 + 19.6t - 4.9t^2$$

BY THE END OF THIS CHAPTER, WE'LL BE ABLE TO CALCULATE THE VELOCITY OF A **FLYING APPLE**...

...AND MUCH, MUCH MORE.

YOU CAN FIND ALL THE TOOLS TOGETHER IN THE GLOSSARY.

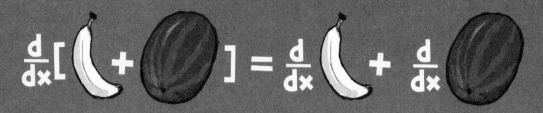

LET'S START WITH THE SUM RULE...
$\frac{d}{dx}$[banana + watermelon] = $\frac{d}{dx}$ banana + $\frac{d}{dx}$ watermelon
...WHICH WE USE WHEN WE'RE ADDING FUNCTIONS.
$\frac{d}{dx}[f(x)+g(x)] = \frac{d}{dx}f(x)+\frac{d}{dx}g(x)$
THE DERIVATIVE OF A SUM...
...EQUALS THE SUM OF THE DERIVATIVES!
YOU CAN GET THE INTUITION BY THINKING ABOUT POSITION AND VELOCITY.
IF I'M ON A FLYING BANANA GOING 10 mph...
$\frac{d}{dx}g(x)=5$
...AND I THROW A WATERMELON FORWARD AT 5 mph...
$\frac{d}{dx}f(x)=10$
...THEN THE WATERMELON WILL BE TRAVELING AT 15 mph.
$\frac{d}{dx}[f(x)+g(x)]=15$

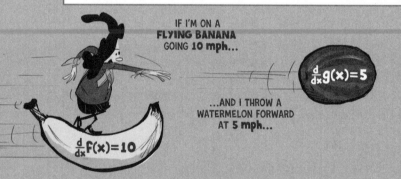

THE SUM RULE ALSO MAKES SENSE IF YOU THINK ABOUT **SLOPES**...

THE SLOPE OF
f(x)+g(x)...

...EQUALS THE
SLOPE OF f(x)...

...PLUS THE SLOPE
OF g(x).

...AND HERE'S HOW WE CAN **PROVE IT USING THE BASIC FORMULA.**

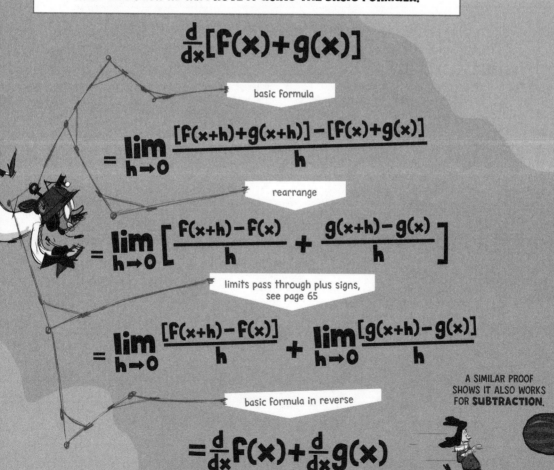

$$\frac{d}{dx}[f(x)+g(x)]$$

basic formula

$$= \lim_{h \to 0} \frac{[f(x+h)+g(x+h)]-[f(x)+g(x)]}{h}$$

rearrange

$$= \lim_{h \to 0} \left[\frac{f(x+h)-f(x)}{h} + \frac{g(x+h)-g(x)}{h} \right]$$

limits pass through plus signs,
see page 65

$$= \lim_{h \to 0} \frac{[f(x+h)-f(x)]}{h} + \lim_{h \to 0} \frac{[g(x+h)-g(x)]}{h}$$

A SIMILAR PROOF
SHOWS IT ALSO WORKS
FOR **SUBTRACTION.**

basic formula in reverse

$$= \frac{d}{dx}f(x) + \frac{d}{dx}g(x)$$

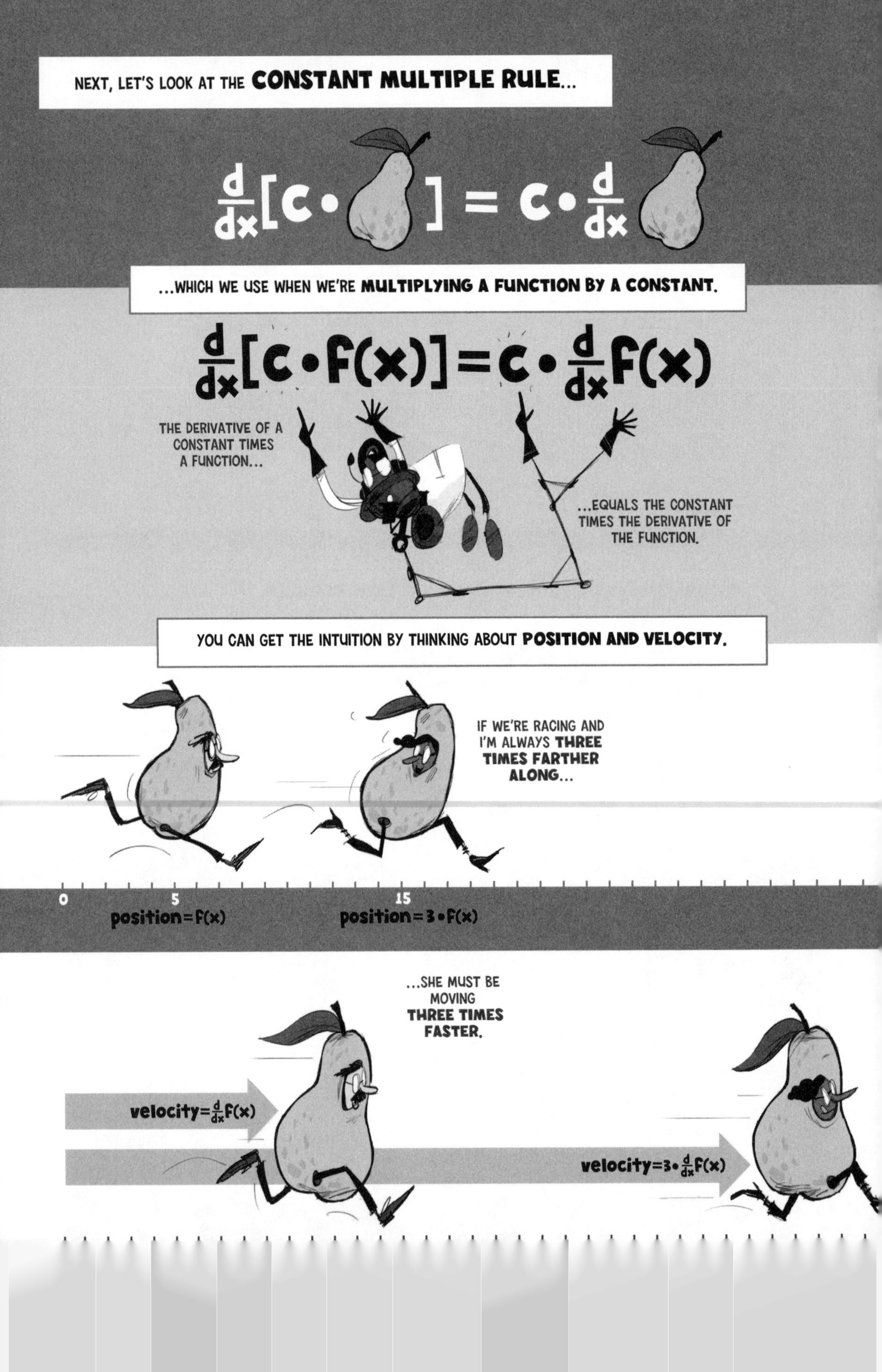
NEXT, LET'S LOOK AT THE **CONSTANT MULTIPLE RULE**...
$\frac{d}{dx}[c \cdot \quad] = c \cdot \frac{d}{dx}$
...WHICH WE USE WHEN WE'RE **MULTIPLYING A FUNCTION BY A CONSTANT.**
$\frac{d}{dx}[c \cdot f(x)] = c \cdot \frac{d}{dx}f(x)$
THE DERIVATIVE OF A CONSTANT TIMES A FUNCTION...
...EQUALS THE CONSTANT TIMES THE DERIVATIVE OF THE FUNCTION.
YOU CAN GET THE INTUITION BY THINKING ABOUT **POSITION AND VELOCITY.**
IF WE'RE RACING AND I'M ALWAYS **THREE TIMES FARTHER ALONG**...
0
5
15
position=f(x)
position=3•f(x)
...SHE MUST BE MOVING **THREE TIMES FASTER.**
velocity=$\frac{d}{dx}$f(x)
velocity=3•$\frac{d}{dx}$f(x)

THE CONSTANT MULTIPLE RULE ALSO MAKES SENSE IF YOU THINK ABOUT **SLOPES...**

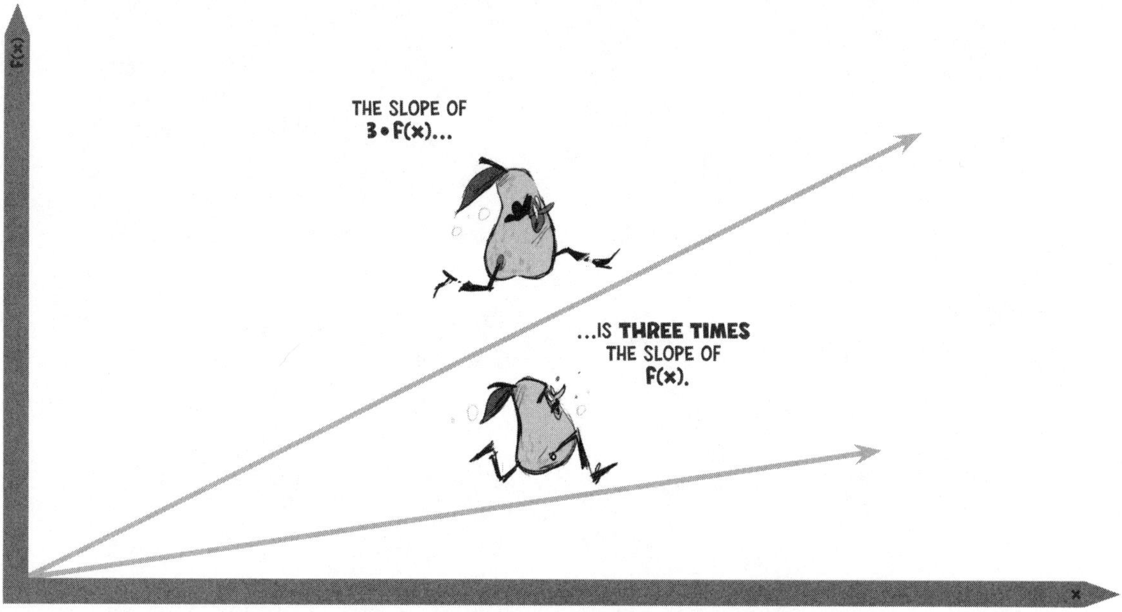
f(x)
THE SLOPE OF
3 • f(x)...
...IS **THREE TIMES**
THE SLOPE OF
f(x).
x

...AND HERE'S HOW WE CAN **PROVE IT USING THE BASIC FORMULA.**

$$\frac{d}{dx}[c \cdot f(x)]$$

basic formula

$$= \lim_{h \to 0} \frac{[c \cdot f(x+h)] - [c \cdot f(x)]}{h}$$

factor out the c

$$= \lim_{h \to 0} c \cdot \frac{[f(x+h) - f(x)]}{h}$$

limits pass through constant multiples, see page 65

$$= c \cdot \lim_{h \to 0} \frac{[f(x+h) - f(x)]}{h}$$

basic formula in reverse

$$= c \cdot \frac{d}{dx} f(x)$$

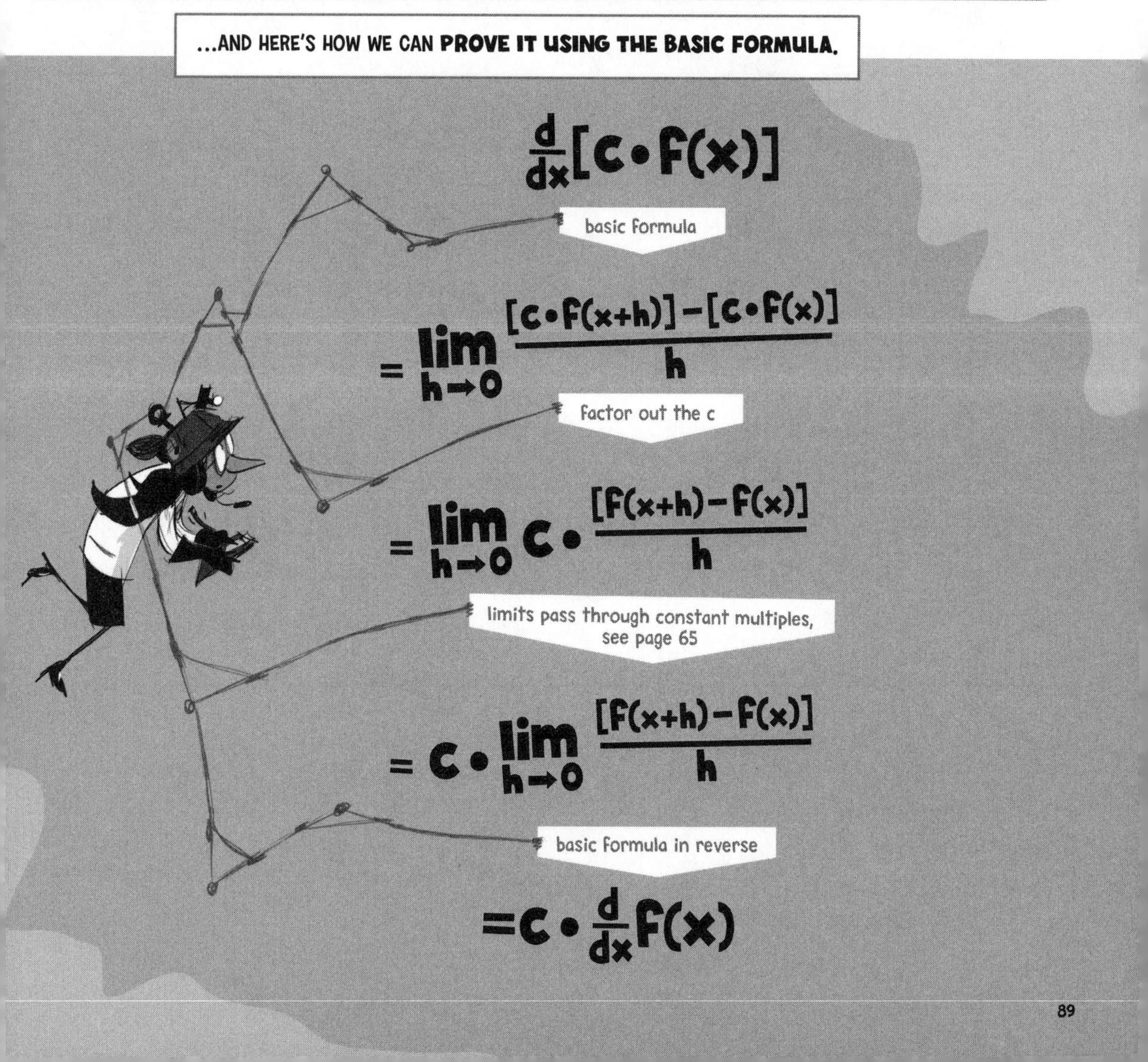

SOMEWHAT MORE COMPLICATED IS THE **PRODUCT RULE**...

...WHICH WE USE WHEN WE'RE **MULTIPLYING TWO FUNCTIONS.**

$$\frac{d}{dx}[f(x) \cdot g(x)] = f(x) \cdot \frac{d}{dx}g(x) + g(x) \cdot \frac{d}{dx}f(x)$$

IT'S ALSO CALLED **LEIBNIZ'S RULE!**

THE BEST **INTUITION** HERE DOESN'T COME FROM POSITION, VELOCITY, OR GRAPHS...

...BUT FROM THINKING ABOUT **AREA.**

—SIGH—

IN PARTICULAR, CONSIDER A RECTANGLE WITH AREA **f(x) · g(x).**

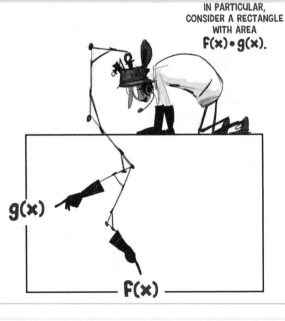

g(x)

f(x)

AS x CHANGES, SO DOES THE AREA OF THE RECTANGLE...
g(x)
f(x)
...AND THE CHANGE IN AREA DEPENDS ON THE **CHANGE IN HEIGHT...**
...AND THE **CHANGE IN WIDTH.**
$\frac{d}{dx}g(x)$ TELLS US HOW FAST THE HEIGHT IS CHANGING...
...SO $f(x)\cdot\frac{d}{dx}g(x)$ TELLS US HOW FAST **THIS** AREA IS CHANGING.
$\frac{d}{dx}g(x)$
g(x)
f(x)
$\frac{d}{dx}f(x)$ TELLS US HOW FAST THE WIDTH IS CHANGING...
...SO $g(x)\cdot\frac{d}{dx}f(x)$ TELLS US HOW FAST **THIS** AREA IS CHANGING.
g(x)
$\frac{d}{dx}f(x)$
f(x)
PUT THEM TOGETHER AND YOU GET THE INTUITION BEHIND THE **PRODUCT RULE.**
THE OVERALL RATE OF CHANGE OF THE AREA, $\frac{d}{dx}[f(x)\cdot g(x)]$...
...CAN BE SPLIT INTO A CHUNK FOR THE HEIGHT, $f(x)\cdot\frac{d}{dx}g(x)$...
...PLUS A CHUNK FOR THE WIDTH, $g(x)\cdot\frac{d}{dx}f(x)$.
=
+

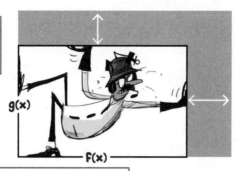

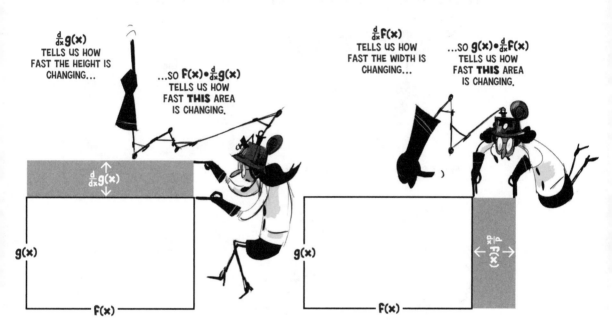

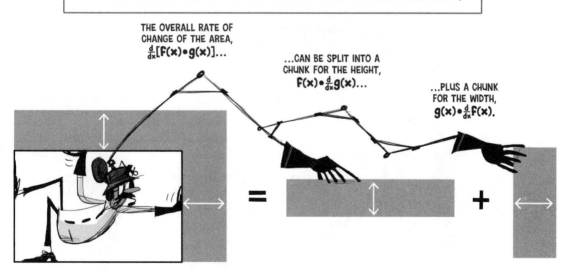

THERE ARE **LOTS** OF OTHER RULES...

CHAIN RULE...

QUOTIENT RULE...

POWER RULE...

...BUT AS A FINAL EXAMPLE, LET'S PROVE THE **POWER RULE** FOR POSITIVE INTEGERS.

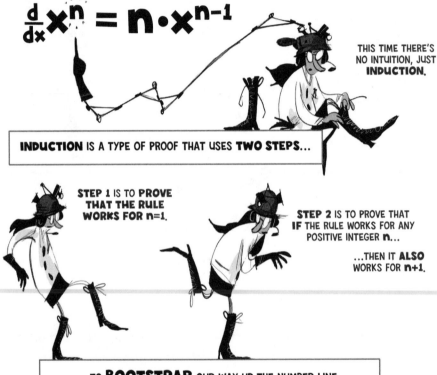

$$\frac{d}{dx} x^n = n \cdot x^{n-1}$$

THIS TIME THERE'S NO INTUITION, JUST **INDUCTION**.

INDUCTION IS A TYPE OF PROOF THAT USES **TWO STEPS**...

STEP 1 IS TO **PROVE THAT THE RULE WORKS FOR n=1.**

STEP 2 IS TO PROVE THAT **IF** THE RULE WORKS FOR ANY POSITIVE INTEGER **n**...

...THEN IT **ALSO** WORKS FOR **n+1**.

...TO **BOOTSTRAP** OUR WAY UP THE NUMBER LINE.

WE KNOW FROM STEP 1 THAT IT WORKS FOR **n=1**...

...SO WE CAN USE STEP 2 OVER AND OVER TO SHOW THAT IT WORKS FOR **n=2**...

...FOR **n=3**...

...AND FOR **ALL OTHER POSITIVE INTEGERS.**

1   2   3   4   5

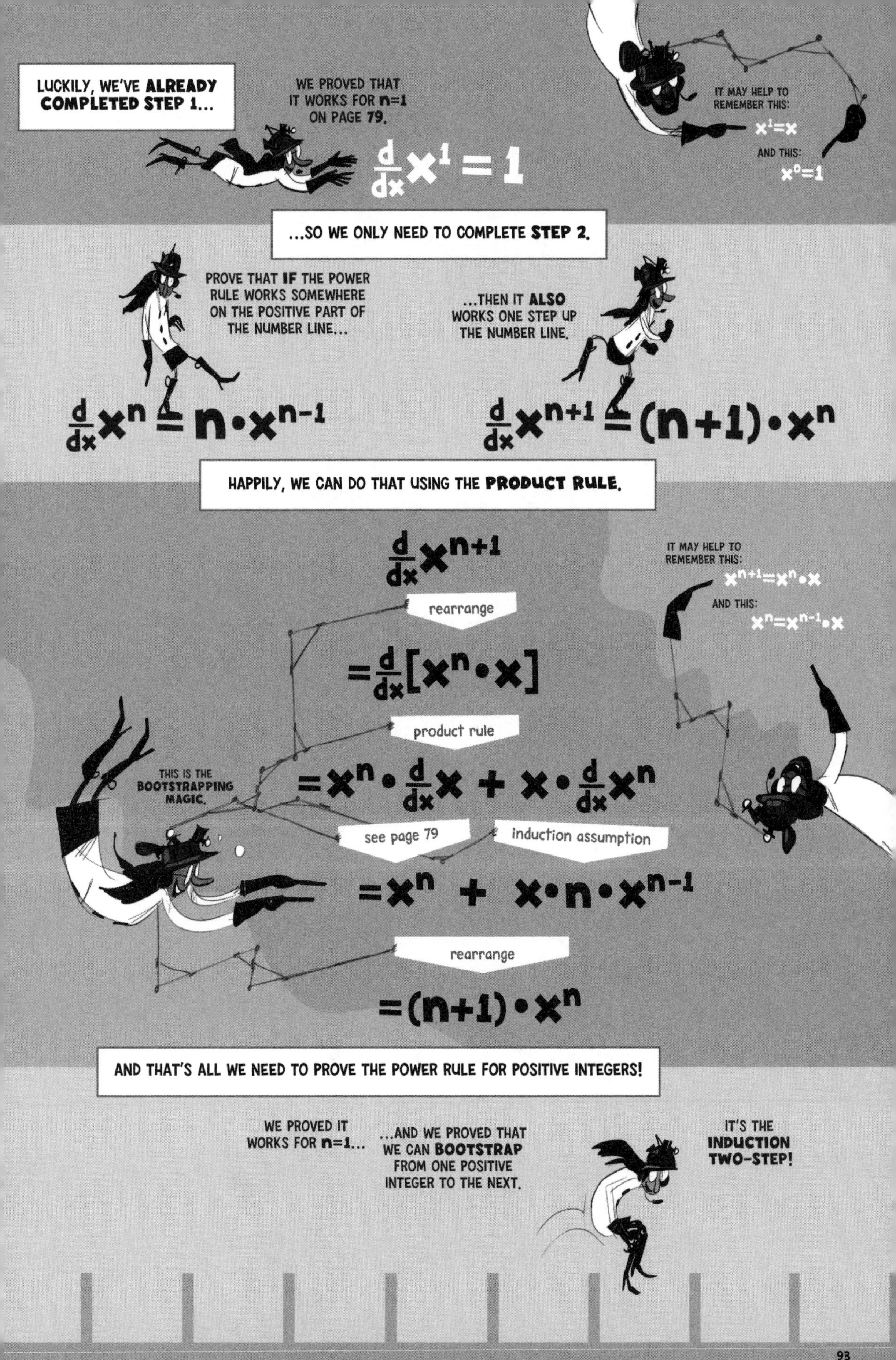

LUCKILY, WE'VE **ALREADY COMPLETED STEP 1...**

WE PROVED THAT IT WORKS FOR **n=1** ON PAGE **79**.

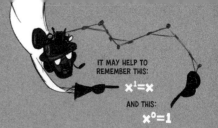

IT MAY HELP TO REMEMBER THIS:
$$x^1 = x$$
AND THIS:
$$x^0 = 1$$

$$\frac{d}{dx}x^1 = 1$$

...SO WE ONLY NEED TO COMPLETE **STEP 2**.

PROVE THAT **IF** THE POWER RULE WORKS SOMEWHERE ON THE POSITIVE PART OF THE NUMBER LINE...

...THEN IT **ALSO** WORKS ONE STEP UP THE NUMBER LINE.

$$\frac{d}{dx}x^n \overset{?}{=} n \cdot x^{n-1}$$

$$\frac{d}{dx}x^{n+1} \overset{?}{=} (n+1) \cdot x^n$$

HAPPILY, WE CAN DO THAT USING THE **PRODUCT RULE**.

$$\frac{d}{dx}x^{n+1}$$

rearrange

IT MAY HELP TO REMEMBER THIS:
$$x^{n+1} = x^n \cdot x$$
AND THIS:
$$x^n = x^{n-1} \cdot x$$

$$= \frac{d}{dx}[x^n \cdot x]$$

product rule

THIS IS THE **BOOTSTRAPPING MAGIC.**

$$= x^n \cdot \frac{d}{dx}x + x \cdot \frac{d}{dx}x^n$$

see page 79   induction assumption

$$= x^n + x \cdot n \cdot x^{n-1}$$

rearrange

$$= (n+1) \cdot x^n$$

AND THAT'S ALL WE NEED TO PROVE THE POWER RULE FOR POSITIVE INTEGERS!

WE PROVED IT WORKS FOR **n=1**...

...AND WE PROVED THAT WE CAN **BOOTSTRAP** FROM ONE POSITIVE INTEGER TO THE NEXT.

IT'S THE **INDUCTION TWO-STEP!**

THANKS TO THE **CALCULUS TOOLKIT**, WE CAN TAKE **COMPLICATED DERIVATIVES**...

LIKE THE VELOCITY OF A **FLYING APPLE**.

...AND TURN THEM INTO **APPLESAUCE**.

$$\frac{d}{dt}[2 + 19.6t - 4.9t^2]$$

sum rule

$$= \frac{d}{dt}2 + \frac{d}{dt}19.6t - \frac{d}{dt}4.9t^2$$

see page 78　　constant multiple rule

$$= 0 + 19.6\frac{d}{dt}t - 4.9\frac{d}{dt}t^2$$

see page 79　　power rule

$$= 19.6 - 4.9 \cdot 2t$$

simplify

$$= 19.6 - 9.8t$$

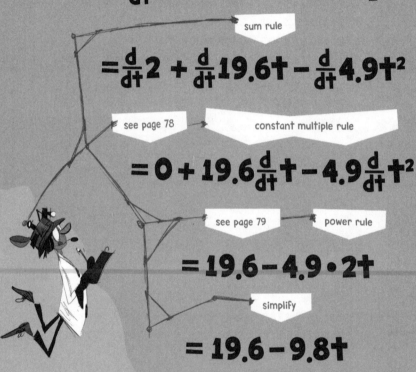

$$\frac{d}{dt}f(t) = 19.6 - 9.8t$$

THAT'S THE VELOCITY OF THE APPLE AT TIME t!

CHAPTER 8
EXTREME VALUES

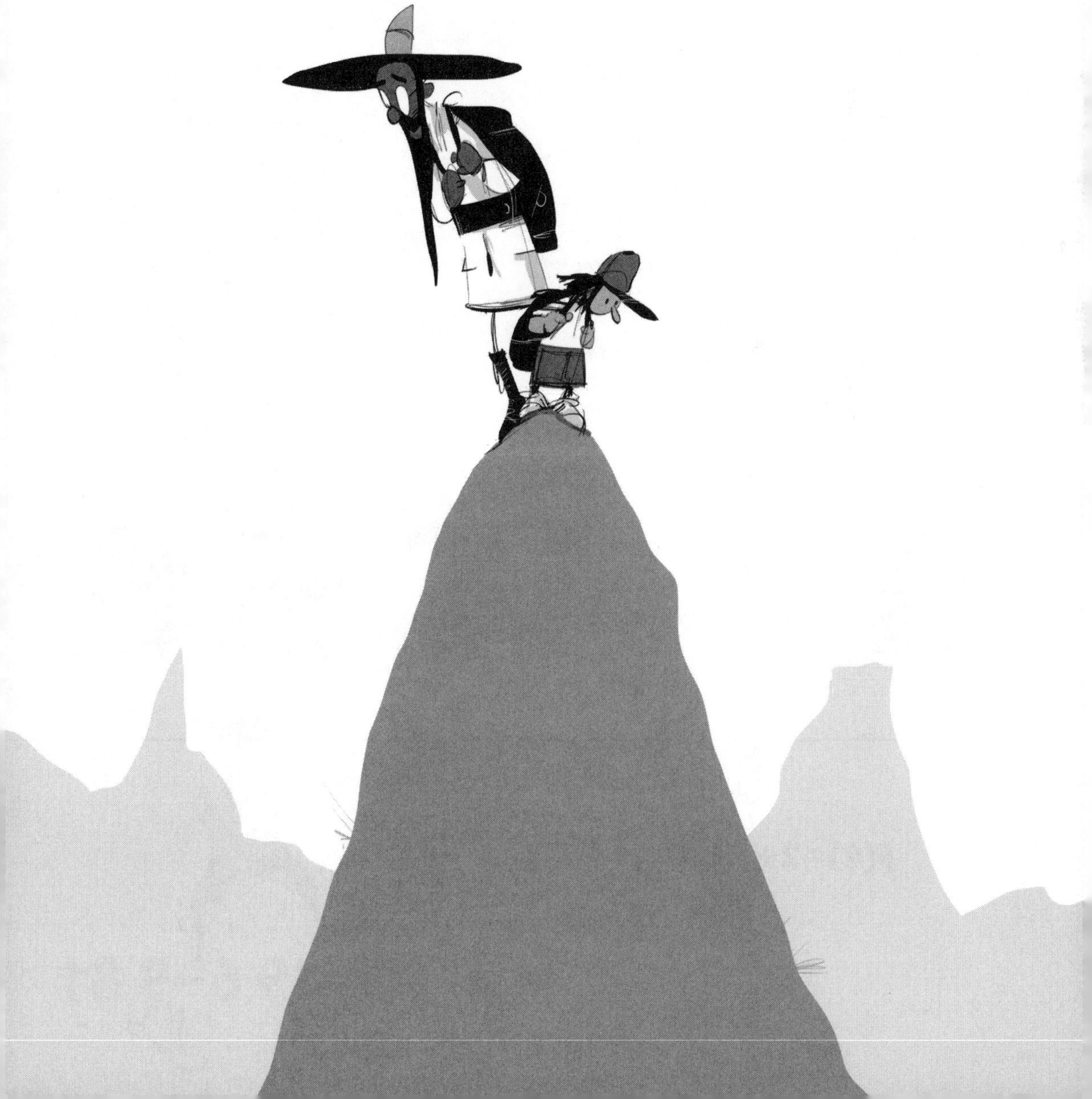

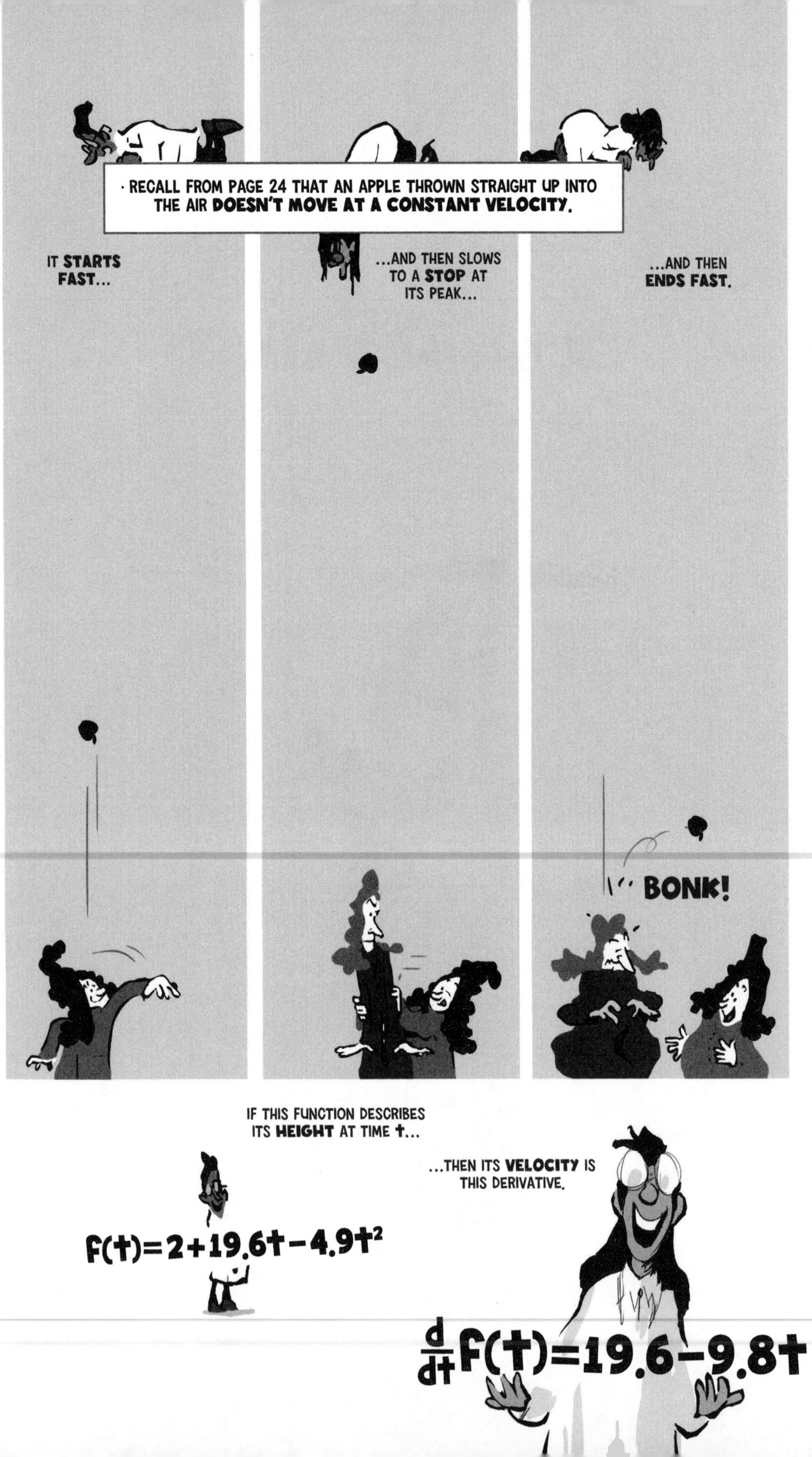

· RECALL FROM PAGE 24 THAT AN APPLE THROWN STRAIGHT UP INTO THE AIR **DOESN'T MOVE AT A CONSTANT VELOCITY.**
IT **STARTS** FAST...
...AND THEN SLOWS TO A **STOP** AT ITS PEAK...
...AND THEN **ENDS FAST.**
BONK!
IF THIS FUNCTION DESCRIBES ITS **HEIGHT** AT TIME **t**...
$f(t) = 2 + 19.6t - 4.9t^2$
...THEN ITS **VELOCITY** IS THIS DERIVATIVE.
$\frac{d}{dt} f(t) = 19.6 - 9.8t$

AND BECAUSE THE APPLE **STOPS FOR AN INSTANT** AT ITS **HIGHEST POINT...**

HEY LOOK, WE HAVE A **VELOCITY** OF **ZERO.**

EEK! IT'S A **TALKING WORM!**

...WE CAN FIND OUT **WHEN** THAT INSTANT OCCURS BY LOOKING AT THE **DERIVATIVE...**

...AND FINDING WHEN IT EQUALS **ZERO!**

$$\frac{d}{dt}f(t) = 0$$

$$19.6 - 9.8t = 0$$

$$t = 2$$

SO THE MAXIMUM HEIGHT WAS AFTER **2 SECONDS...**

...AT A HEIGHT OF f(2)=21.6 **METERS.**

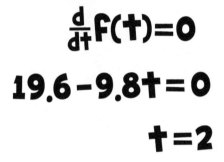

THAT'S AN EXAMPLE OF USING CALCULUS TO **FIND EXTREME VALUES.**

I WANT TO LEARN **MORE!**

TAKE A HIKE, KID!

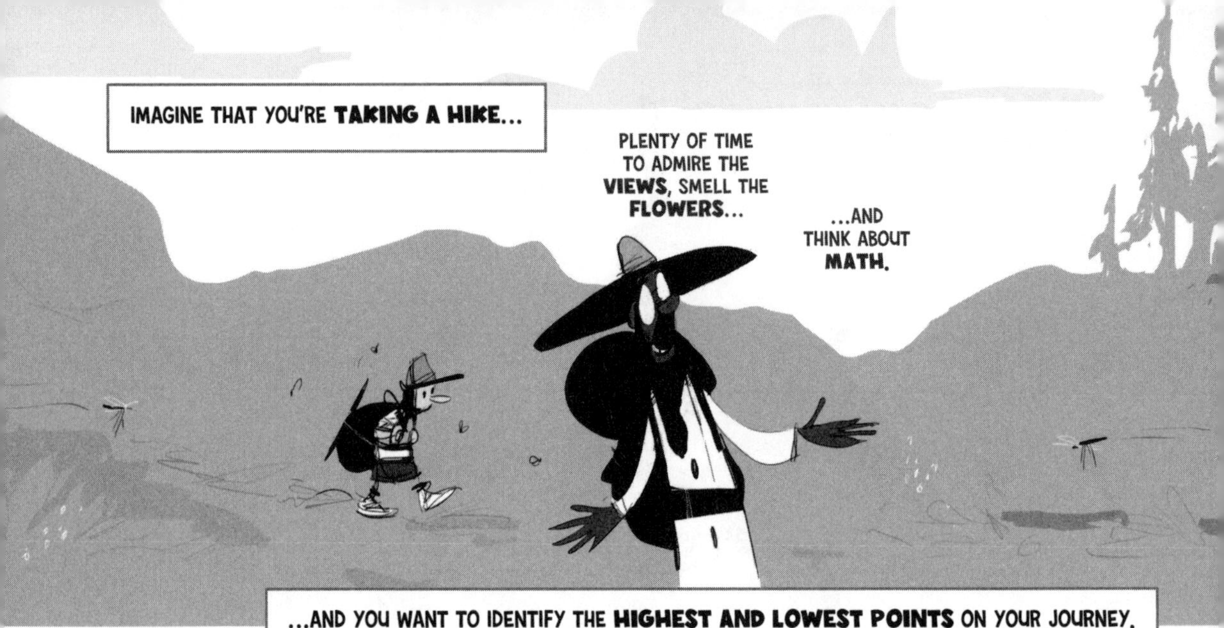
IMAGINE THAT YOU'RE **TAKING A HIKE**...
PLENTY OF TIME TO ADMIRE THE **VIEWS**, SMELL THE **FLOWERS**...
...AND THINK ABOUT **MATH**.
...AND YOU WANT TO IDENTIFY THE **HIGHEST AND LOWEST POINTS** ON YOUR JOURNEY.

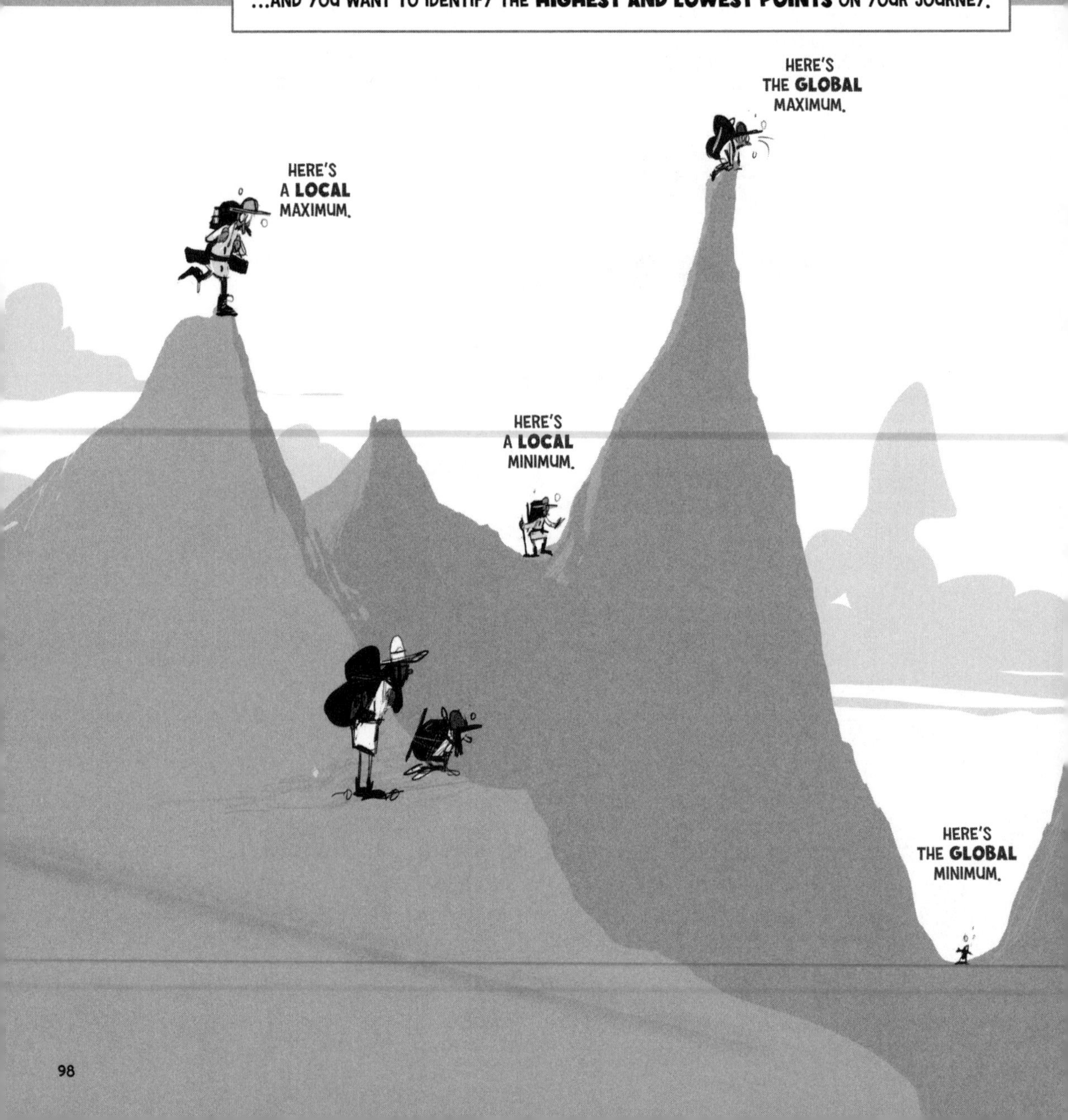
HERE'S A **LOCAL** MAXIMUM.
HERE'S THE **GLOBAL** MAXIMUM.
HERE'S A **LOCAL** MINIMUM.
HERE'S THE **GLOBAL** MINIMUM.

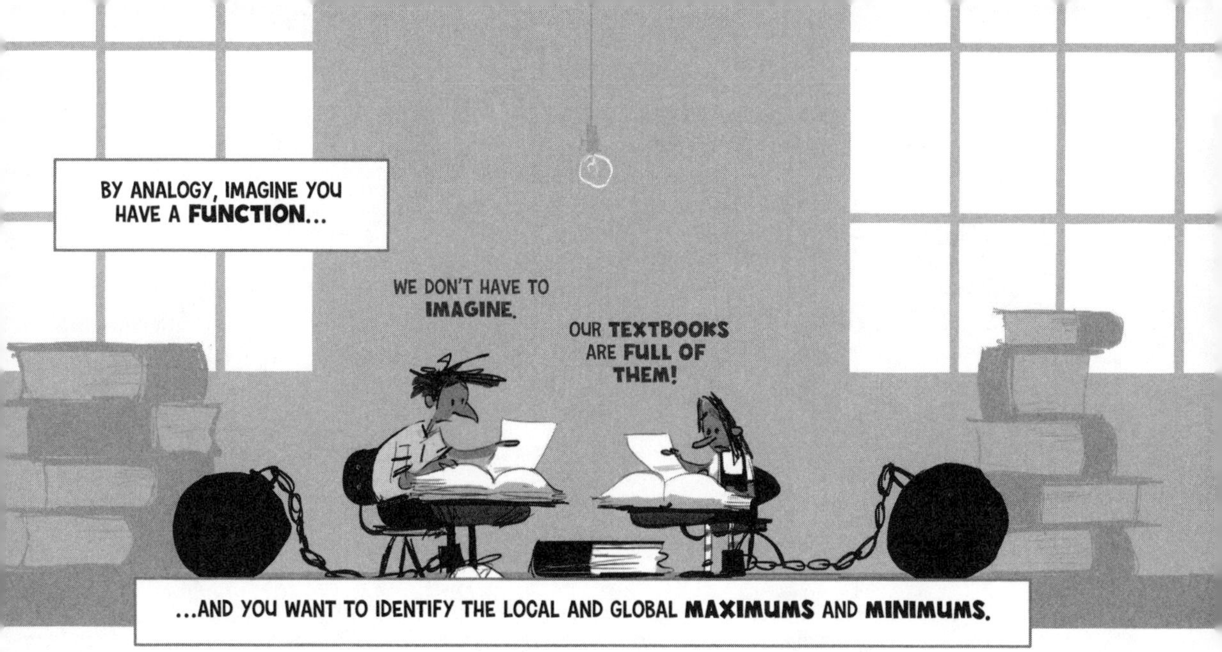
BY ANALOGY, IMAGINE YOU HAVE A **FUNCTION**...
WE DON'T HAVE TO **IMAGINE.**
OUR **TEXTBOOKS** ARE **FULL OF THEM!**
...AND YOU WANT TO IDENTIFY THE LOCAL AND GLOBAL **MAXIMUMS** AND **MINIMUMS.**

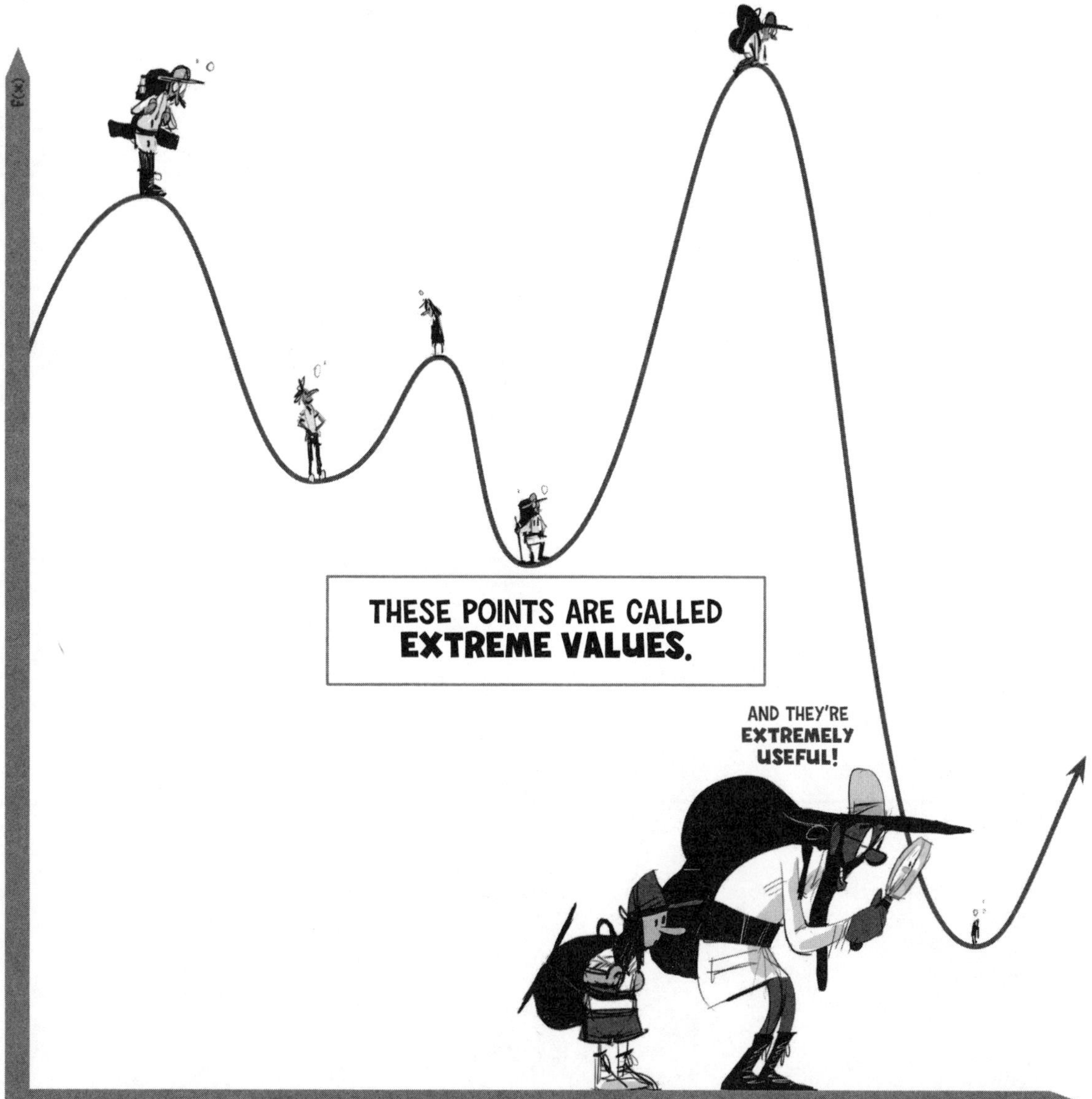
F(x)
THESE POINTS ARE CALLED **EXTREME VALUES.**
AND THEY'RE **EXTREMELY USEFUL!**
x

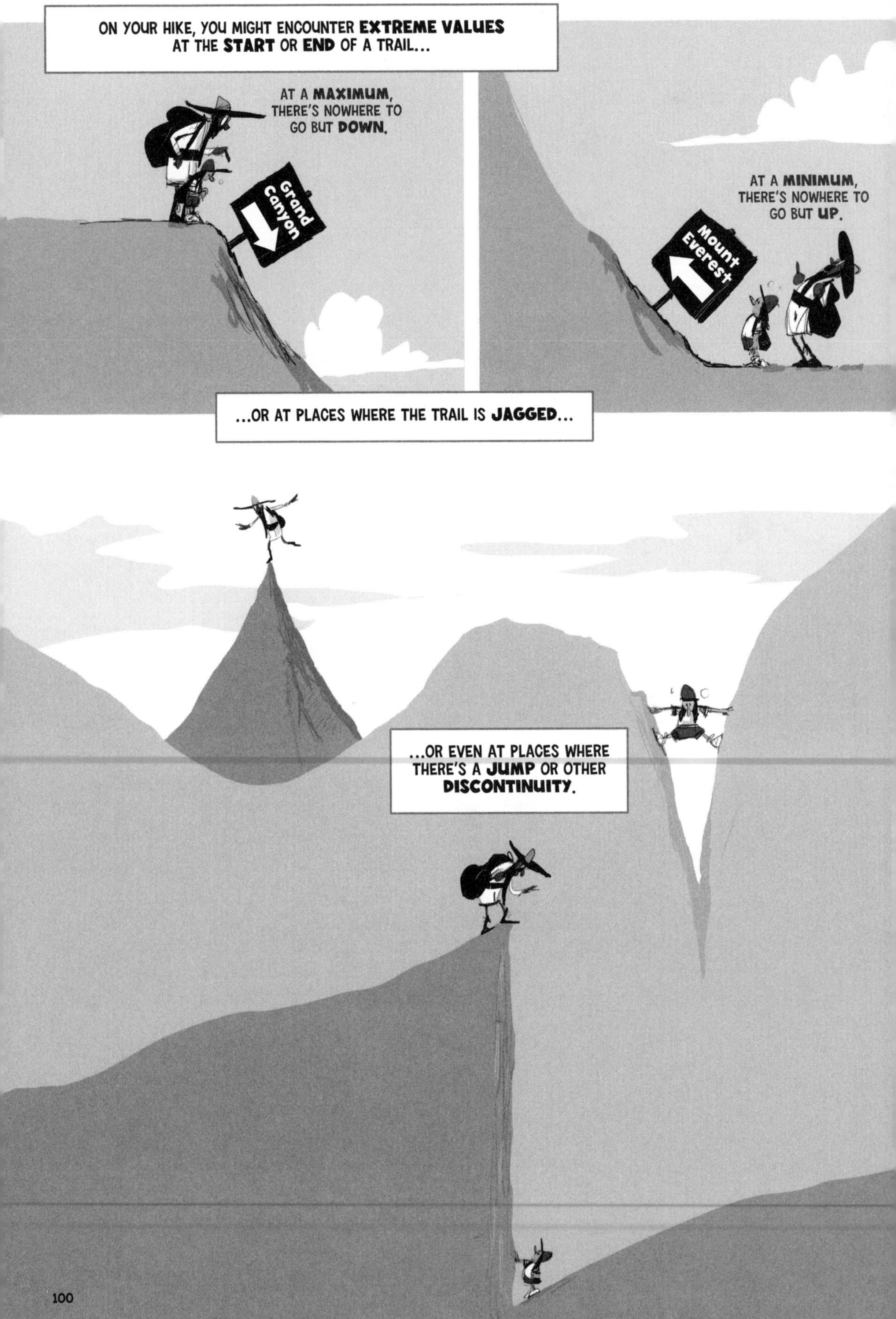
ON YOUR HIKE, YOU MIGHT ENCOUNTER EXTREME VALUES AT THE START OR END OF A TRAIL...
AT A MAXIMUM, THERE'S NOWHERE TO GO BUT DOWN.
Grand Canyon
AT A MINIMUM, THERE'S NOWHERE TO GO BUT UP.
Mount Everest
...OR AT PLACES WHERE THE TRAIL IS JAGGED...
...OR EVEN AT PLACES WHERE THERE'S A JUMP OR OTHER DISCONTINUITY.

SIMILARLY, THE **EXTREME VALUES OF A FUNCTION** CAN OCCUR WHERE THE FUNCTION **STARTS** OR **ENDS**...

LIKE THIS FUNCTION, WHICH STARTS AT **x=1**...

...AND ENDS AT **x=5**.

WE CALL THOSE **CORNER SOLUTIONS.**

...OR AT POINTS WHERE THE FUNCTION IS **JAGGED**...

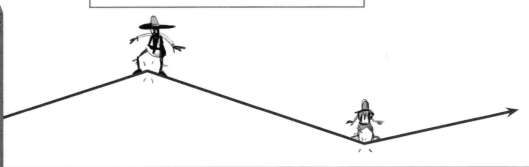

...OR EVEN WHERE IT **ISN'T CONTINUOUS.**

SADLY, CALCULUS **CAN'T HELP US** FIND THESE PARTICULAR EXTREME VALUES.

YOU NEED TO **CHECK BY HAND**...

...TO SEE IF THERE ARE EXTREME VALUES AT **NON-DIFFERENTIABLE** POINTS.

FORTUNATELY, THEY'RE **PRETTY RARE.**

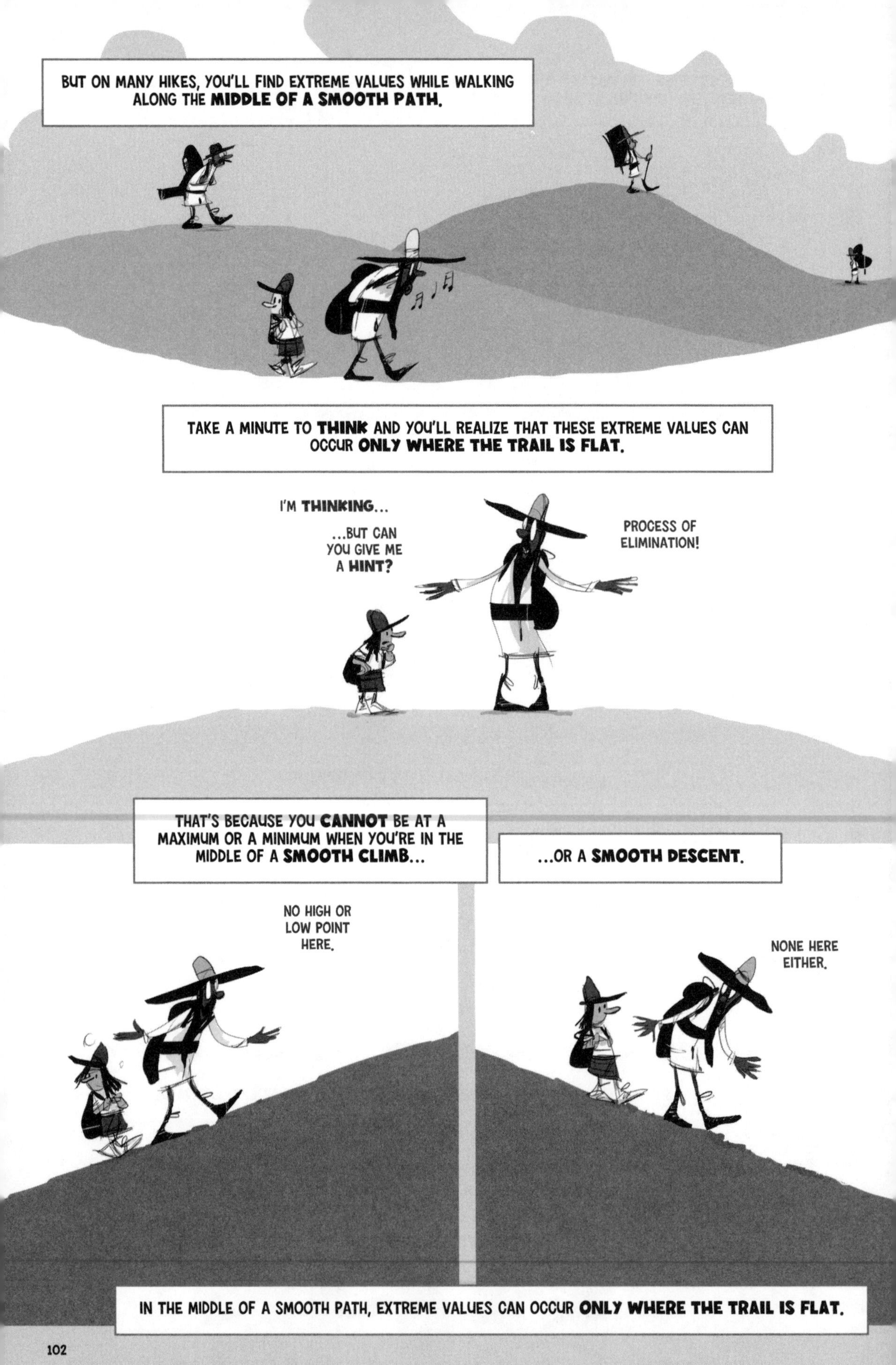
BUT ON MANY HIKES, YOU'LL FIND EXTREME VALUES WHILE WALKING ALONG THE **MIDDLE OF A SMOOTH PATH.**
TAKE A MINUTE TO **THINK** AND YOU'LL REALIZE THAT THESE EXTREME VALUES CAN OCCUR **ONLY WHERE THE TRAIL IS FLAT.**
I'M **THINKING**...
...BUT CAN YOU GIVE ME A **HINT?**
PROCESS OF ELIMINATION!
THAT'S BECAUSE YOU **CANNOT** BE AT A MAXIMUM OR A MINIMUM WHEN YOU'RE IN THE MIDDLE OF A **SMOOTH CLIMB**...
NO HIGH OR LOW POINT HERE.
...OR A **SMOOTH DESCENT.**
NONE HERE EITHER.
IN THE MIDDLE OF A SMOOTH PATH, EXTREME VALUES CAN OCCUR **ONLY WHERE THE TRAIL IS FLAT.**

SIMILARLY, MANY FUNCTIONS HAVE EXTREME VALUES AT **DIFFERENTIABLE POINTS**.

AT DIFFERENTIABLE POINTS, THE FUNCTION IS **SMOOTH**...

...WHICH MEANS THAT THE DERIVATIVE **EXISTS**.

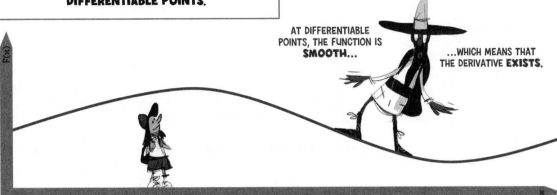

THE **BIG IDEA** OF THIS CHAPTER IS THAT THESE EXTREME VALUES CAN OCCUR **ONLY WHERE THE DERIVATIVE EQUALS ZERO**.

YOU GET THE **SAME HINT:**

PROCESS OF ELIMINATION!

THAT'S BECAUSE THESE EXTREME VALUES **CANNOT OCCUR** AT POINTS WHERE THE DERIVATIVE IS **POSITIVE**...

...OR **NEGATIVE**.

NO MAXIMUM OR MINIMUM VALUE HERE.

NONE HERE EITHER.

WHERE A FUNCTION IS DIFFERENTIABLE, **EXTREME VALUES CAN OCCUR ONLY WHERE THE DERIVATIVE IS ZERO**.

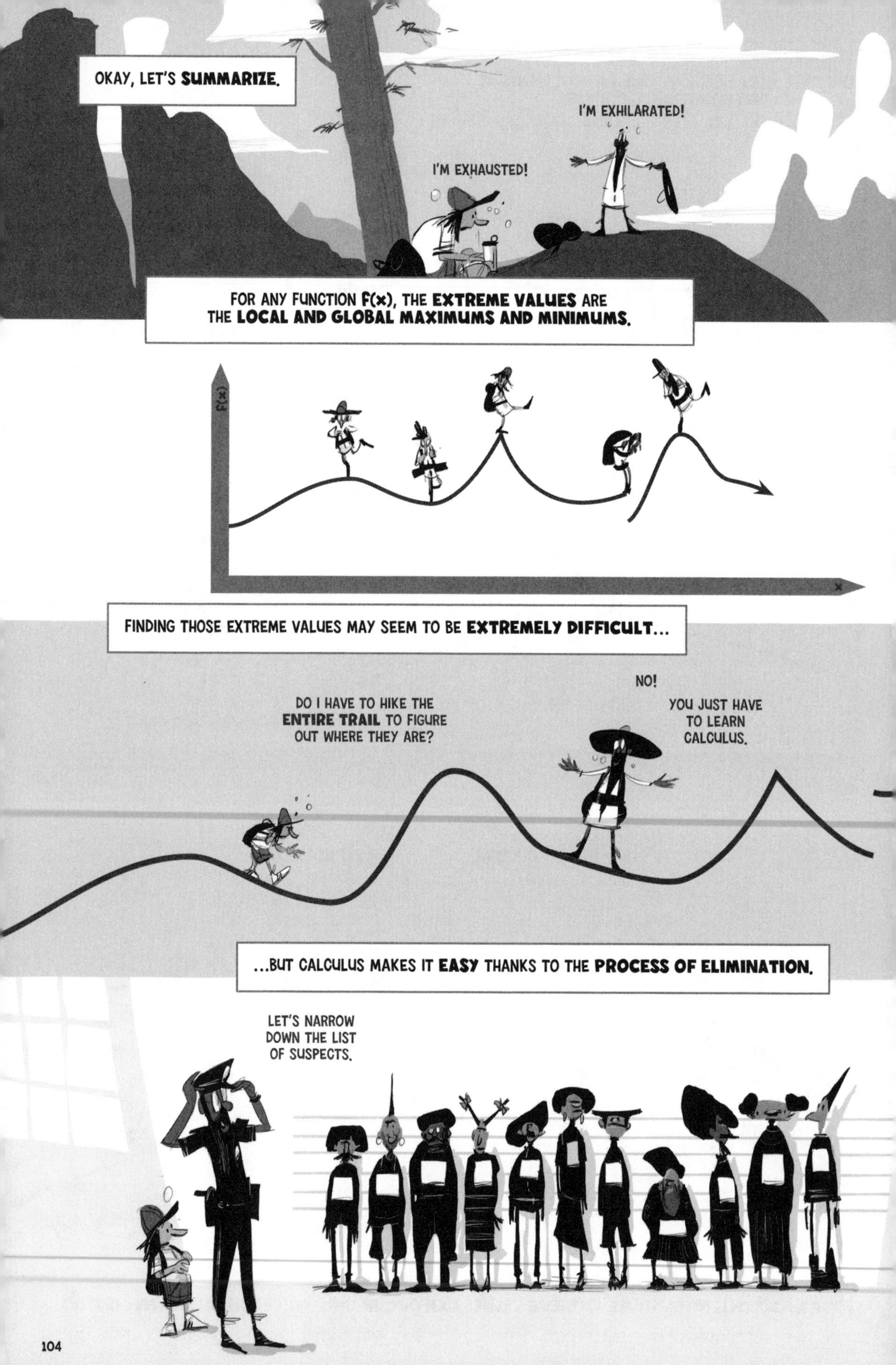
OKAY, LET'S **SUMMARIZE.**
I'M EXHAUSTED!
I'M EXHILARATED!
FOR ANY FUNCTION f(x), THE **EXTREME VALUES** ARE THE **LOCAL AND GLOBAL MAXIMUMS AND MINIMUMS.**
f(x)
x
FINDING THOSE EXTREME VALUES MAY SEEM TO BE **EXTREMELY DIFFICULT...**
DO I HAVE TO HIKE THE **ENTIRE TRAIL** TO FIGURE OUT WHERE THEY ARE?
NO!
YOU JUST HAVE TO LEARN CALCULUS.
...BUT CALCULUS MAKES IT **EASY** THANKS TO THE **PROCESS OF ELIMINATION.**
LET'S NARROW DOWN THE LIST OF SUSPECTS.

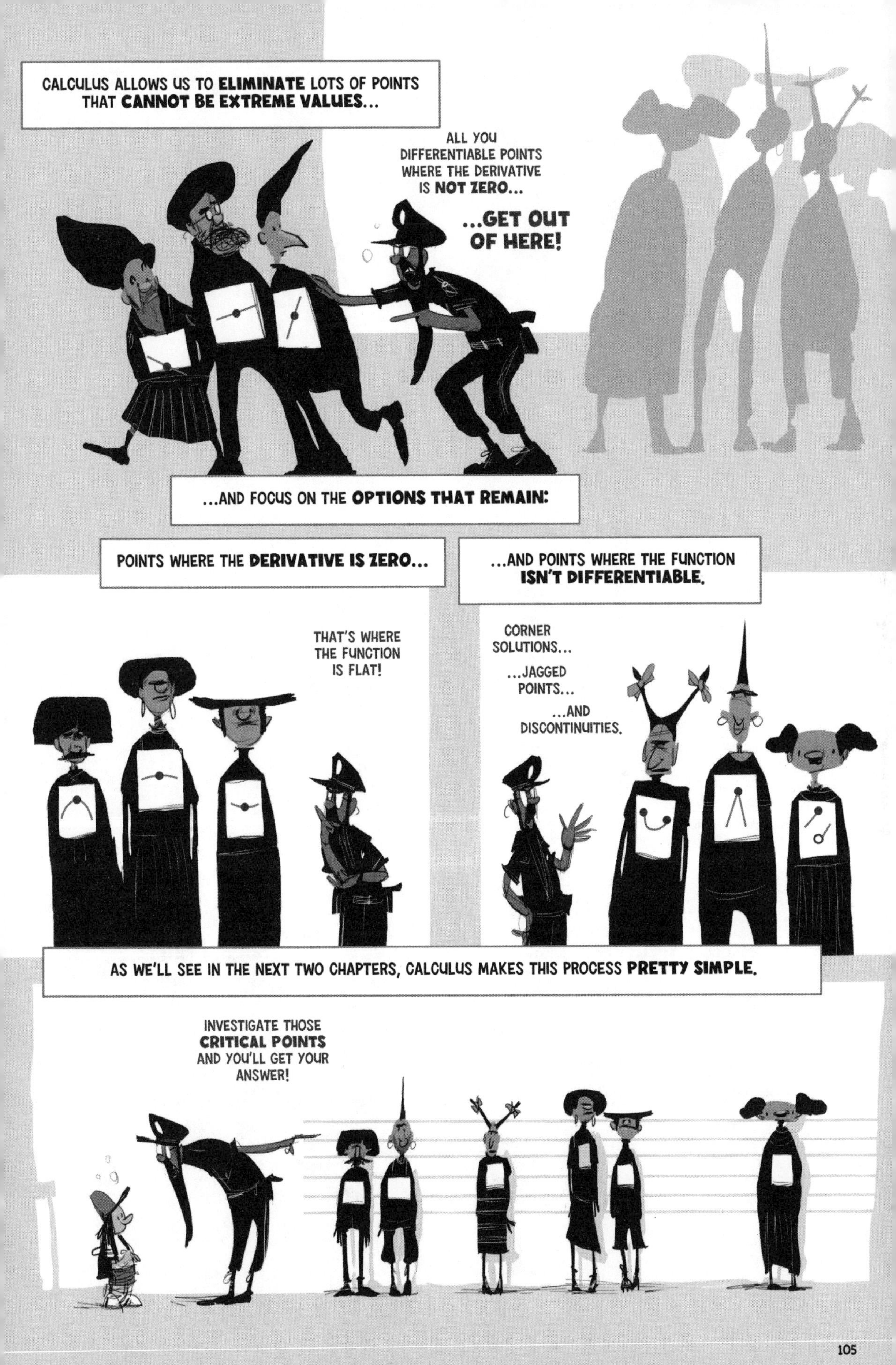

CALCULUS ALLOWS US TO ELIMINATE LOTS OF POINTS THAT CANNOT BE EXTREME VALUES...
ALL YOU DIFFERENTIABLE POINTS WHERE THE DERIVATIVE IS NOT ZERO...
...GET OUT OF HERE!
...AND FOCUS ON THE OPTIONS THAT REMAIN:
POINTS WHERE THE DERIVATIVE IS ZERO...
...AND POINTS WHERE THE FUNCTION ISN'T DIFFERENTIABLE.
THAT'S WHERE THE FUNCTION IS FLAT!
CORNER SOLUTIONS...
...JAGGED POINTS...
...AND DISCONTINUITIES.
AS WE'LL SEE IN THE NEXT TWO CHAPTERS, CALCULUS MAKES THIS PROCESS PRETTY SIMPLE.
INVESTIGATE THOSE CRITICAL POINTS AND YOU'LL GET YOUR ANSWER!

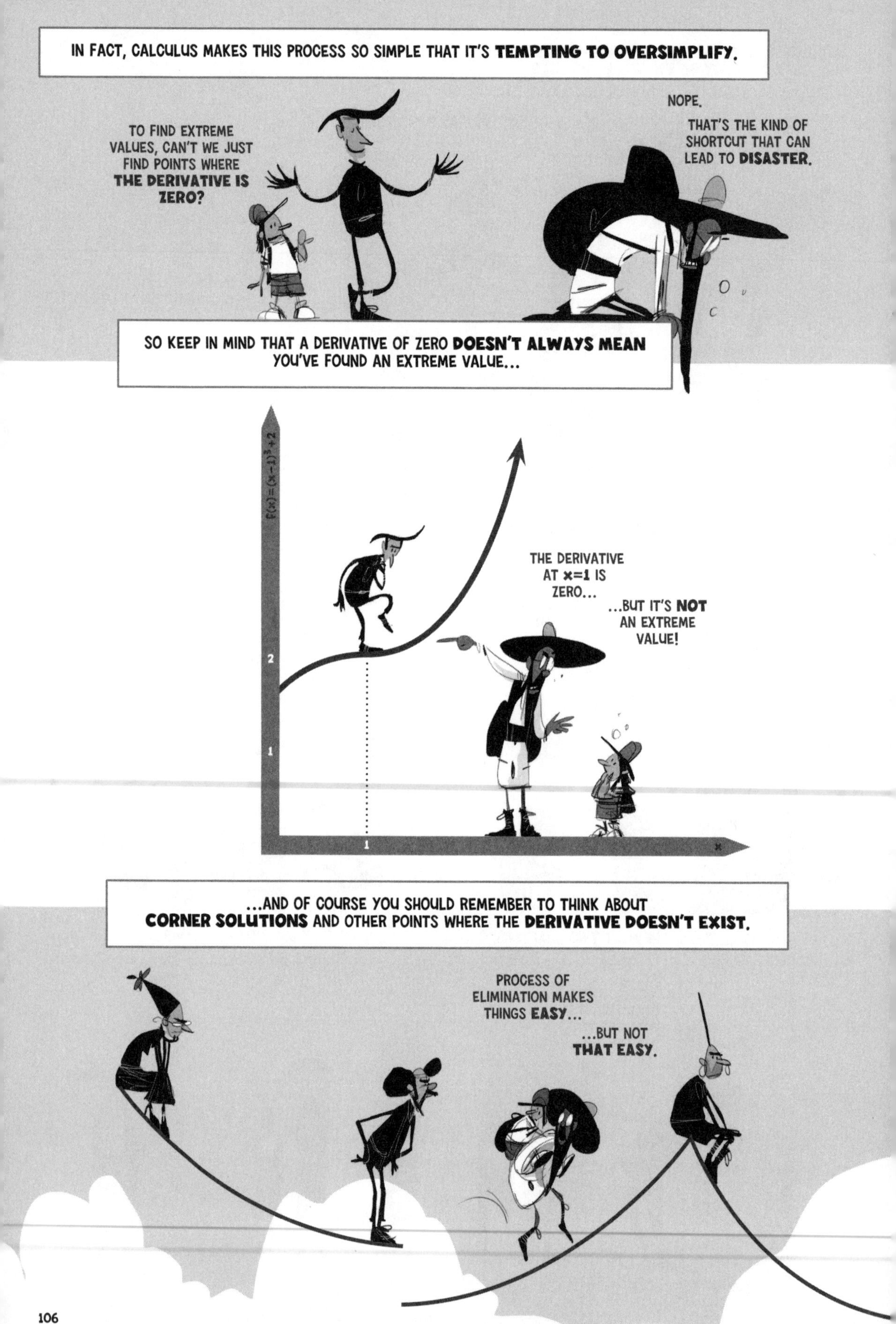
IN FACT, CALCULUS MAKES THIS PROCESS SO SIMPLE THAT IT'S **TEMPTING TO OVERSIMPLIFY.**
TO FIND EXTREME VALUES, CAN'T WE JUST FIND POINTS WHERE **THE DERIVATIVE IS ZERO?**
NOPE.
THAT'S THE KIND OF SHORTCUT THAT CAN LEAD TO **DISASTER.**
SO KEEP IN MIND THAT A DERIVATIVE OF ZERO **DOESN'T ALWAYS MEAN** YOU'VE FOUND AN EXTREME VALUE...
$f(x)=(x-1)^3+2$
2
1
1
x
THE DERIVATIVE AT **x=1** IS ZERO...
...BUT IT'S **NOT** AN EXTREME VALUE!
...AND OF COURSE YOU SHOULD REMEMBER TO THINK ABOUT **CORNER SOLUTIONS** AND OTHER POINTS WHERE THE **DERIVATIVE DOESN'T EXIST.**
PROCESS OF ELIMINATION MAKES THINGS **EASY...**
...BUT NOT **THAT EASY.**

CHAPTER 9
OPTIMIZATION

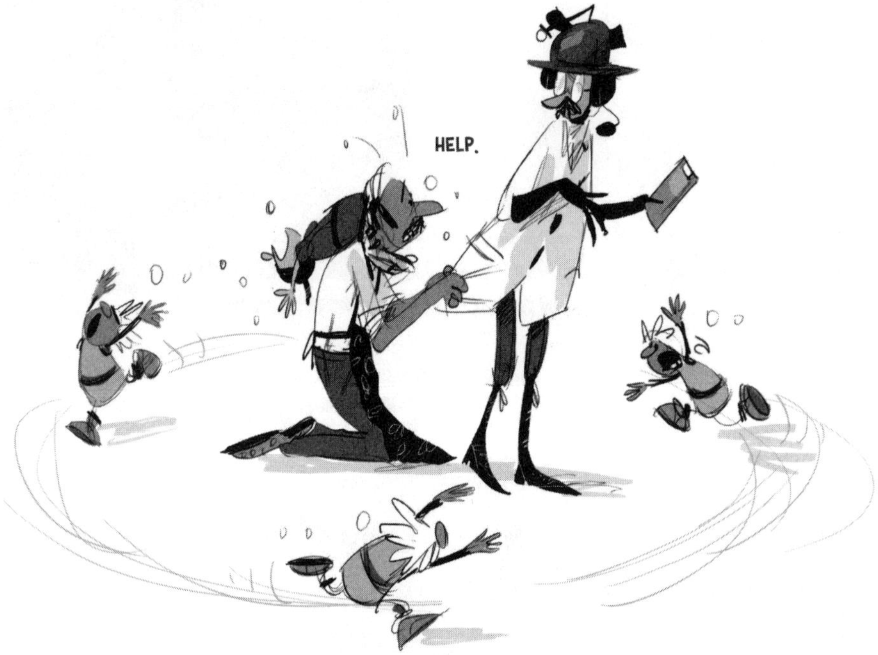
HELP.

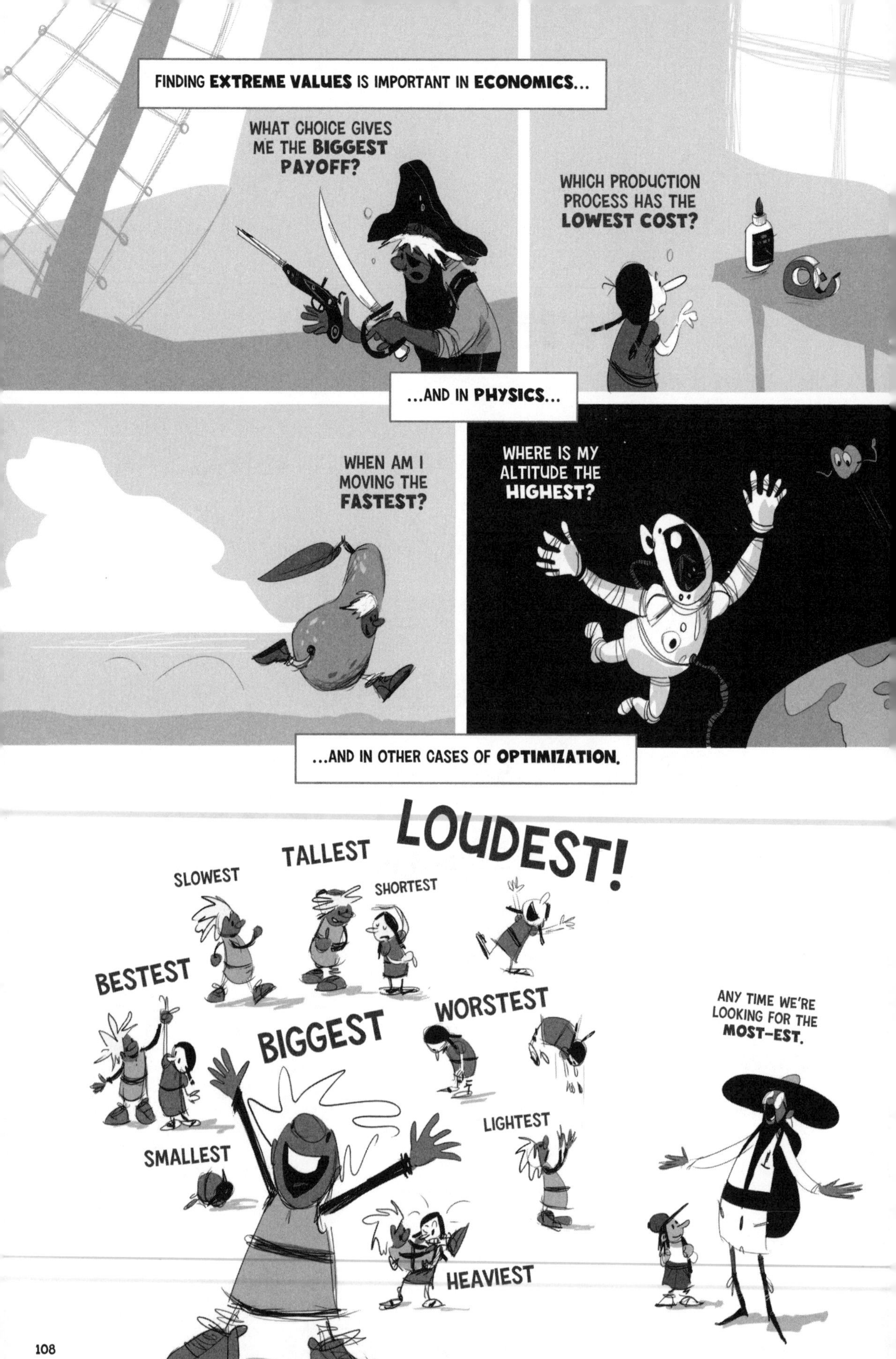
FINDING **EXTREME VALUES** IS IMPORTANT IN **ECONOMICS**...
WHAT CHOICE GIVES ME THE **BIGGEST PAYOFF?**
WHICH PRODUCTION PROCESS HAS THE **LOWEST COST?**
...AND IN **PHYSICS**...
WHEN AM I MOVING THE **FASTEST?**
WHERE IS MY ALTITUDE THE **HIGHEST?**
...AND IN OTHER CASES OF **OPTIMIZATION.**
LOUDEST!
TALLEST
SLOWEST
SHORTEST
BESTEST
WORSTEST
BIGGEST
LIGHTEST
SMALLEST
HEAVIEST
ANY TIME WE'RE LOOKING FOR THE **MOST-EST.**

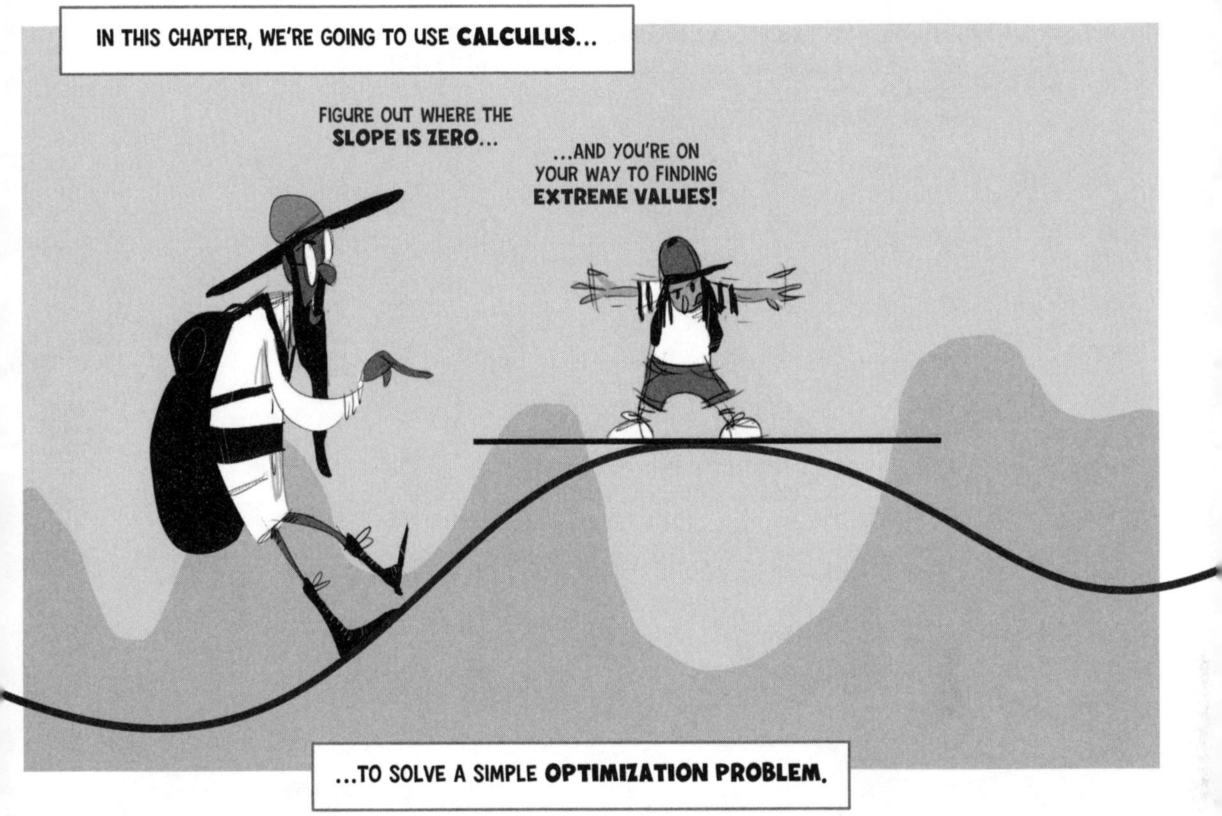
IN THIS CHAPTER, WE'RE GOING TO USE **CALCULUS**...
FIGURE OUT WHERE THE **SLOPE IS ZERO**...
...AND YOU'RE ON YOUR WAY TO FINDING **EXTREME VALUES!**
...TO SOLVE A SIMPLE **OPTIMIZATION PROBLEM.**

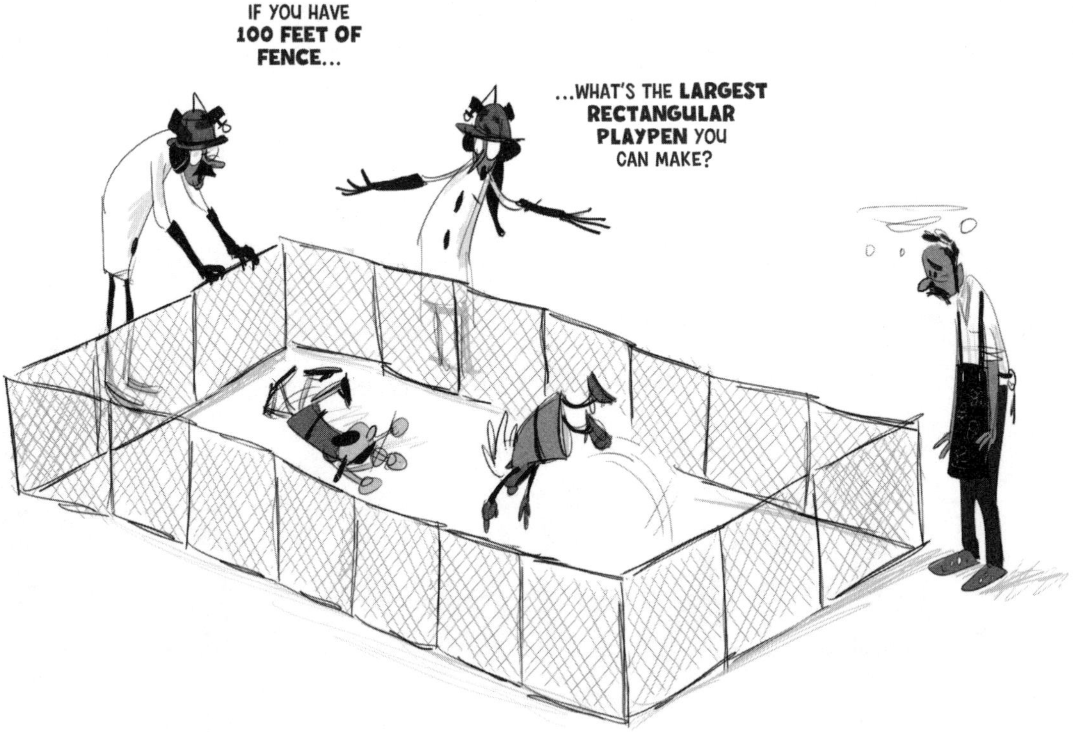
IF YOU HAVE **100 FEET OF FENCE**...
...WHAT'S THE **LARGEST RECTANGULAR PLAYPEN** YOU CAN MAKE?

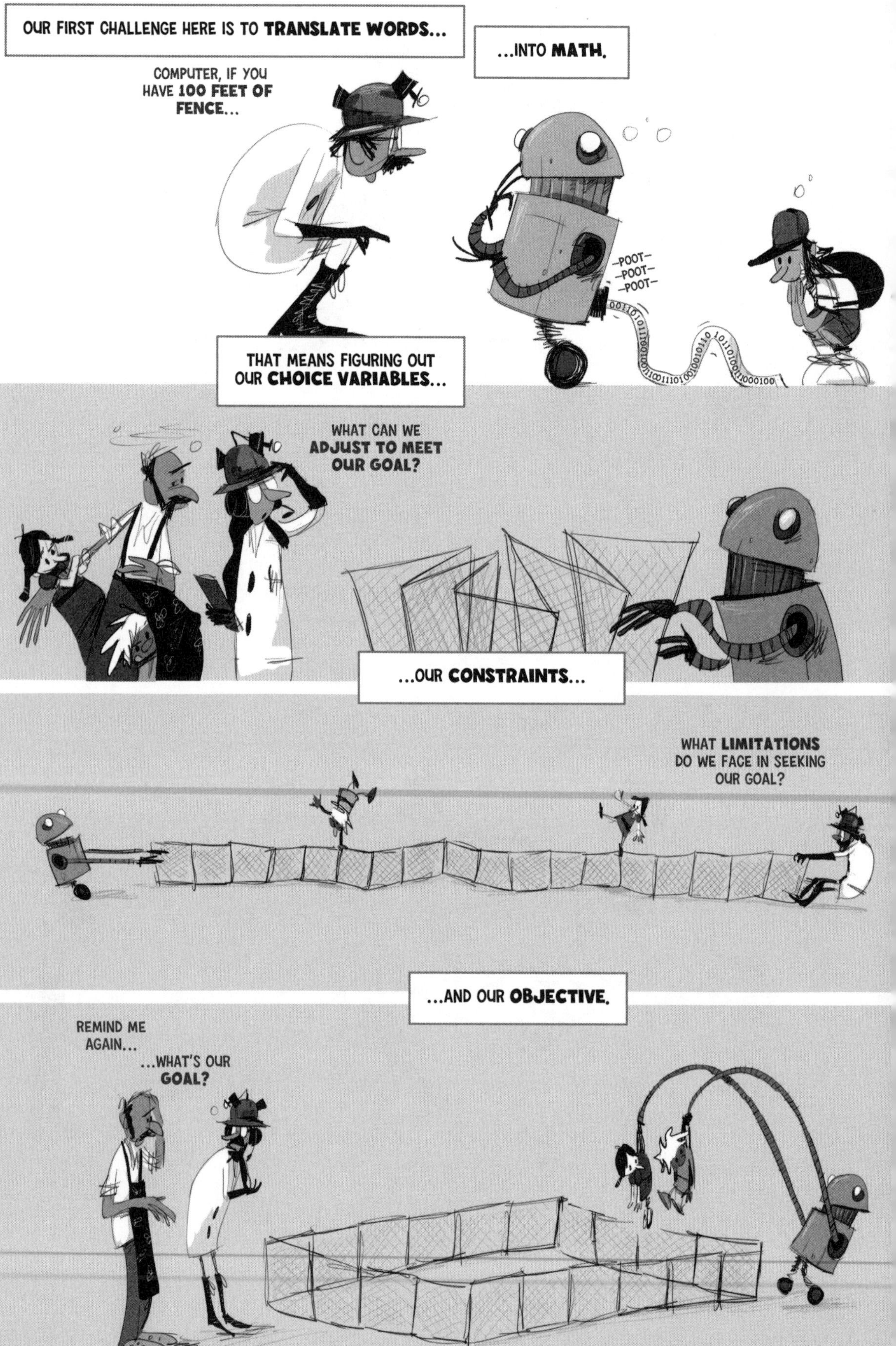

OUR FIRST CHALLENGE HERE IS TO **TRANSLATE WORDS...**
...INTO **MATH.**
COMPUTER, IF YOU HAVE **100 FEET OF FENCE...**
—POOT— —POOT— —POOT—
THAT MEANS FIGURING OUT OUR **CHOICE VARIABLES...**
WHAT CAN WE **ADJUST TO MEET OUR GOAL?**
...OUR **CONSTRAINTS...**
WHAT **LIMITATIONS** DO WE FACE IN SEEKING OUR GOAL?
...AND OUR **OBJECTIVE.**
REMIND ME AGAIN...
...WHAT'S OUR **GOAL?**

FOR OUR PROBLEM, IT'S EASIEST TO FOCUS ON ONLY ONE CHOICE VARIABLE: **LENGTH L.**

ONCE WE CHOOSE THE LENGTH, **L**, WE KNOW THAT THE WIDTH, **W**, HAS TO BE 50-L.

$$2L + 2W = 100$$
$$\Rightarrow L + W = 50$$
$$\Rightarrow W = 50 - L$$

THE **CONSTRAINT** WE FACE IS THAT **L CAN'T BE LESS THAN ZERO OR GREATER THAN 50.**

THE LENGTH **L** AND THE WIDTH **W=50-L** BOTH HAVE TO BE **NON-NEGATIVE...**

...SO $0 \le L \le 50$.

AND OUR OBJECTIVE IS TO **MAXIMIZE THE AREA OF THE RECTANGLE.**

AREA IS **LENGTH TIMES WIDTH.**

BIGGER IS BETTER!

$$A = L \bullet W$$
$$= L(50 - L)$$
$$= 50L - L^2$$

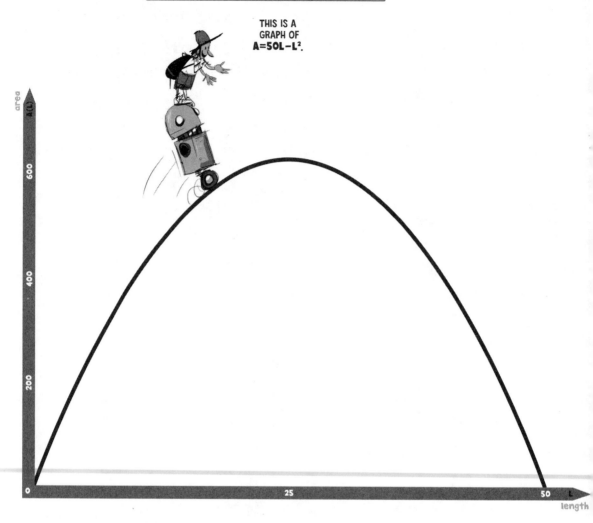

IT OFTEN HELPS TO SKETCH A **GRAPH**...
THIS IS A GRAPH OF **A=50L−L²**.
area
A(L)
600
400
200
0
25
50
L
length
AT ONE EXTREME (L=0) WE HAVE AN AREA OF ZERO...
...BECAUSE OUR RECTANGLE HAS **ZERO LENGTH**...
...AND AT THE OTHER EXTREME (L=50) WE HAVE AN AREA OF ZERO...
...BECAUSE OUR RECTANGLE HAS **ZERO WIDTH**.

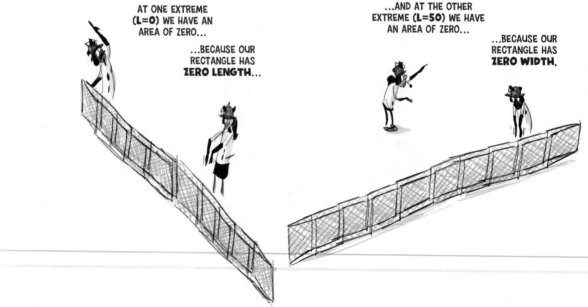

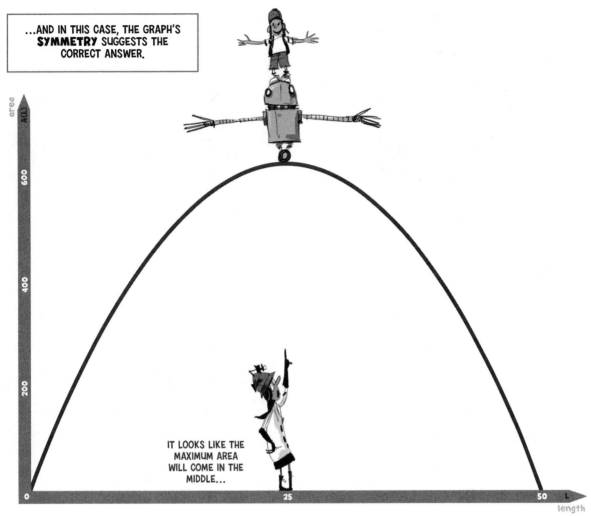

...AND IN THIS CASE, THE GRAPH'S **SYMMETRY** SUGGESTS THE CORRECT ANSWER.

area
A(L)

600

400

200

0

25

50

L

length

IT LOOKS LIKE THE MAXIMUM AREA WILL COME IN THE MIDDLE...

...WHERE **L=25** AND WE HAVE A SQUARE.

BUT LET'S USE **CALCULUS** TO FIND OUT **FOR SURE.**

IN CHAPTER 8, WE LEARNED THAT **EXTREME VALUES** CAN BE **CORNER SOLUTIONS**...

THOSE ARE AT THE **START** AND THE **END**...

...SO WE NEED TO THINK ABOUT **L=0** AND **L=50**.

...OR OTHER POINTS WHERE THE **DERIVATIVE DOESN'T EXIST**...

THAT'S NOT AN ISSUE IN OUR PROBLEM...

...BECAUSE OUR GRAPH IS A **SMOOTH PATH**.

...OR POINTS THAT ARE IN THE MIDDLE OF A SMOOTH PATH AND **HAVE A DERIVATIVE OF ZERO**.

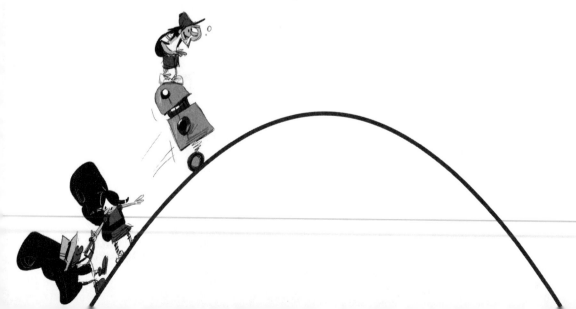

LET'S FIND **THOSE POINTS!**

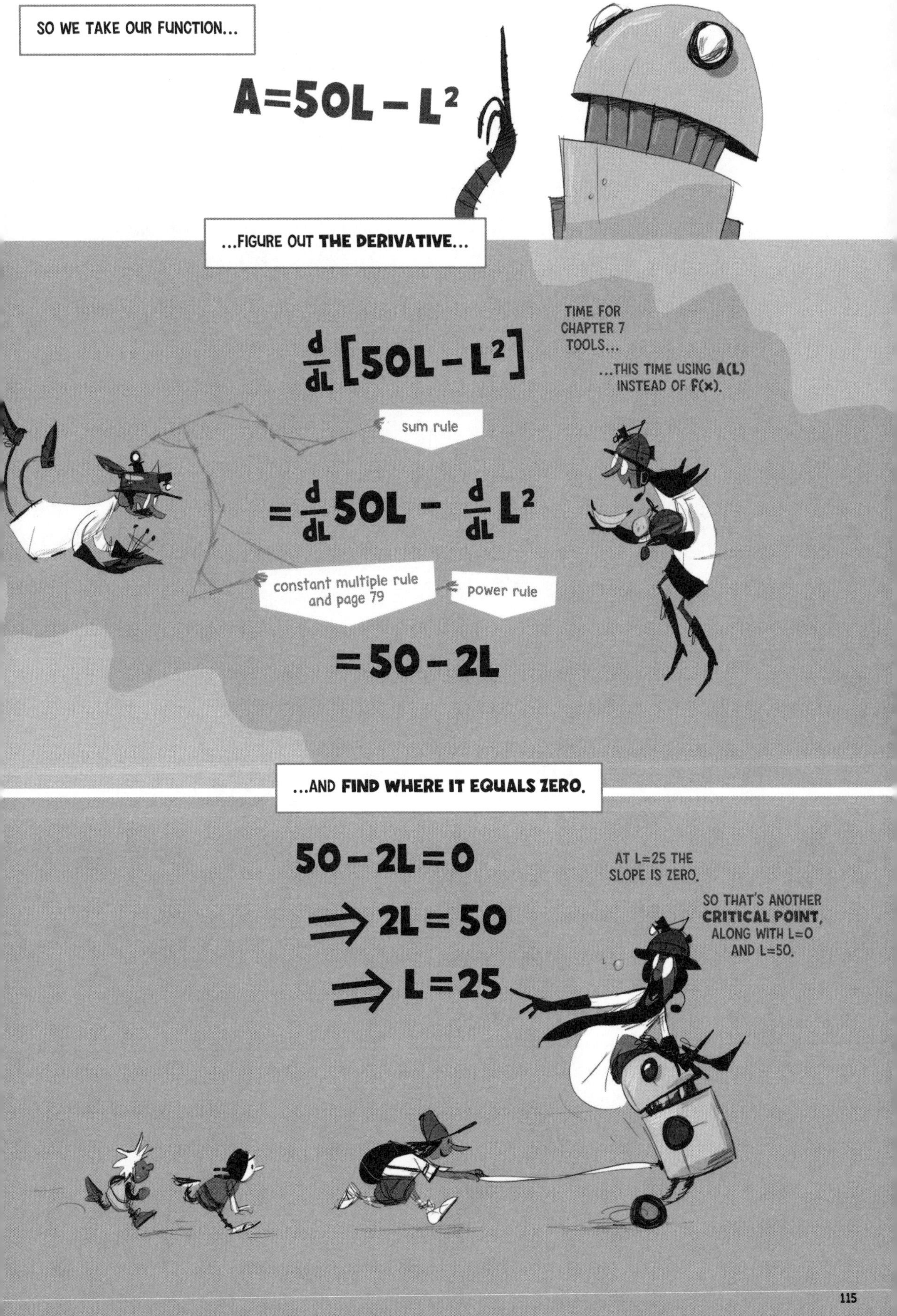

$$A = 50L - L^2$$

...FIGURE OUT **THE DERIVATIVE**...

TIME FOR
CHAPTER 7
TOOLS...

...THIS TIME USING **A(L)**
INSTEAD OF **f(x)**.

$$\frac{d}{dL}\left[50L - L^2\right]$$

sum rule

$$= \frac{d}{dL}50L - \frac{d}{dL}L^2$$

constant multiple rule
and page 79

power rule

$$= 50 - 2L$$

...AND **FIND WHERE IT EQUALS ZERO.**

$$50 - 2L = 0$$

$$\Rightarrow 2L = 50$$

$$\Rightarrow L = 25$$

AT L=25 THE
SLOPE IS ZERO.

SO THAT'S ANOTHER
CRITICAL POINT,
ALONG WITH L=0
AND L=50.

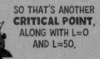

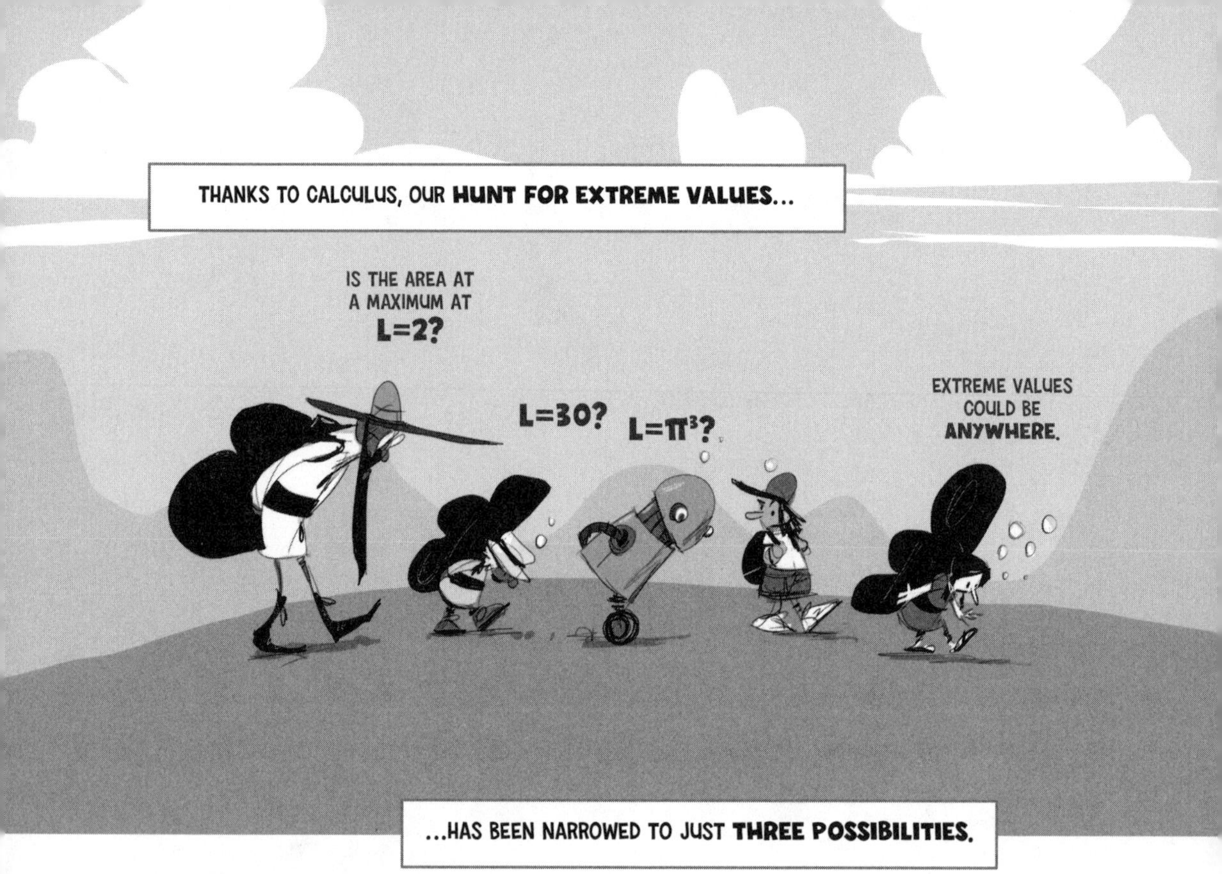
THANKS TO CALCULUS, OUR **HUNT FOR EXTREME VALUES...**
IS THE AREA AT A MAXIMUM AT **L=2?**
L=30?
L=π^3?
EXTREME VALUES COULD BE **ANYWHERE.**
...HAS BEEN NARROWED TO JUST **THREE POSSIBILITIES.**

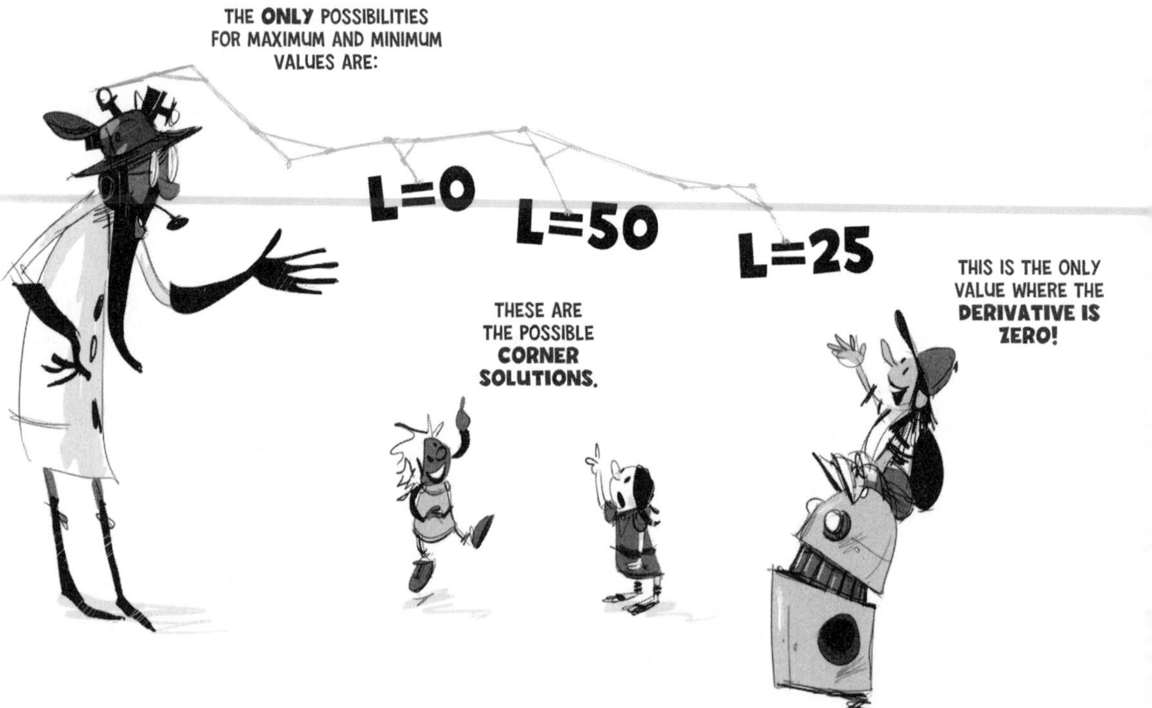
THE **ONLY** POSSIBILITIES FOR MAXIMUM AND MINIMUM VALUES ARE:
L=0
L=50
L=25
THESE ARE THE POSSIBLE **CORNER SOLUTIONS.**
THIS IS THE ONLY VALUE WHERE THE **DERIVATIVE IS ZERO!**

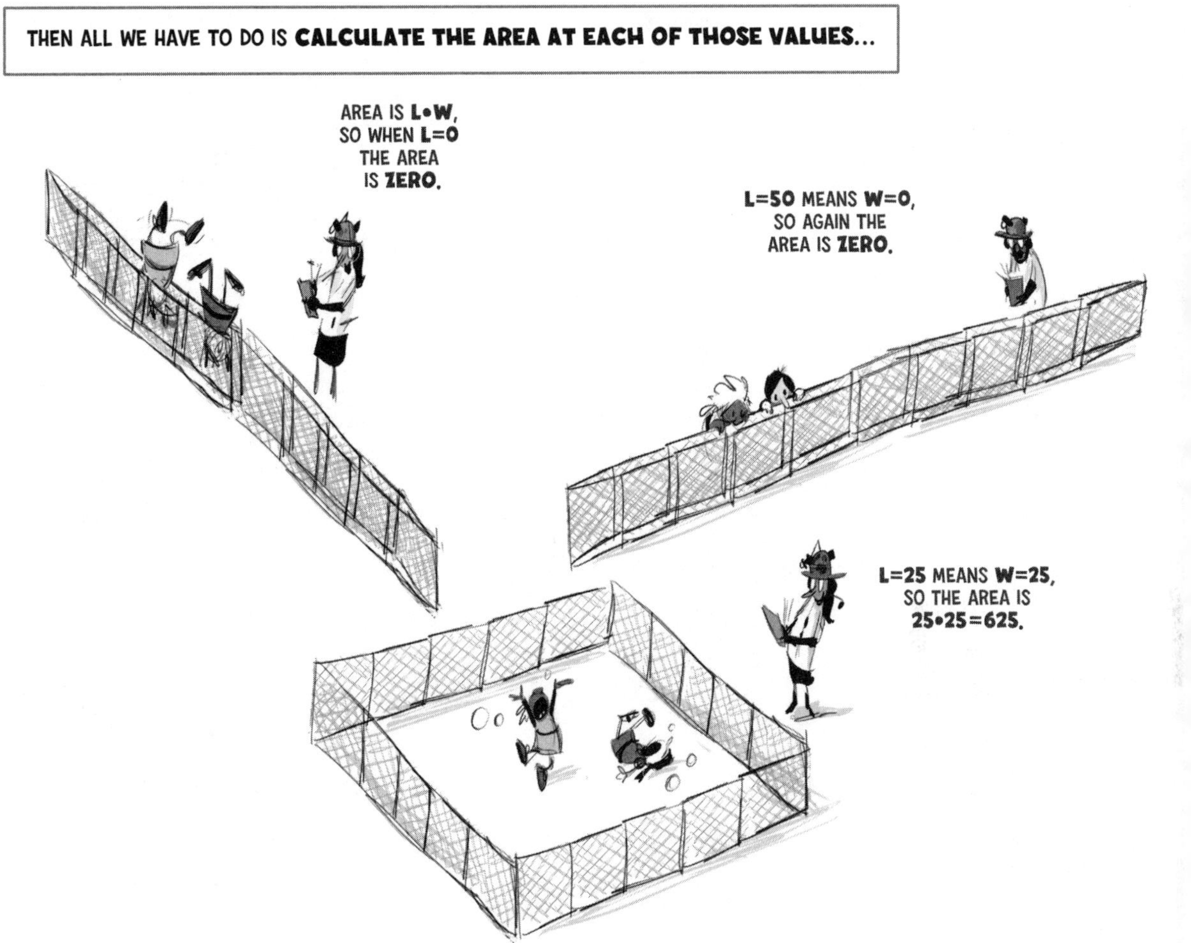
THEN ALL WE HAVE TO DO IS **CALCULATE THE AREA AT EACH OF THOSE VALUES...**
AREA IS **L•W**, SO WHEN **L=0** THE AREA IS **ZERO**.
L=50 MEANS **W=0**, SO AGAIN THE AREA IS **ZERO**.
L=25 MEANS **W=25**, SO THE AREA IS **25•25=625**.

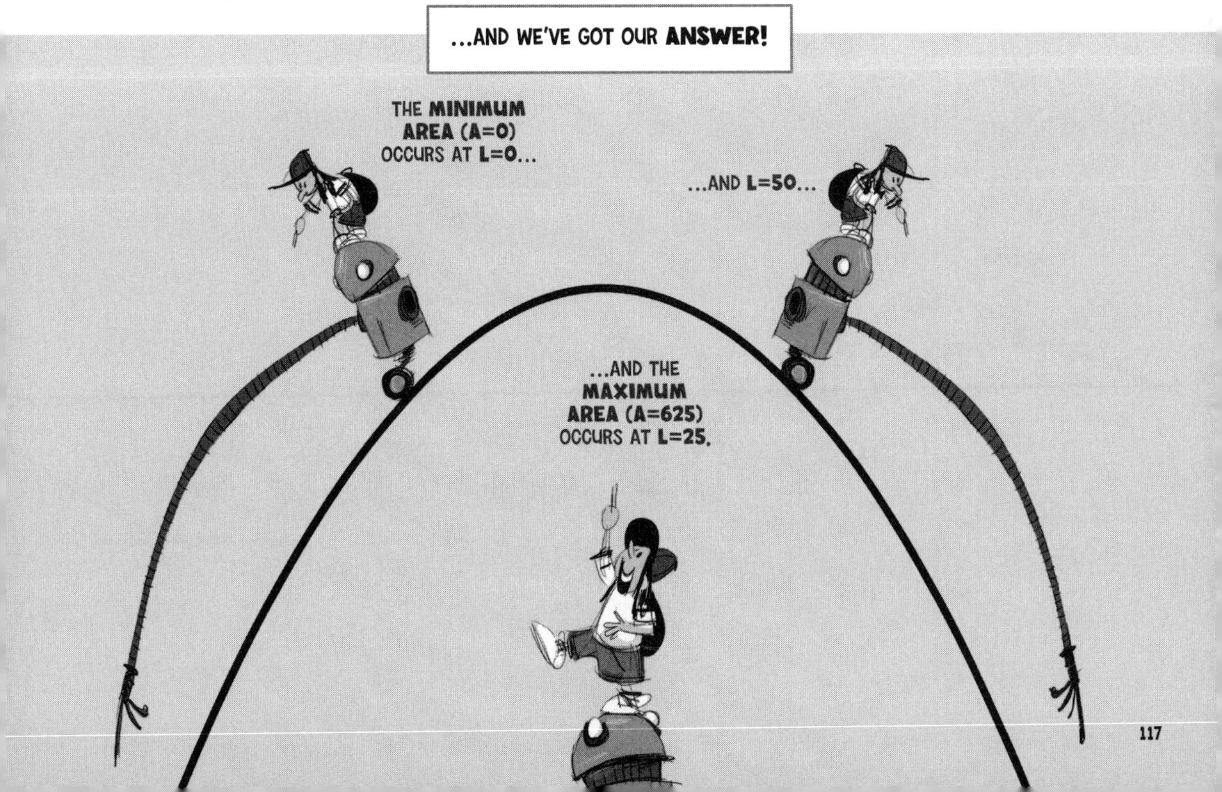
...AND WE'VE GOT OUR **ANSWER!**
THE **MINIMUM AREA (A=0)** OCCURS AT **L=0**...
...AND **L=50**...
...AND THE **MAXIMUM AREA (A=625)** OCCURS AT **L=25**.

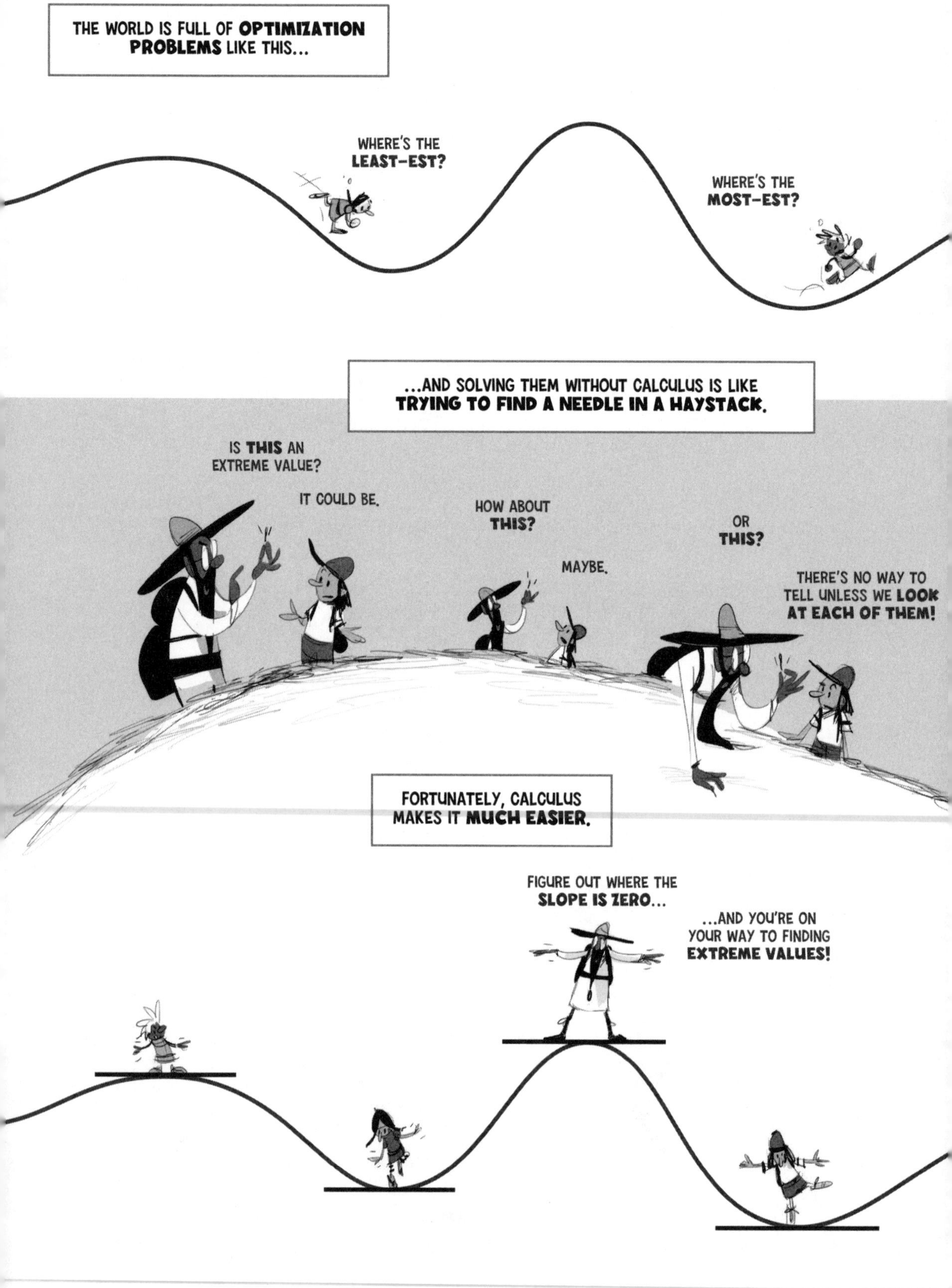
THE WORLD IS FULL OF **OPTIMIZATION PROBLEMS** LIKE THIS...
WHERE'S THE **LEAST-EST?**
WHERE'S THE **MOST-EST?**
...AND SOLVING THEM WITHOUT CALCULUS IS LIKE **TRYING TO FIND A NEEDLE IN A HAYSTACK.**
IS **THIS** AN EXTREME VALUE?
IT COULD BE.
HOW ABOUT **THIS?**
MAYBE.
OR **THIS?**
THERE'S NO WAY TO TELL UNLESS WE **LOOK AT EACH OF THEM!**
FORTUNATELY, CALCULUS MAKES IT **MUCH EASIER.**
FIGURE OUT WHERE THE **SLOPE IS ZERO...**
...AND YOU'RE ON YOUR WAY TO FINDING **EXTREME VALUES!**

CHAPTER 10

ECONOMICS

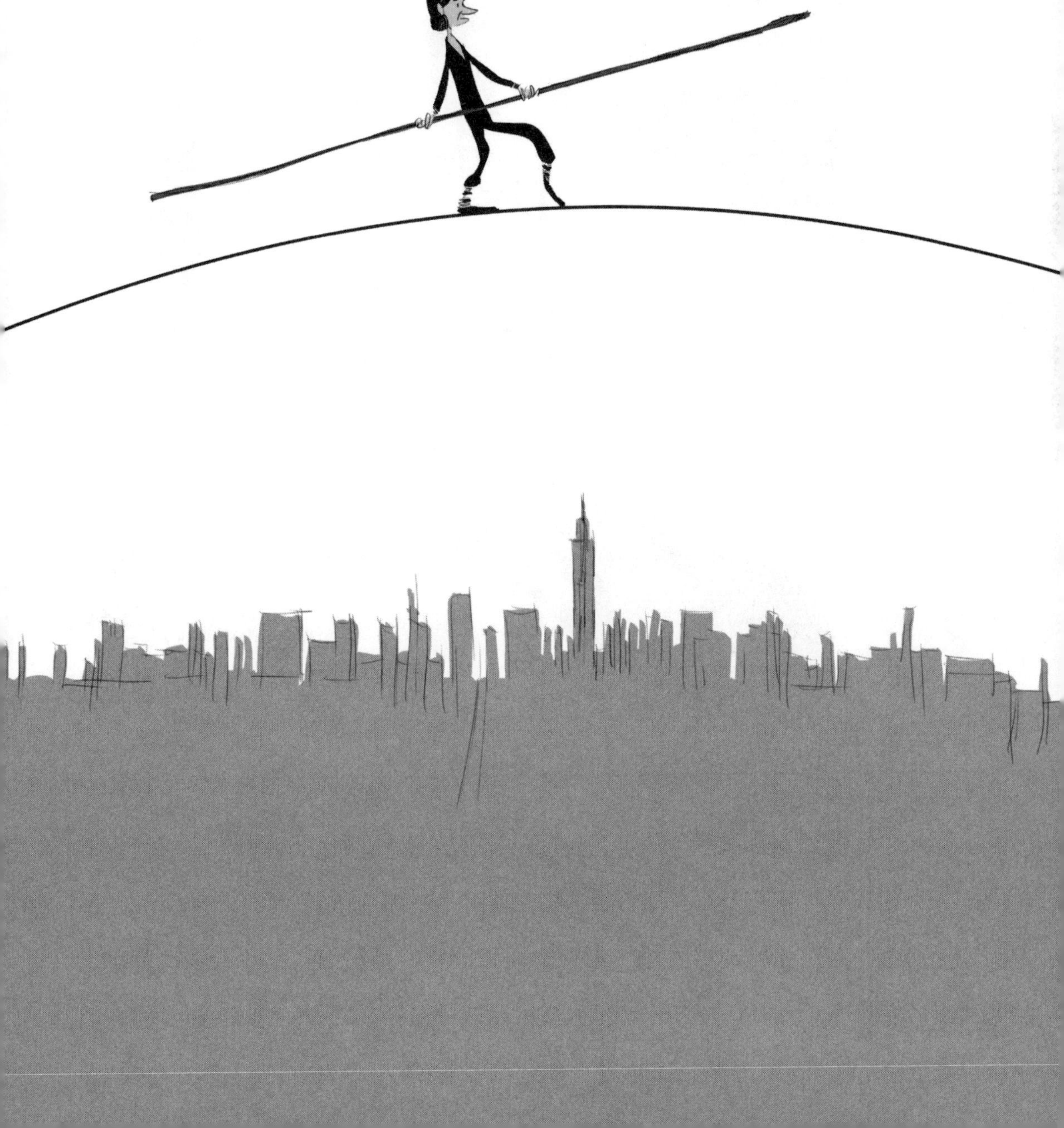

ECONOMICS IS ABOUT THE **ACTIONS AND INTERACTIONS OF OPTIMIZING INDIVIDUALS...**

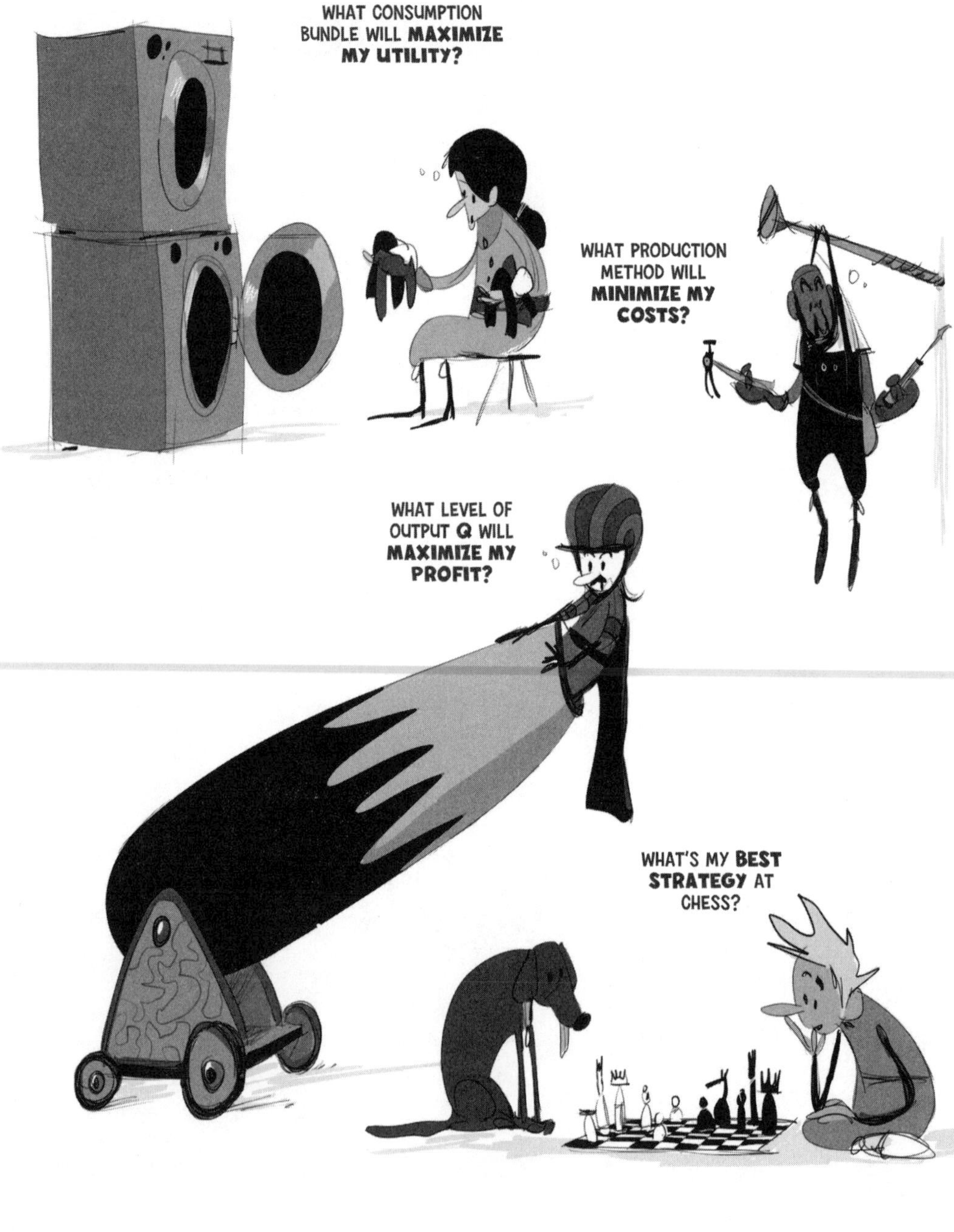
WHAT CONSUMPTION BUNDLE WILL MAXIMIZE MY UTILITY?
WHAT PRODUCTION METHOD WILL MINIMIZE MY COSTS?
WHAT LEVEL OF OUTPUT Q WILL MAXIMIZE MY PROFIT?
WHAT'S MY BEST STRATEGY AT CHESS?

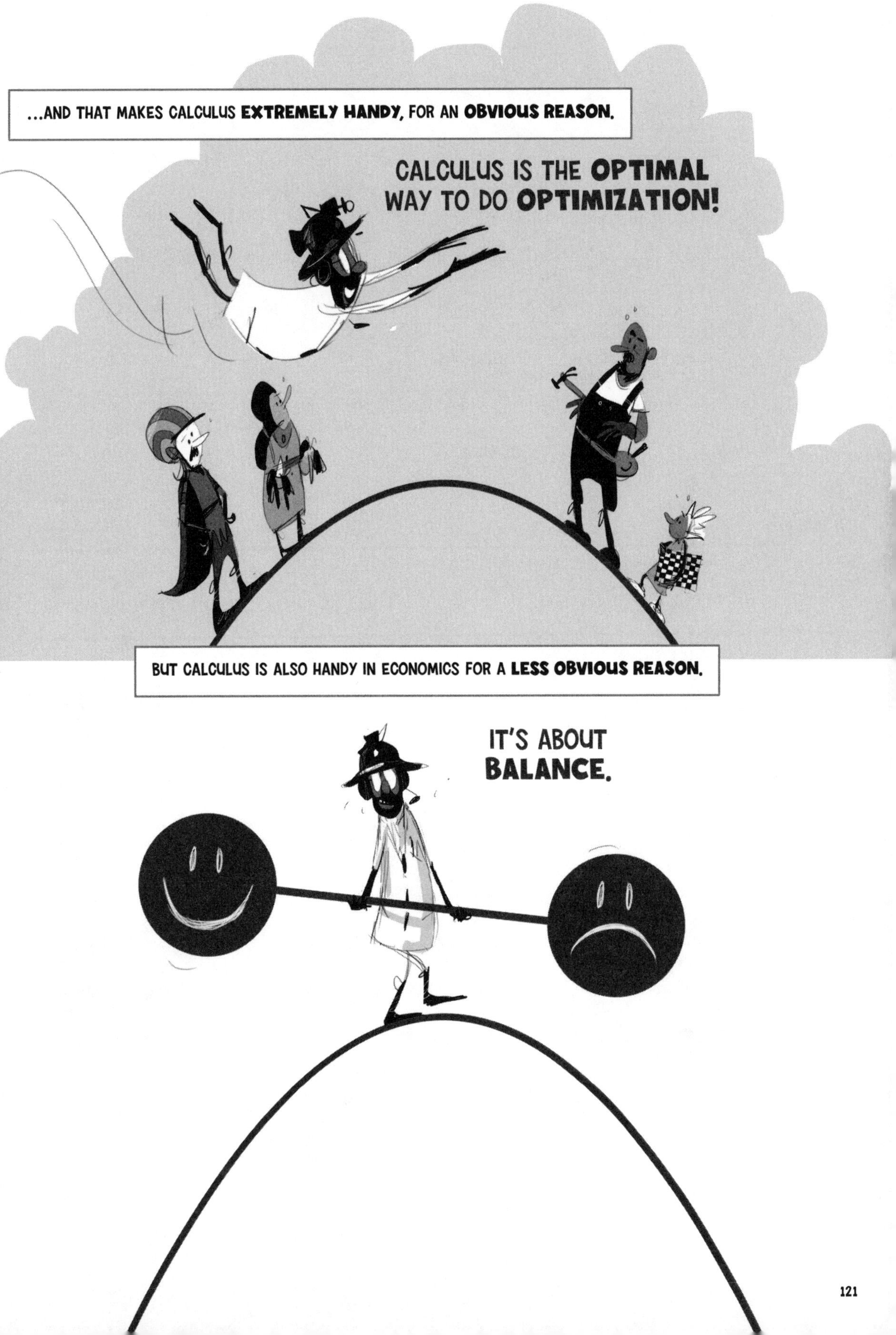
...AND THAT MAKES CALCULUS **EXTREMELY HANDY**, FOR AN **OBVIOUS REASON.**
CALCULUS IS THE **OPTIMAL** WAY TO DO **OPTIMIZATION!**
BUT CALCULUS IS ALSO HANDY IN ECONOMICS FOR A **LESS OBVIOUS REASON.**
IT'S ABOUT **BALANCE.**

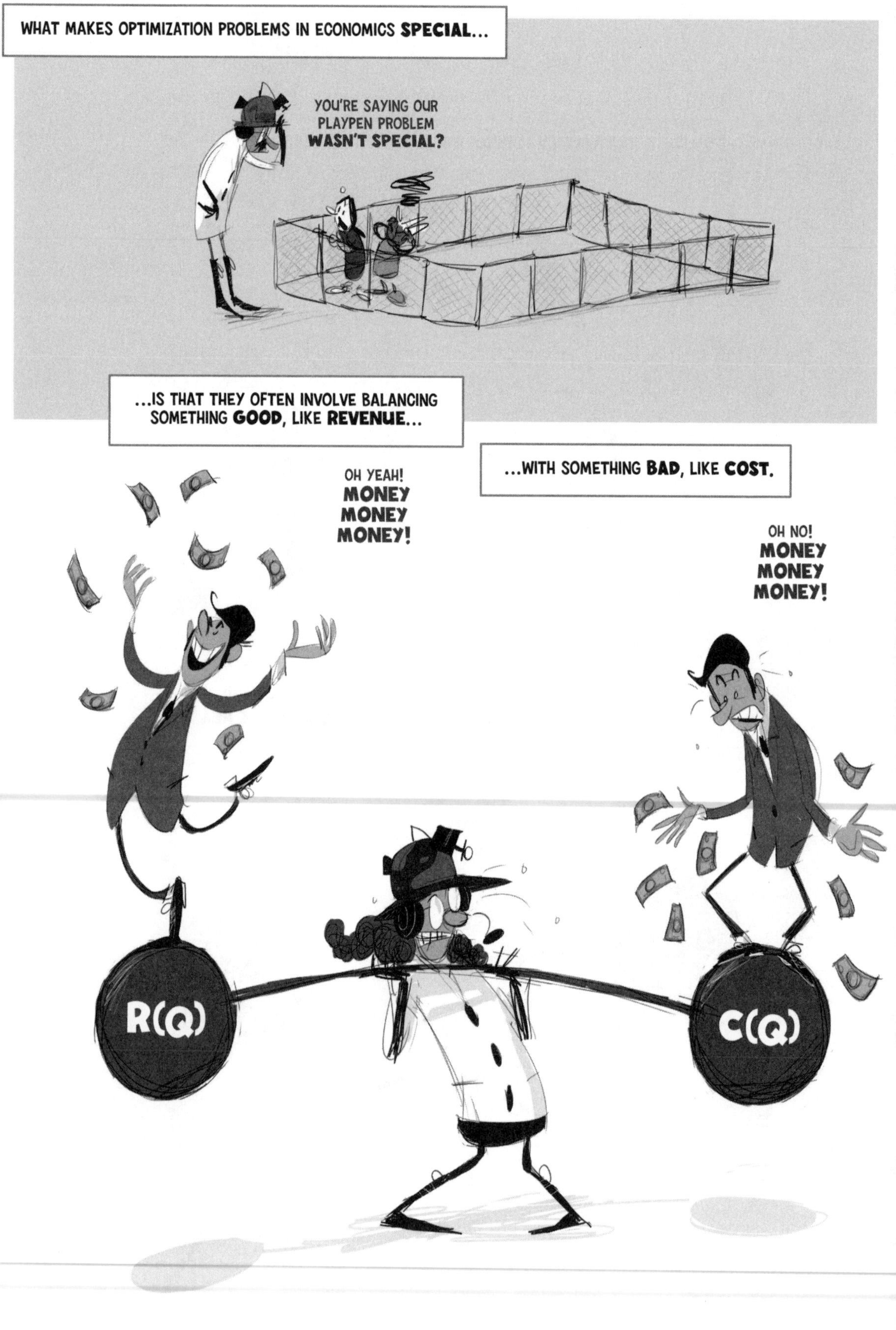
WHAT MAKES OPTIMIZATION PROBLEMS IN ECONOMICS **SPECIAL**...
YOU'RE SAYING OUR PLAYPEN PROBLEM **WASN'T SPECIAL?**
...IS THAT THEY OFTEN INVOLVE BALANCING SOMETHING **GOOD**, LIKE **REVENUE**...
OH YEAH! **MONEY MONEY MONEY!**
...WITH SOMETHING **BAD**, LIKE **COST**.
OH NO! **MONEY MONEY MONEY!**
R(Q)
C(Q)

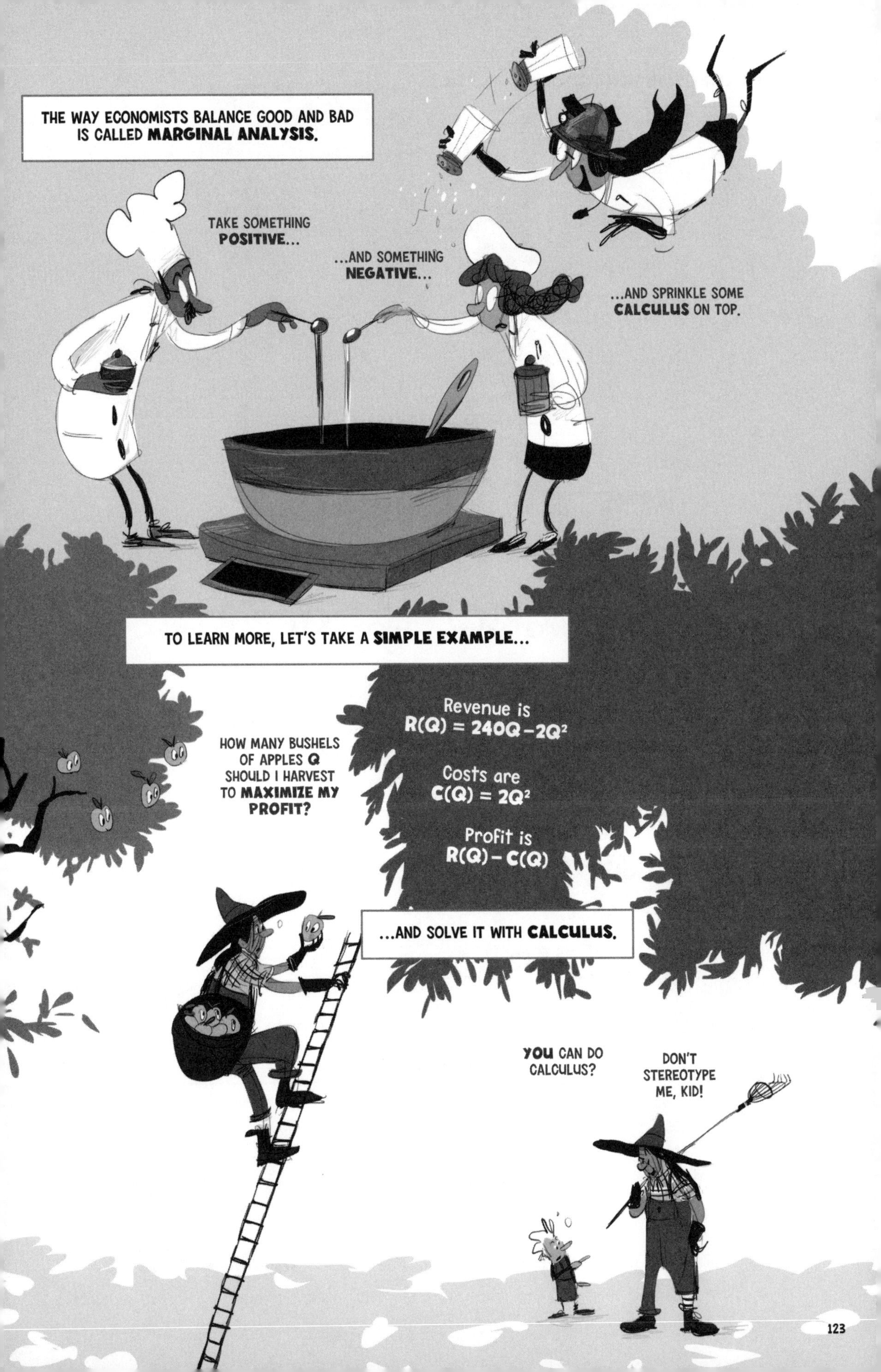
THE WAY ECONOMISTS BALANCE GOOD AND BAD IS CALLED **MARGINAL ANALYSIS.**
TAKE SOMETHING **POSITIVE...**
...AND SOMETHING **NEGATIVE...**
...AND SPRINKLE SOME **CALCULUS** ON TOP.
TO LEARN MORE, LET'S TAKE A **SIMPLE EXAMPLE...**
HOW MANY BUSHELS OF APPLES Q SHOULD I HARVEST TO **MAXIMIZE MY PROFIT?**
Revenue is $R(Q) = 240Q - 2Q^2$
Costs are $C(Q) = 2Q^2$
Profit is $R(Q) - C(Q)$
...AND SOLVE IT WITH **CALCULUS.**
YOU CAN DO CALCULUS?
DON'T STEREOTYPE ME, KID!

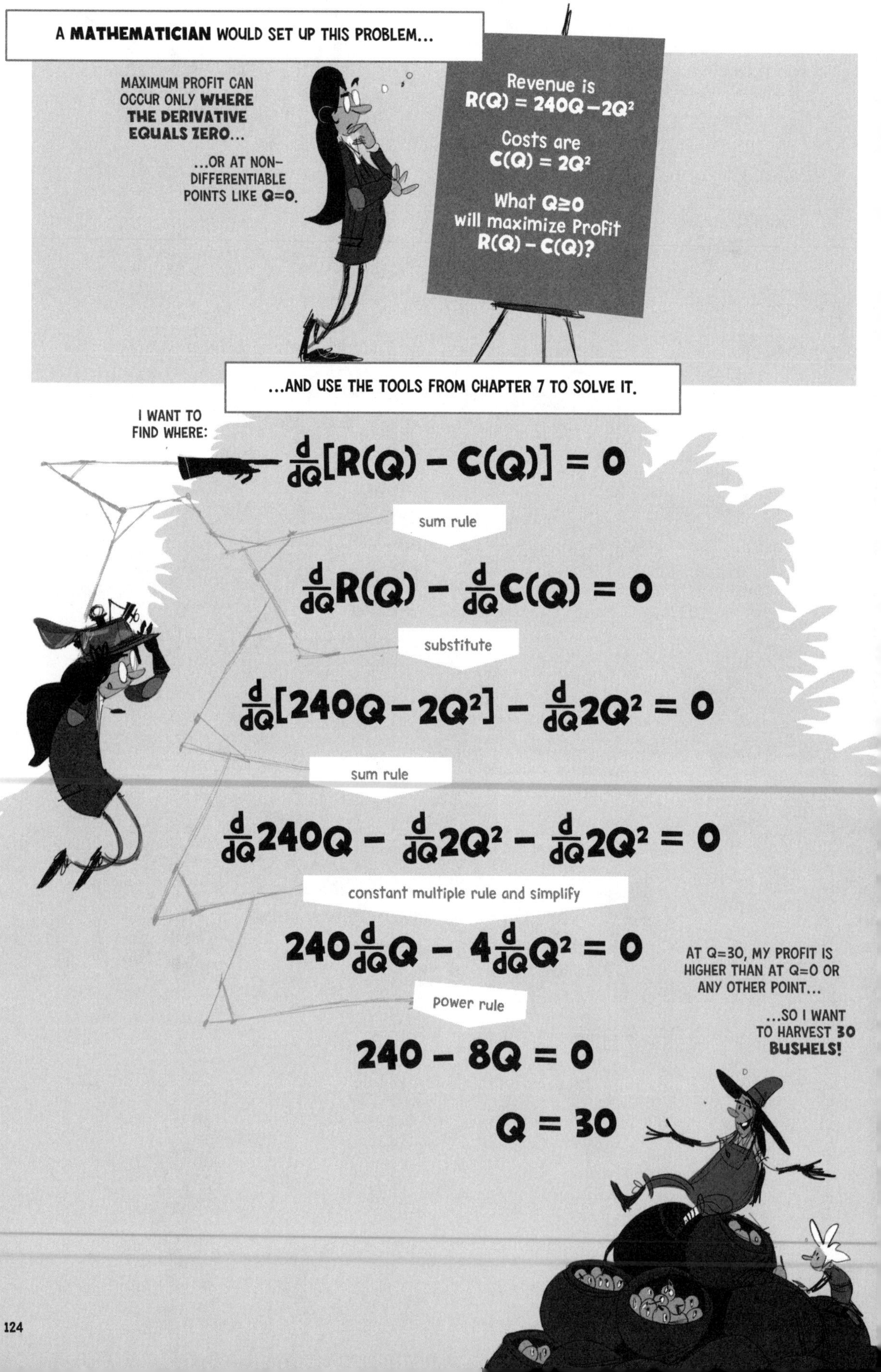
A MATHEMATICIAN WOULD SET UP THIS PROBLEM...
MAXIMUM PROFIT CAN OCCUR ONLY WHERE THE DERIVATIVE EQUALS ZERO...
...OR AT NON-DIFFERENTIABLE POINTS LIKE Q=0.
Revenue is $R(Q) = 240Q - 2Q^2$
Costs are $C(Q) = 2Q^2$
What $Q \geq 0$ will maximize Profit $R(Q) - C(Q)$?
...AND USE THE TOOLS FROM CHAPTER 7 TO SOLVE IT.
I WANT TO FIND WHERE:
$\frac{d}{dQ}[R(Q) - C(Q)] = 0$
sum rule
$\frac{d}{dQ}R(Q) - \frac{d}{dQ}C(Q) = 0$
substitute
$\frac{d}{dQ}[240Q - 2Q^2] - \frac{d}{dQ}2Q^2 = 0$
sum rule
$\frac{d}{dQ}240Q - \frac{d}{dQ}2Q^2 - \frac{d}{dQ}2Q^2 = 0$
constant multiple rule and simplify
$240\frac{d}{dQ}Q - 4\frac{d}{dQ}Q^2 = 0$
power rule
$240 - 8Q = 0$
$Q = 30$
AT Q=30, MY PROFIT IS HIGHER THAN AT Q=0 OR ANY OTHER POINT...
...SO I WANT TO HARVEST 30 BUSHELS!

AN **ECONOMIST** WOULD HEAD FOR THE SAME DESTINATION...

SEE YOU AT Q=30?

Revenue is
$R(Q) = 240Q - 2Q^2$

Costs are
$C(Q) = 2Q^2$

What $Q \geq 0$
will maximize Profit
$R(Q) - C(Q)$?

YES, BUT I'M GOING A **DIFFERENT WAY.**

...AND WOULD START DOWN THE SAME PATH...

I WANT TO FIND WHERE:

$$\frac{d}{dQ}[R(Q) - C(Q)] = 0$$

sum rule

$$\frac{d}{dQ}R(Q) - \frac{d}{dQ}C(Q) = 0$$

...BUT WOULD TAKE THE **SCENIC ROUTE**...

rearrange

$$\frac{d}{dQ}R(Q) = \frac{d}{dQ}C(Q)$$

...IN SEARCH OF ECONOMICS **ENLIGHTENMENT.**

AHA! **MARGINAL REVENUE...** ...EQUALS **MARGINAL COST!**

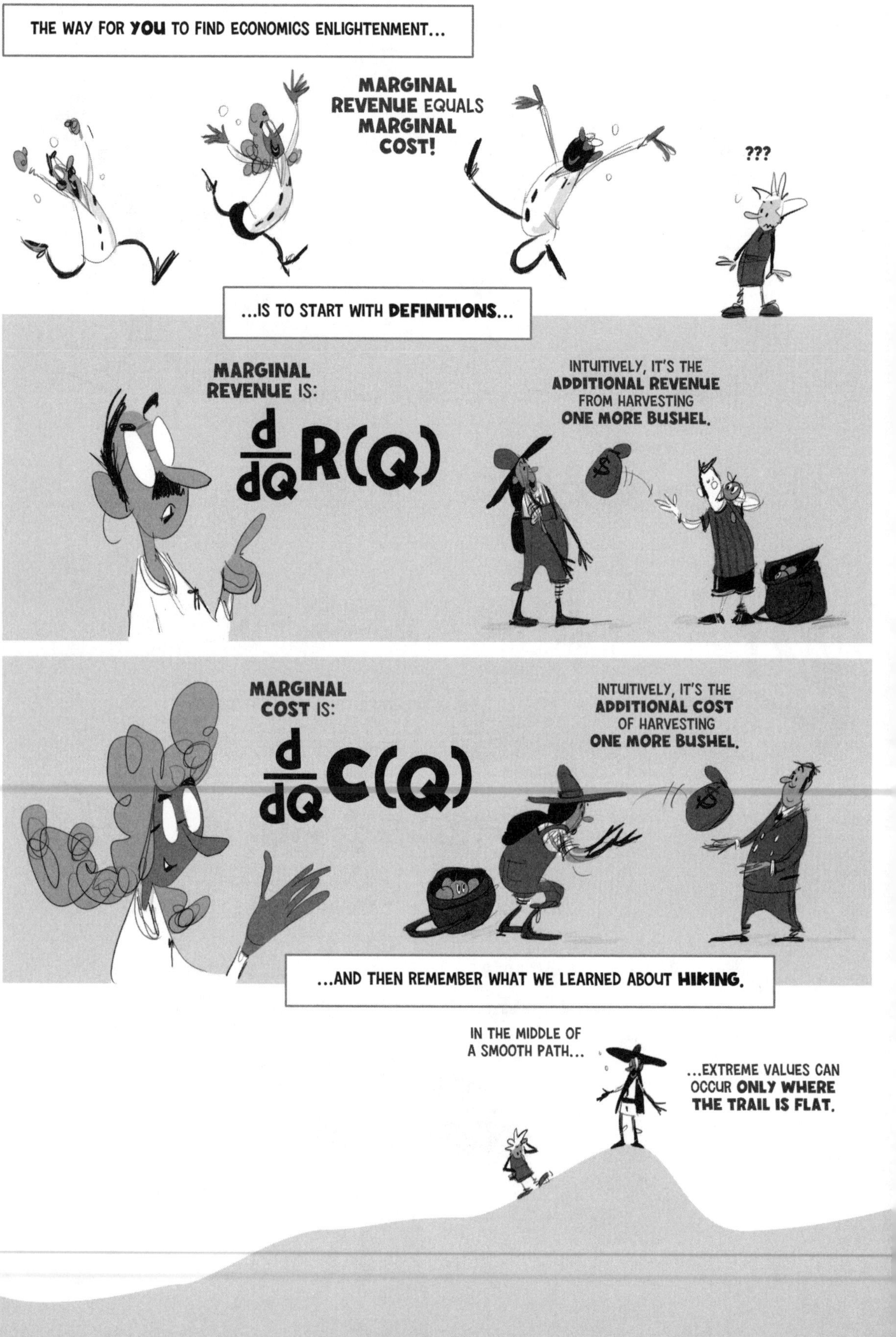
THE WAY FOR **YOU** TO FIND ECONOMICS ENLIGHTENMENT...
MARGINAL REVENUE EQUALS **MARGINAL COST!**
???
...IS TO START WITH **DEFINITIONS...**
MARGINAL REVENUE IS:
$\frac{d}{dQ}R(Q)$
INTUITIVELY, IT'S THE **ADDITIONAL REVENUE** FROM HARVESTING **ONE MORE BUSHEL.**
MARGINAL COST IS:
$\frac{d}{dQ}C(Q)$
INTUITIVELY, IT'S THE **ADDITIONAL COST** OF HARVESTING **ONE MORE BUSHEL.**
...AND THEN REMEMBER WHAT WE LEARNED ABOUT **HIKING.**
IN THE MIDDLE OF A SMOOTH PATH...
...EXTREME VALUES CAN OCCUR **ONLY WHERE THE TRAIL IS FLAT.**

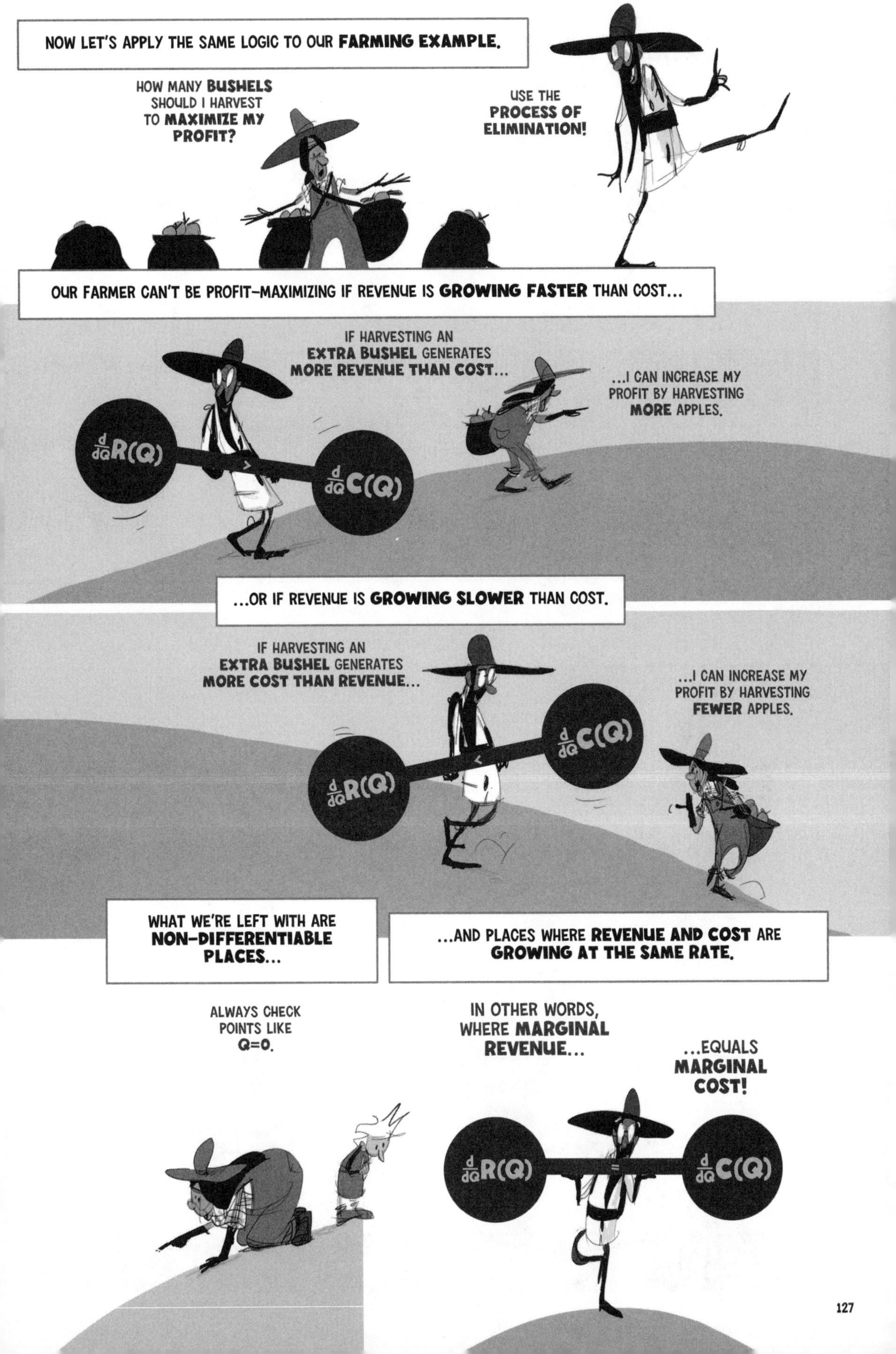
NOW LET'S APPLY THE SAME LOGIC TO OUR FARMING EXAMPLE.
HOW MANY BUSHELS SHOULD I HARVEST TO MAXIMIZE MY PROFIT?
USE THE PROCESS OF ELIMINATION!
OUR FARMER CAN'T BE PROFIT-MAXIMIZING IF REVENUE IS GROWING FASTER THAN COST...
IF HARVESTING AN EXTRA BUSHEL GENERATES MORE REVENUE THAN COST...
...I CAN INCREASE MY PROFIT BY HARVESTING MORE APPLES.
$\frac{d}{dQ}R(Q)$ > $\frac{d}{dQ}C(Q)$
...OR IF REVENUE IS GROWING SLOWER THAN COST.
IF HARVESTING AN EXTRA BUSHEL GENERATES MORE COST THAN REVENUE...
...I CAN INCREASE MY PROFIT BY HARVESTING FEWER APPLES.
$\frac{d}{dQ}R(Q)$ < $\frac{d}{dQ}C(Q)$
WHAT WE'RE LEFT WITH ARE NON-DIFFERENTIABLE PLACES...
...AND PLACES WHERE REVENUE AND COST ARE GROWING AT THE SAME RATE.
ALWAYS CHECK POINTS LIKE Q=0.
IN OTHER WORDS, WHERE MARGINAL REVENUE...
...EQUALS MARGINAL COST!
$\frac{d}{dQ}R(Q)$ = $\frac{d}{dQ}C(Q)$

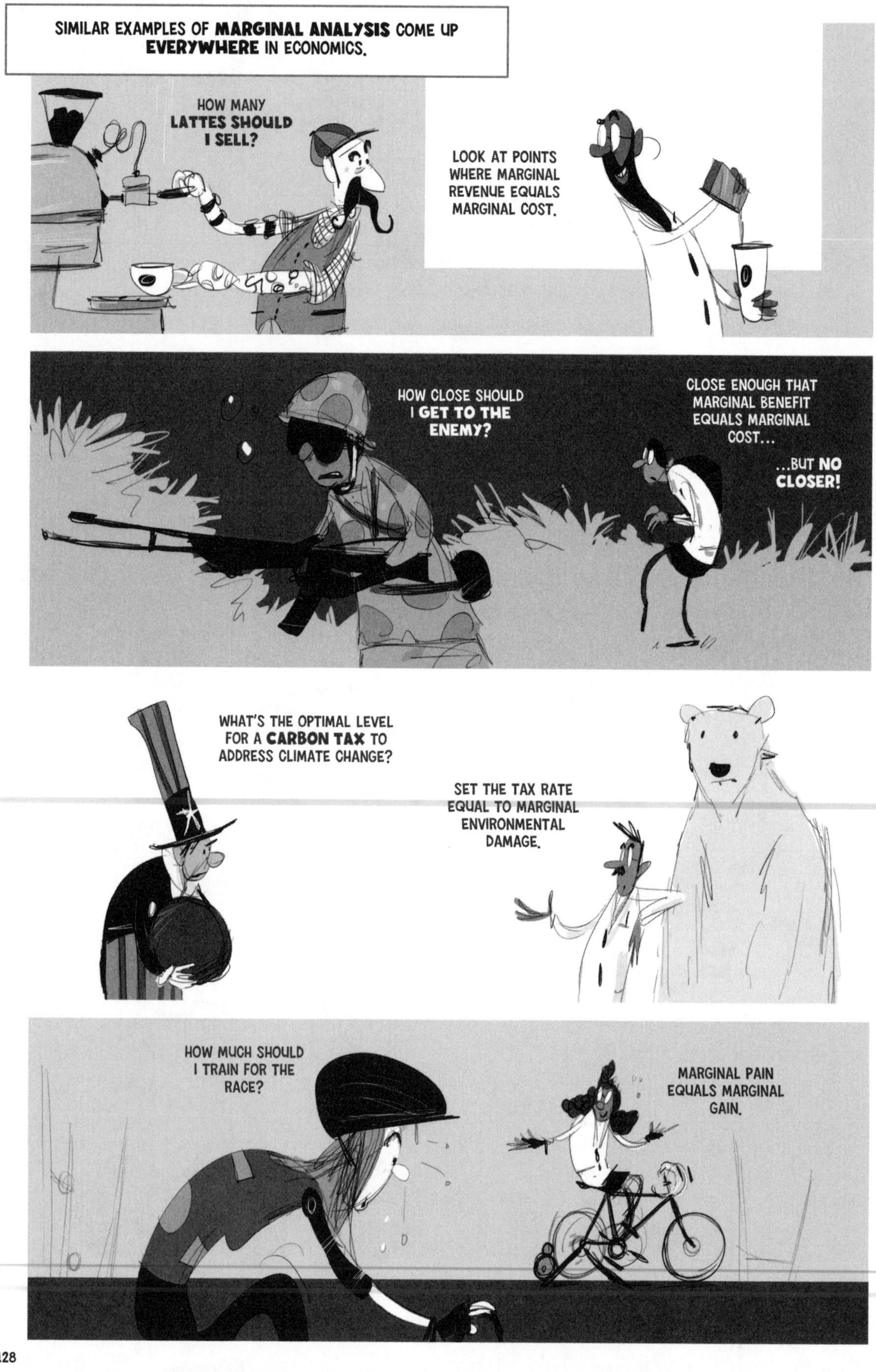
SIMILAR EXAMPLES OF **MARGINAL ANALYSIS** COME UP **EVERYWHERE** IN ECONOMICS.
HOW MANY **LATTES SHOULD I SELL?**
LOOK AT POINTS WHERE MARGINAL REVENUE EQUALS MARGINAL COST.
HOW CLOSE SHOULD I **GET TO THE ENEMY?**
CLOSE ENOUGH THAT MARGINAL BENEFIT EQUALS MARGINAL COST...
...BUT **NO CLOSER!**
WHAT'S THE OPTIMAL LEVEL FOR A **CARBON TAX** TO ADDRESS CLIMATE CHANGE?
SET THE TAX RATE EQUAL TO MARGINAL ENVIRONMENTAL DAMAGE.
HOW MUCH SHOULD I TRAIN FOR THE RACE?
MARGINAL PAIN EQUALS MARGINAL GAIN.

OF COURSE, WE SHOULDN'T FORGET TO CHECK **CORNER SOLUTIONS** AND OTHER **NON-DIFFERENTIABLE POINTS...**

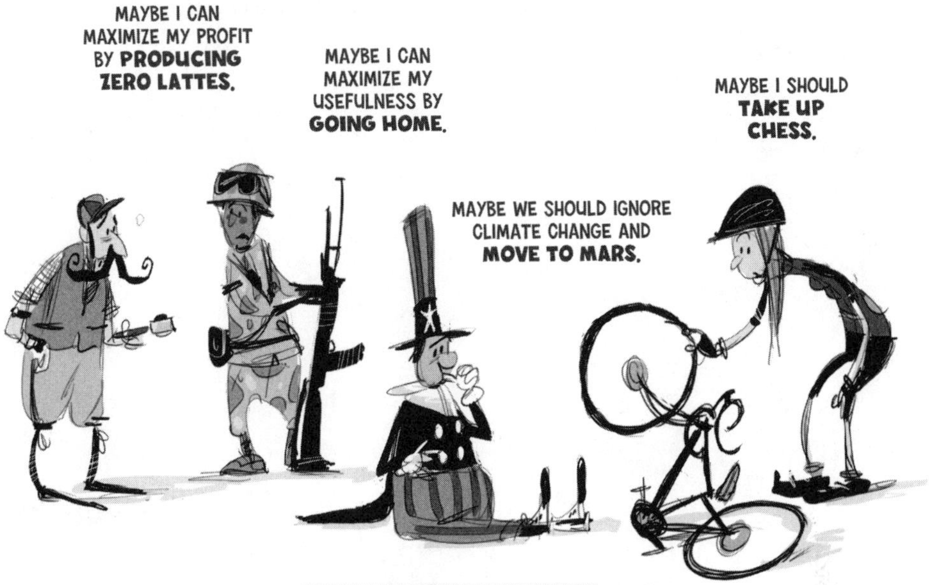
MAYBE I CAN MAXIMIZE MY PROFIT BY **PRODUCING ZERO LATTES.**
MAYBE I CAN MAXIMIZE MY USEFULNESS BY **GOING HOME.**
MAYBE WE SHOULD IGNORE CLIMATE CHANGE AND **MOVE TO MARS.**
MAYBE I SHOULD **TAKE UP CHESS.**

...BUT CALCULUS NONETHELESS GIVES US A **KEY FOCAL POINT.**

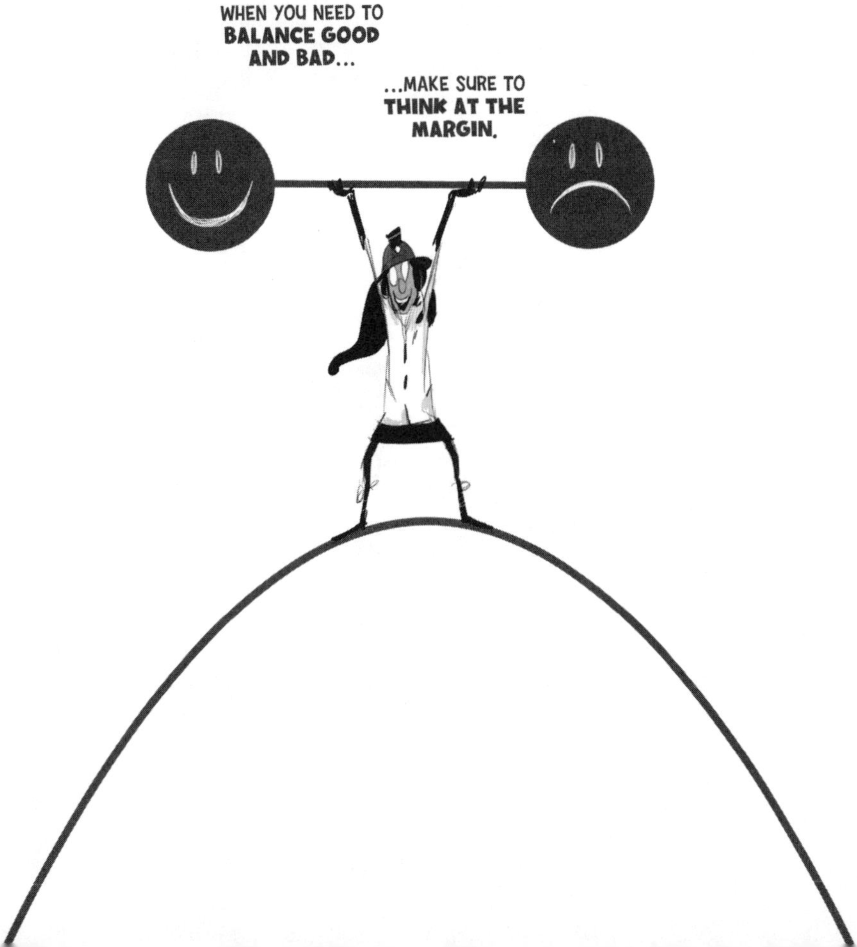
WHEN YOU NEED TO **BALANCE GOOD AND BAD...**
...MAKE SURE TO **THINK AT THE MARGIN.**

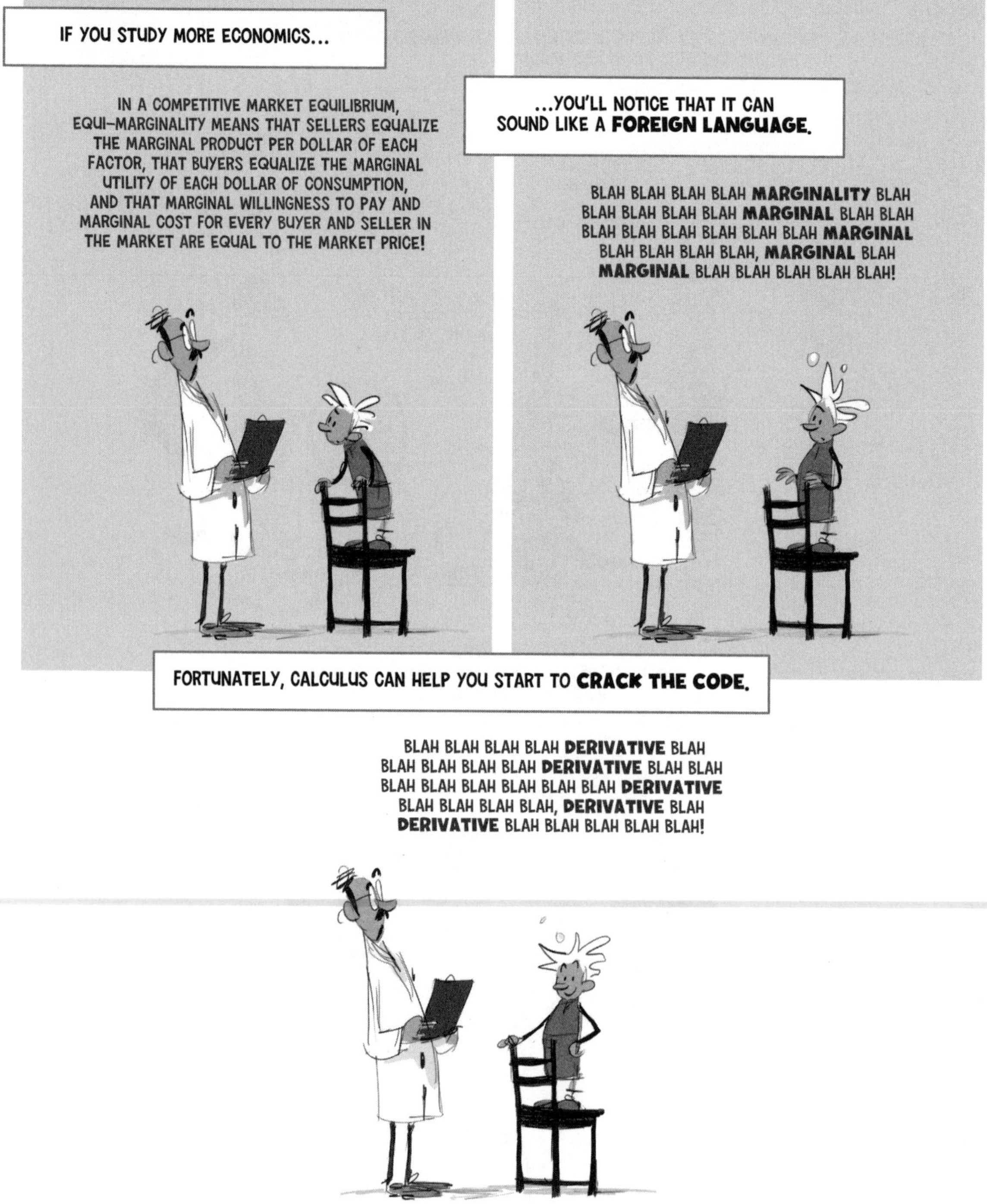
IF YOU STUDY MORE ECONOMICS...
IN A COMPETITIVE MARKET EQUILIBRIUM, EQUI-MARGINALITY MEANS THAT SELLERS EQUALIZE THE MARGINAL PRODUCT PER DOLLAR OF EACH FACTOR, THAT BUYERS EQUALIZE THE MARGINAL UTILITY OF EACH DOLLAR OF CONSUMPTION, AND THAT MARGINAL WILLINGNESS TO PAY AND MARGINAL COST FOR EVERY BUYER AND SELLER IN THE MARKET ARE EQUAL TO THE MARKET PRICE!
...YOU'LL NOTICE THAT IT CAN SOUND LIKE A **FOREIGN LANGUAGE.**
BLAH BLAH BLAH BLAH **MARGINALITY** BLAH BLAH BLAH BLAH BLAH **MARGINAL** BLAH BLAH BLAH BLAH BLAH BLAH BLAH BLAH **MARGINAL** BLAH BLAH BLAH BLAH, **MARGINAL** BLAH **MARGINAL** BLAH BLAH BLAH BLAH BLAH!
FORTUNATELY, CALCULUS CAN HELP YOU START TO **CRACK THE CODE.**
BLAH BLAH BLAH BLAH **DERIVATIVE** BLAH BLAH BLAH BLAH BLAH **DERIVATIVE** BLAH BLAH BLAH BLAH BLAH BLAH BLAH BLAH **DERIVATIVE** BLAH BLAH BLAH BLAH, **DERIVATIVE** BLAH **DERIVATIVE** BLAH BLAH BLAH BLAH BLAH!

PART THREE
INTEGRALS

CHAPTER 11
INTEGRATION, THE HARD WAY

A JOURNEY OF
A THOUSAND MILES
BEGINS WITH A
SINGLE STEP.

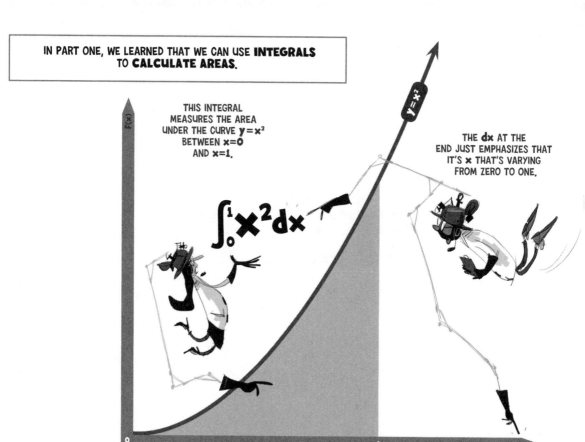

IN PART ONE, WE LEARNED THAT WE CAN USE **INTEGRALS** TO **CALCULATE AREAS.**

THIS INTEGRAL MEASURES THE AREA UNDER THE CURVE $y=x^2$ BETWEEN $x=0$ AND $x=1$.

THE **dx** AT THE END JUST EMPHASIZES THAT IT'S **x** THAT'S VARYING FROM ZERO TO ONE.

$$\int_0^1 x^2 dx$$

$y=x^2$

WE ALSO LEARNED THAT THE **EASY WAY TO CLIMB MT. INTEGRAL** IS TO TAKE THE **ZIPLINE FROM MT. DERIVATIVE...**

BOO!

THE **EASY WAY** IS FOR **WIMPS!**

...BUT WE'RE NOT GOING TO DO THAT UNTIL THE **NEXT CHAPTER.**

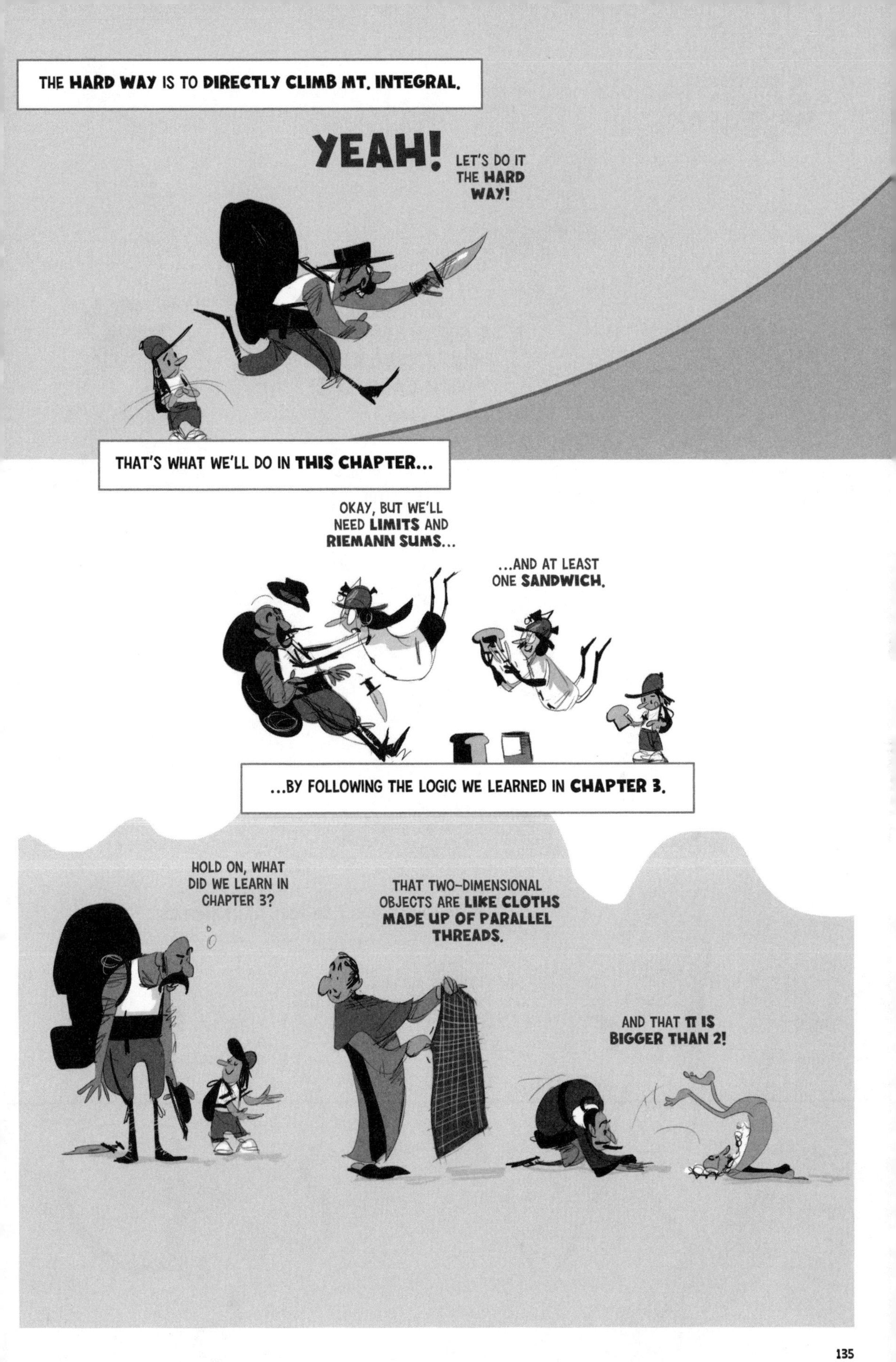
THE **HARD WAY** IS TO **DIRECTLY CLIMB MT. INTEGRAL.**
YEAH!
LET'S DO IT THE **HARD WAY!**
THAT'S WHAT WE'LL DO IN **THIS CHAPTER...**
OKAY, BUT WE'LL NEED **LIMITS** AND **RIEMANN SUMS...**
...AND AT LEAST ONE **SANDWICH.**
...BY FOLLOWING THE LOGIC WE LEARNED IN **CHAPTER 3.**
HOLD ON, WHAT DID WE LEARN IN CHAPTER 3?
THAT TWO-DIMENSIONAL OBJECTS ARE **LIKE CLOTHS MADE UP OF PARALLEL THREADS.**
AND THAT **π IS BIGGER THAN 2!**

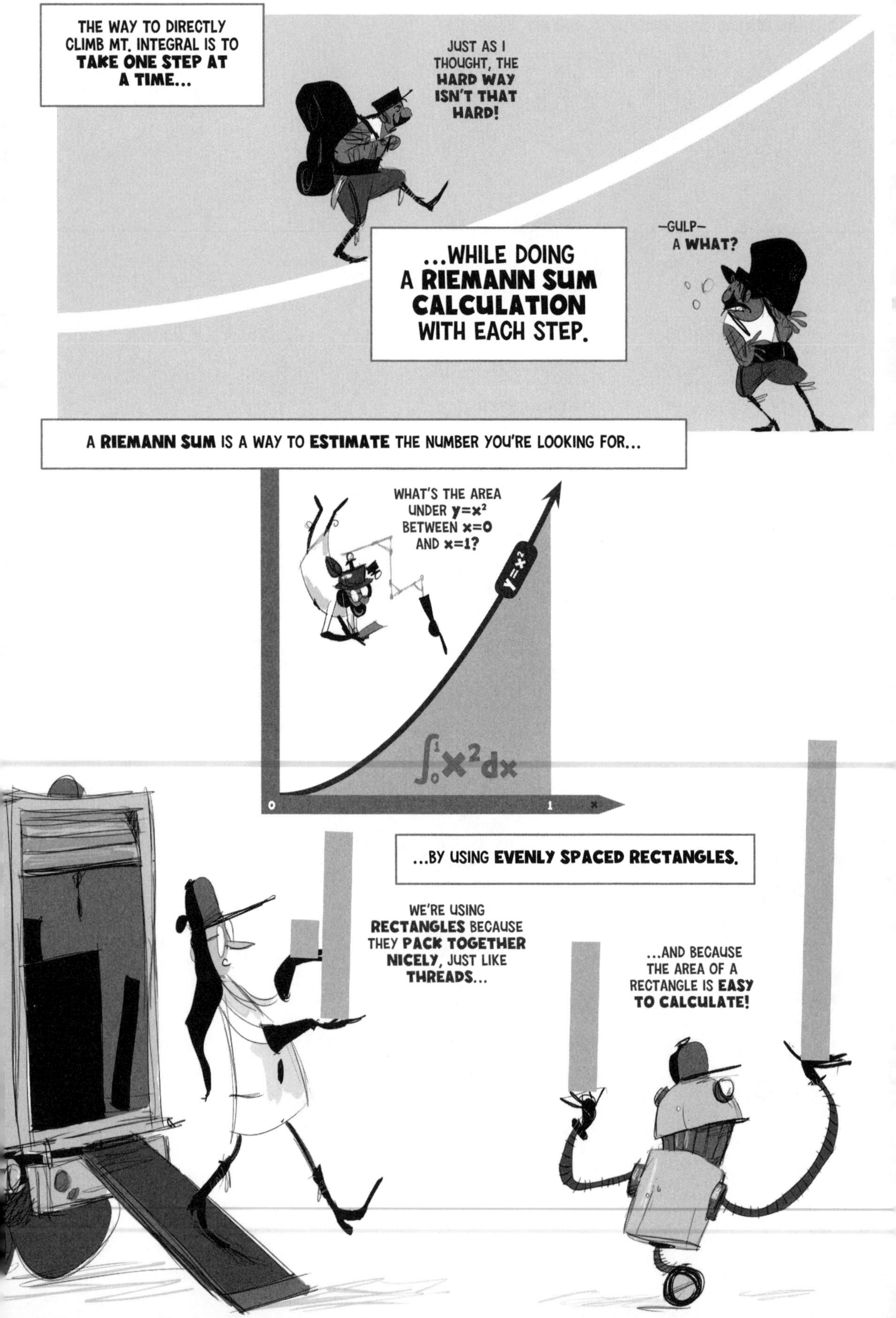
THE WAY TO DIRECTLY CLIMB MT. INTEGRAL IS TO TAKE ONE STEP AT A TIME...
JUST AS I THOUGHT, THE HARD WAY ISN'T THAT HARD!
...WHILE DOING A RIEMANN SUM CALCULATION WITH EACH STEP.
—GULP— A WHAT?
A RIEMANN SUM IS A WAY TO ESTIMATE THE NUMBER YOU'RE LOOKING FOR...
WHAT'S THE AREA UNDER $y=x^2$ BETWEEN $x=0$ AND $x=1$?
$y=x^2$
$\int_0^1 x^2 dx$
0
1
x
...BY USING EVENLY SPACED RECTANGLES.
WE'RE USING RECTANGLES BECAUSE THEY PACK TOGETHER NICELY, JUST LIKE THREADS...
...AND BECAUSE THE AREA OF A RECTANGLE IS EASY TO CALCULATE!

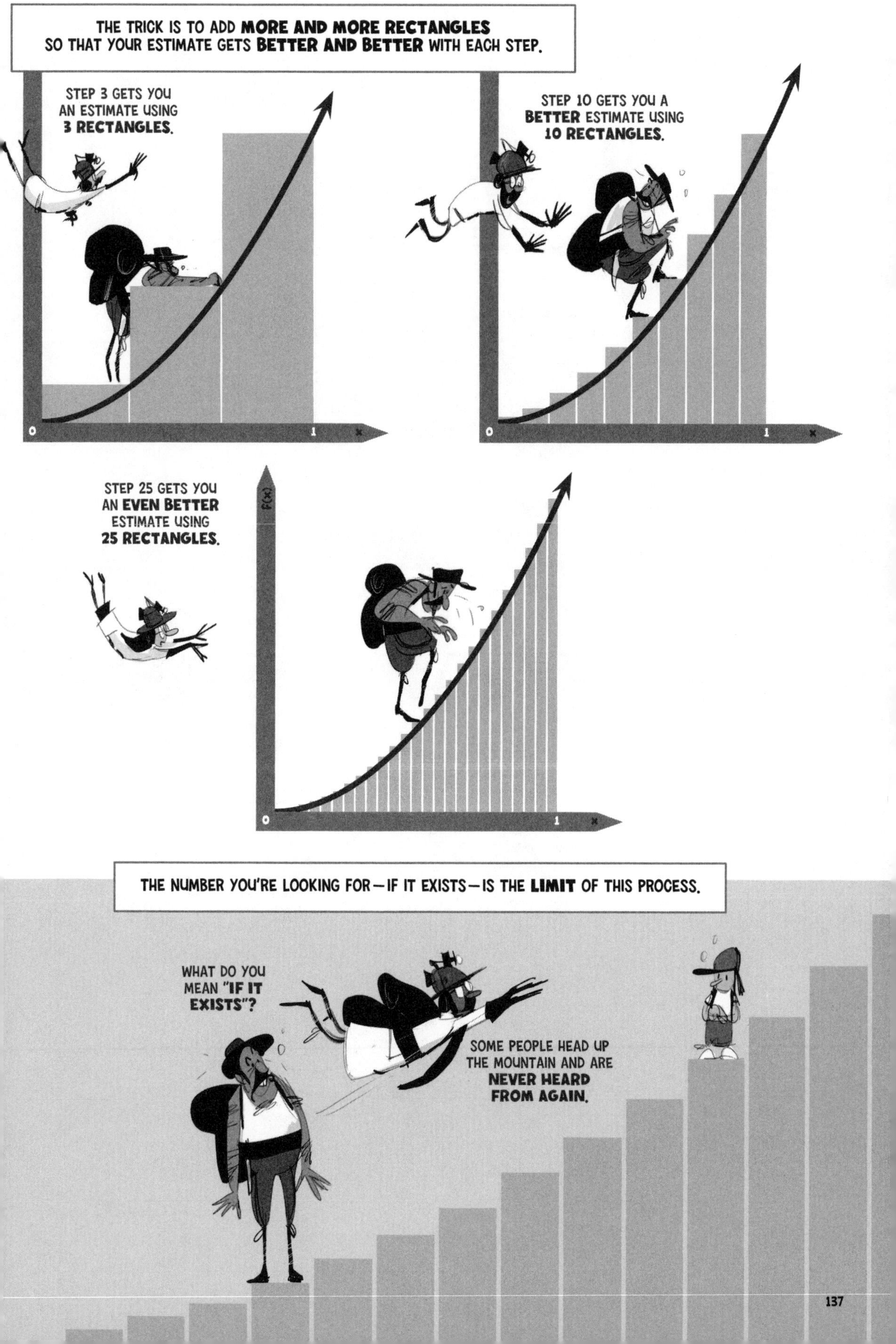
THE TRICK IS TO ADD MORE AND MORE RECTANGLES
SO THAT YOUR ESTIMATE GETS BETTER AND BETTER WITH EACH STEP.
STEP 3 GETS YOU AN ESTIMATE USING 3 RECTANGLES.
0
1
x
STEP 10 GETS YOU A BETTER ESTIMATE USING 10 RECTANGLES.
0
1
x
STEP 25 GETS YOU AN EVEN BETTER ESTIMATE USING 25 RECTANGLES.
f(x)
0
1
x
THE NUMBER YOU'RE LOOKING FOR—IF IT EXISTS—IS THE LIMIT OF THIS PROCESS.
WHAT DO YOU MEAN "IF IT EXISTS"?
SOME PEOPLE HEAD UP THE MOUNTAIN AND ARE NEVER HEARD FROM AGAIN.

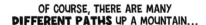

OF COURSE, THERE ARE MANY **DIFFERENT PATHS** UP A MOUNTAIN...

TO GET TO THE TOP OF **MT. EVEREST,** YOU CAN USE THE **SOUTHEAST RIDGE ROUTE**...

...OR THE **NORTH RIDGE ROUTE**...

...OR THE TREACHEROUS **KANGSHUNG FACE!**

...BUT UNLESS **DISASTER STRIKES**...

...THEY ALL END UP AT THE **SAME PLACE.**

THE SUMMIT!

SIMILARLY, THERE ARE MANY **DIFFERENT WAYS** TO ESTIMATE AN AREA USING RIEMANN SUMS...

LIKE **UPPER** RIEMANN SUMS...

...WHICH **OVERESTIMATE** THE ACTUAL AREA.

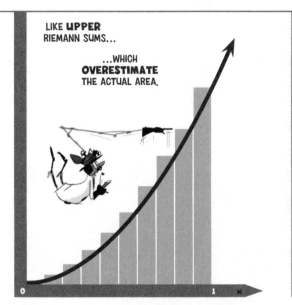

AND **LOWER** RIEMANN SUMS...

...WHICH **UNDERESTIMATE** THE ACTUAL AREA.

...BUT UNLESS **DISASTER STRIKES**...

WHAT HAPPENED TO YOU?

I FELL HEADFIRST INTO AN $f(x)$ THAT EQUALS ONE IF x IS RATIONAL AND ZERO IF x IS IRRATIONAL!

LOOKS LIKE WE NEED A **LEBESGUE MEASURE**.

...IN THE **LIMIT** THEY ALL END UP AT THE **SAME PLACE**.

FANCY MEETING YOU HERE!

AS AN EXAMPLE, LET'S RETURN TO THE INTEGRAL WE LOOKED AT EARLIER...

WE'RE CALCULATING THE AREA UNDER THE CURVE $y=x^2$ BETWEEN $x=0$ AND $x=1$.

$$\int_0^1 x^2\, dx$$

...AND LET'S BEGIN WITH **UPPER RIEMANN SUM** ESTIMATES.

THE RECTANGLES ARE OF **EQUAL WIDTH**...

...AND THE **HEIGHT** OF EACH RECTANGLE IS THE **LARGEST VALUE** OF $F(x)$ IN THAT RECTANGLE.

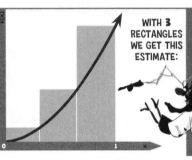

WITH **3** RECTANGLES WE GET THIS ESTIMATE:

WIDTH   HEIGHT

$$\frac{1}{3}\left(\frac{1}{3}\right)^2 + \frac{1}{3}\left(\frac{2}{3}\right)^2 + \frac{1}{3}\left(\frac{3}{3}\right)^2$$

$$= 0.518518$$

WITH **10** RECTANGLES WE GET THIS ESTIMATE:

$$\frac{1}{10}\left(\frac{1}{10}\right)^2 + \frac{1}{10}\left(\frac{2}{10}\right)^2 + \ldots + \frac{1}{10}\left(\frac{10}{10}\right)^2$$

$$= 0.385$$

WITH **25** RECTANGLES WE GET THIS ESTIMATE:

$$\frac{1}{25}\left(\frac{1}{25}\right)^2 + \frac{1}{25}\left(\frac{2}{25}\right)^2 + \ldots + \frac{1}{25}\left(\frac{25}{25}\right)^2$$

$$= 0.3536$$

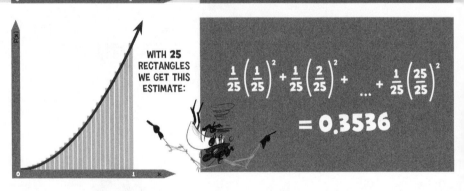

WITH **1,000** RECTANGLES WE GET THIS ESTIMATE:

$$\frac{1}{1,000}\left(\frac{1}{1,000}\right)^2 + \frac{1}{1,000}\left(\frac{2}{1,000}\right)^2 + \ldots + \frac{1}{1,000}\left(\frac{1,000}{1,000}\right)^2$$

$$= \text{Heart Attack!}$$

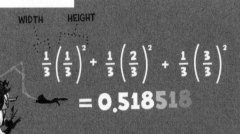

THIS EXAMPLE TURNS OUT TO BE **LUCKY** BECAUSE WE CAN ADD UP ALL THOSE RECTANGLES...

WITH **n** RECTANGLES WE GET THIS ESTIMATE:

$$\frac{1}{n}\left(\frac{1}{n}\right)^2 + \frac{1}{n}\left(\frac{2}{n}\right)^2 + ... + \frac{1}{n}\left(\frac{n}{n}\right)^2$$

...WITH A **HANDY FORMULA.**

$$= \frac{1}{3} + \frac{1}{2n} + \frac{1}{6n^2}$$

YOU CAN PROVE THIS **SUM OF SQUARES** FORMULA USING **INDUCTION.**

THIS FORMULA HELPS US **AVOID HEART ATTACKS**...

PHEW.

$$\frac{1}{1,000}\left(\frac{1}{1,000}\right)^2 + \frac{1}{1,000}\left(\frac{2}{1,000}\right)^2 + ... + \frac{1}{1,000}\left(\frac{1,000}{1,000}\right)^2$$

$$= \frac{1}{3} + \frac{1}{2,000} + \frac{1}{6,000,000}$$

...AND IT ALSO HELPS US **FIND THE LIMIT.**

THE LIMIT OF ALL THOSE UPPER RIEMANN SUM ESTIMATES IS:

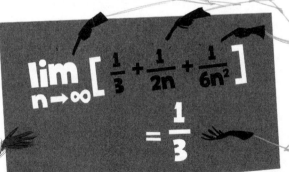

$$\lim_{n \to \infty}\left[\frac{1}{3} + \frac{1}{2n} + \frac{1}{6n^2}\right] = \frac{1}{3}$$

AS **n** GETS LARGER AND LARGER...

... $\frac{1}{2n}$ GETS CLOSER AND CLOSER TO ZERO...

...AND $\frac{1}{6n^2}$ GETS EVEN CLOSER TO ZERO...

...SO THE RIEMANN SUM ESTIMATES GET CLOSER AND CLOSER TO $\frac{1}{3}$.

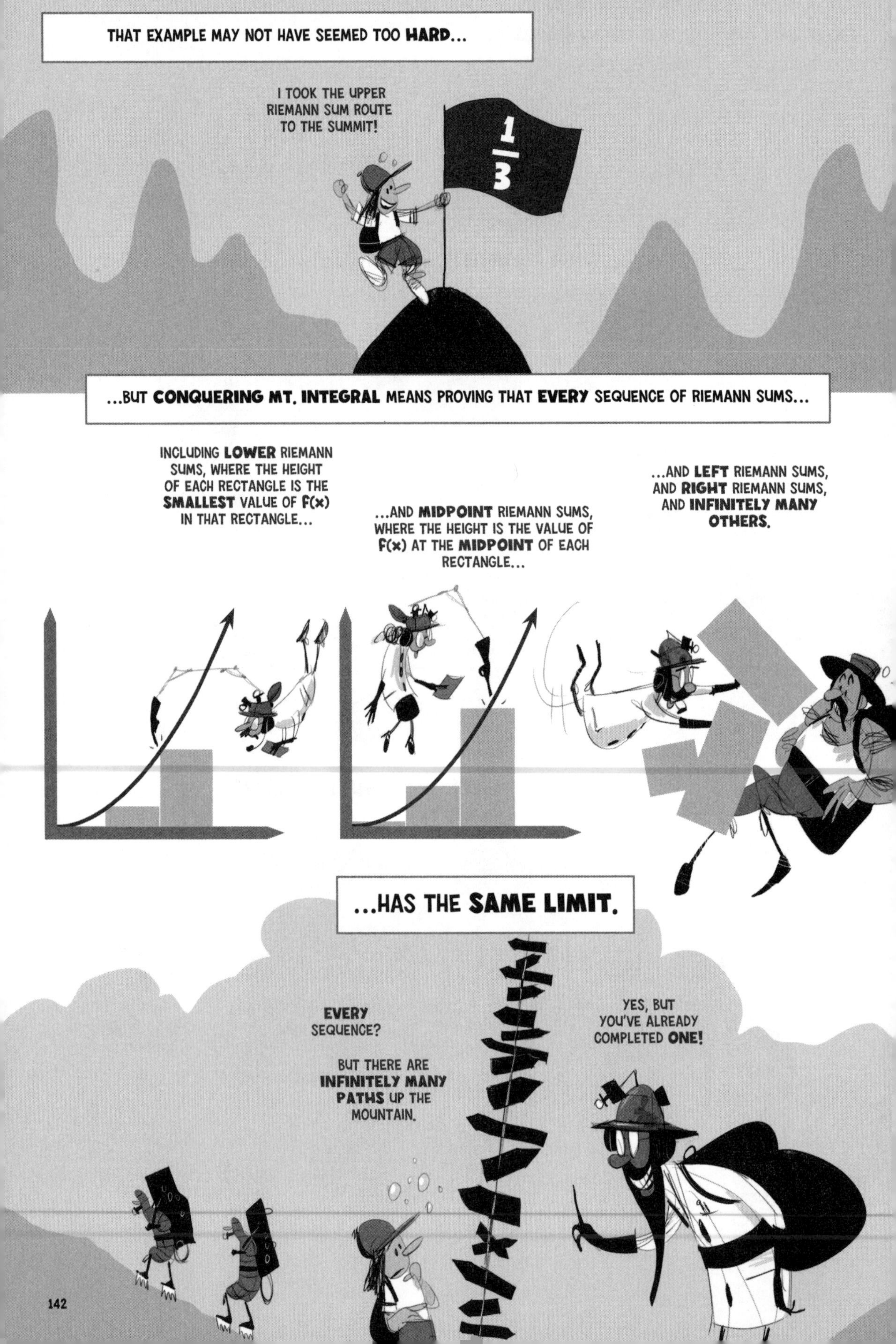
THAT EXAMPLE MAY NOT HAVE SEEMED TOO **HARD**...
I TOOK THE UPPER RIEMANN SUM ROUTE TO THE SUMMIT!
1/3
...BUT **CONQUERING MT. INTEGRAL** MEANS PROVING THAT **EVERY** SEQUENCE OF RIEMANN SUMS...
INCLUDING **LOWER** RIEMANN SUMS, WHERE THE HEIGHT OF EACH RECTANGLE IS THE **SMALLEST** VALUE OF F(x) IN THAT RECTANGLE...
...AND **MIDPOINT** RIEMANN SUMS, WHERE THE HEIGHT IS THE VALUE OF F(x) AT THE **MIDPOINT** OF EACH RECTANGLE...
...AND **LEFT** RIEMANN SUMS, AND **RIGHT** RIEMANN SUMS, AND **INFINITELY MANY OTHERS.**
...HAS THE **SAME LIMIT.**
EVERY SEQUENCE?
BUT THERE ARE **INFINITELY MANY PATHS** UP THE MOUNTAIN.
YES, BUT YOU'VE ALREADY COMPLETED **ONE!**

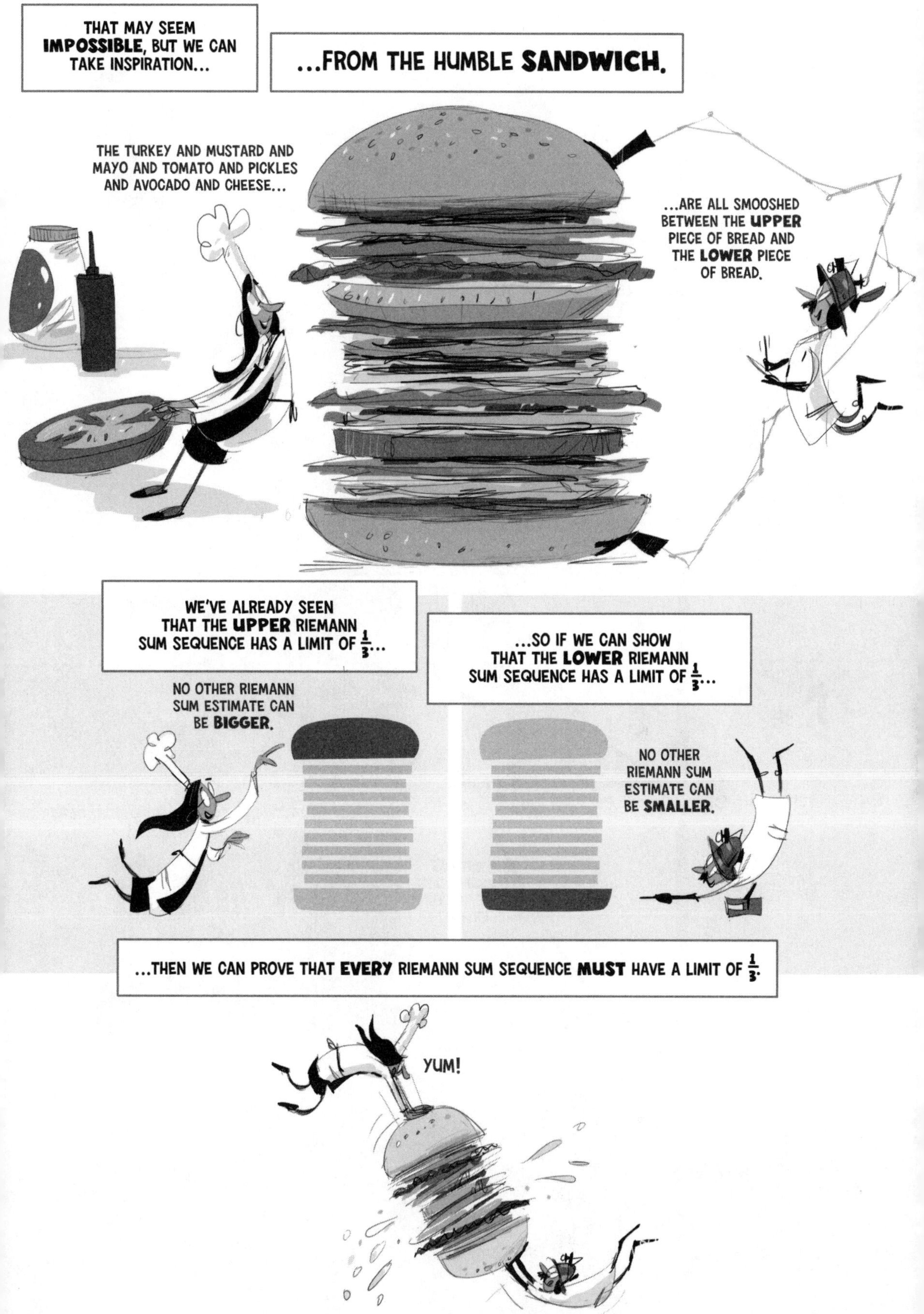

THAT MAY SEEM IMPOSSIBLE, BUT WE CAN TAKE INSPIRATION...
...FROM THE HUMBLE SANDWICH.
THE TURKEY AND MUSTARD AND MAYO AND TOMATO AND PICKLES AND AVOCADO AND CHEESE...
...ARE ALL SMOOSHED BETWEEN THE UPPER PIECE OF BREAD AND THE LOWER PIECE OF BREAD.
WE'VE ALREADY SEEN THAT THE UPPER RIEMANN SUM SEQUENCE HAS A LIMIT OF $\frac{1}{3}$...
NO OTHER RIEMANN SUM ESTIMATE CAN BE BIGGER.
...SO IF WE CAN SHOW THAT THE LOWER RIEMANN SUM SEQUENCE HAS A LIMIT OF $\frac{1}{3}$...
NO OTHER RIEMANN SUM ESTIMATE CAN BE SMALLER.
...THEN WE CAN PROVE THAT EVERY RIEMANN SUM SEQUENCE MUST HAVE A LIMIT OF $\frac{1}{3}$.
YUM!

SO LET'S RETURN TO THE **SAME MOUNTAIN** AND FINISH OUR **PROOF BY SANDWICH**...

ONE STEP AT A TIME...

...FOR THE **SECOND TIME!**

$\int_0^1 x^2 dx$

...WITH **LOWER RIEMANN SUM** ESTIMATES.

THE RECTANGLES ARE OF **EQUAL WIDTH**...

...AND THE **HEIGHT** OF EACH RECTANGLE IS THE **SMALLEST VALUE** OF F(x) IN THAT RECTANGLE.

WITH **3** RECTANGLES WE GET THIS ESTIMATE:

WIDTH   HEIGHT

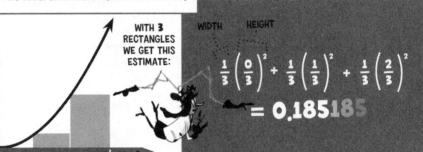

$$\frac{1}{3}\left(\frac{0}{3}\right)^2 + \frac{1}{3}\left(\frac{1}{3}\right)^2 + \frac{1}{3}\left(\frac{2}{3}\right)^2$$

$$= 0.185185$$

WITH **10** RECTANGLES WE GET THIS ESTIMATE:

$$\frac{1}{10}\left(\frac{0}{10}\right)^2 + \frac{1}{10}\left(\frac{1}{10}\right)^2 + \ldots + \frac{1}{10}\left(\frac{9}{10}\right)^2$$

$$= 0.285$$

WITH **25** RECTANGLES WE GET THIS ESTIMATE:

$$\frac{1}{25}\left(\frac{0}{25}\right)^2 + \frac{1}{25}\left(\frac{1}{25}\right)^2 + \ldots + \frac{1}{25}\left(\frac{24}{25}\right)^2$$

$$= 0.3136$$

WITH **1,000** RECTANGLES WE GET THIS ESTIMATE:

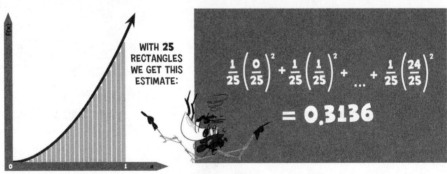

$$\frac{1}{1,000}\left(\frac{0}{1,000}\right)^2 + \frac{1}{1,000}\left(\frac{1}{1,000}\right)^2 + \ldots \frac{1}{1,000}\left(\frac{999}{1,000}\right)^2$$

$$= \textbf{BRAIN FREEZE!}$$

ONCE AGAIN WE'RE **LUCKY** BECAUSE WE CAN ADD UP ALL THOSE RECTANGLES...

WITH **n** RECTANGLES WE GET THIS ESTIMATE:

$$\frac{1}{n}\left(\frac{0}{n}\right)^2 + \frac{1}{n}\left(\frac{1}{n}\right)^2 + ... + \frac{1}{n}\left(\frac{n-1}{n}\right)^2$$

...WITH A **HANDY FORMULA.**

INDUCTION AGAIN!

$$= \frac{1}{3} - \frac{1}{2n} + \frac{1}{6n^2}$$

THIS FORMULA HELPS US **AVOID BRAIN FREEZES**...

PHEW!

$$\frac{1}{1,000}\left(\frac{0}{1,000}\right)^2 + \frac{1}{1,000}\left(\frac{1}{1,000}\right)^2 + ... + \frac{1}{1,000}\left(\frac{999}{1,000}\right)^2$$

$$= \frac{1}{3} - \frac{1}{2,000} + \frac{1}{6,000,000}$$

...AND IT ALSO ALLOWS US TO **FIND THE LIMIT.**

AS **n** GETS LARGER AND LARGER...

$\frac{1}{2n}$ GETS CLOSER AND CLOSER TO ZERO...

...AND $\frac{1}{6n^2}$ GETS EVEN CLOSER TO ZERO...

...SO THE RIEMANN SUM ESTIMATES GET CLOSER AND CLOSER TO $\frac{1}{3}$.

THE LIMIT OF THE **LOWER** RIEMANN SUMS IS THE **SAME** AS THE LIMIT OF THE **UPPER** RIEMANN SUMS!

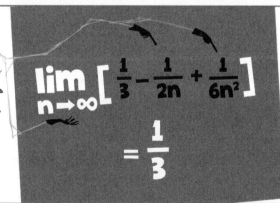

$$\lim_{n \to \infty}\left[\frac{1}{3} - \frac{1}{2n} + \frac{1}{6n^2}\right]$$

$$= \frac{1}{3}$$

OUR **PROOF BY SANDWICH** IS COMPLETE:
FOR $\int_0^1 x^2 dx$, **EVERY** SEQUENCE OF RIEMANN SUMS MUST HAVE THE **SAME LIMIT.**

THAT'S WHAT IT MEANS TO BE **INTEGRABLE.**

THIS PROOF IS **DELICIOUS.**

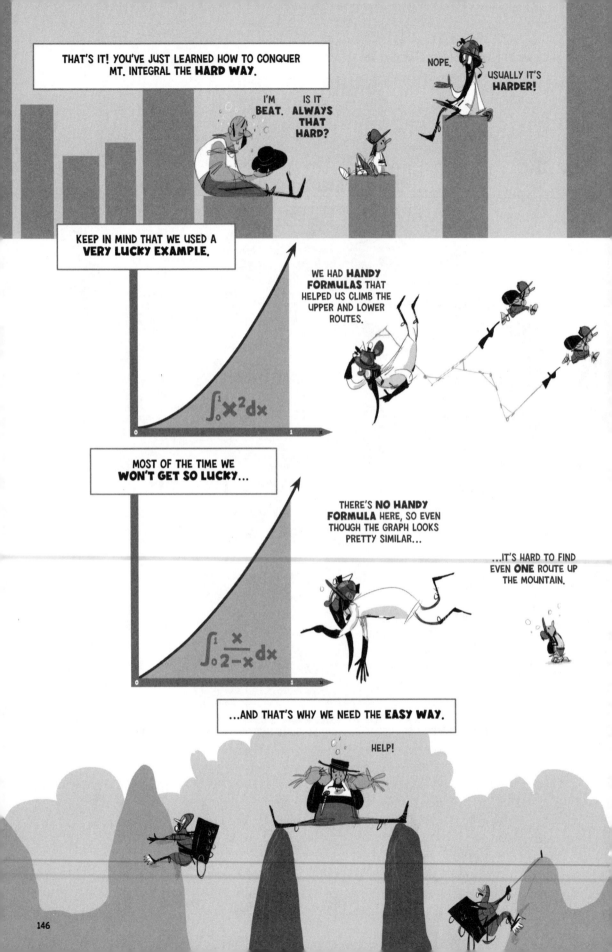

THAT'S IT! YOU'VE JUST LEARNED HOW TO CONQUER MT. INTEGRAL THE **HARD WAY.**

NOPE.

USUALLY IT'S **HARDER!**

I'M **BEAT.**

IS IT **ALWAYS THAT HARD?**

KEEP IN MIND THAT WE USED A **VERY LUCKY EXAMPLE.**

WE HAD **HANDY FORMULAS** THAT HELPED US CLIMB THE UPPER AND LOWER ROUTES.

$$\int_0^1 x^2 \, dx$$

MOST OF THE TIME WE **WON'T GET SO LUCKY…**

THERE'S **NO HANDY FORMULA** HERE, SO EVEN THOUGH THE GRAPH LOOKS PRETTY SIMILAR…

…IT'S HARD TO FIND EVEN **ONE** ROUTE UP THE MOUNTAIN.

$$\int_0^1 \frac{x}{2-x} \, dx$$

…AND THAT'S WHY WE NEED THE **EASY WAY.**

HELP!

CHAPTER 12
INTEGRATION, THE EASY WAY

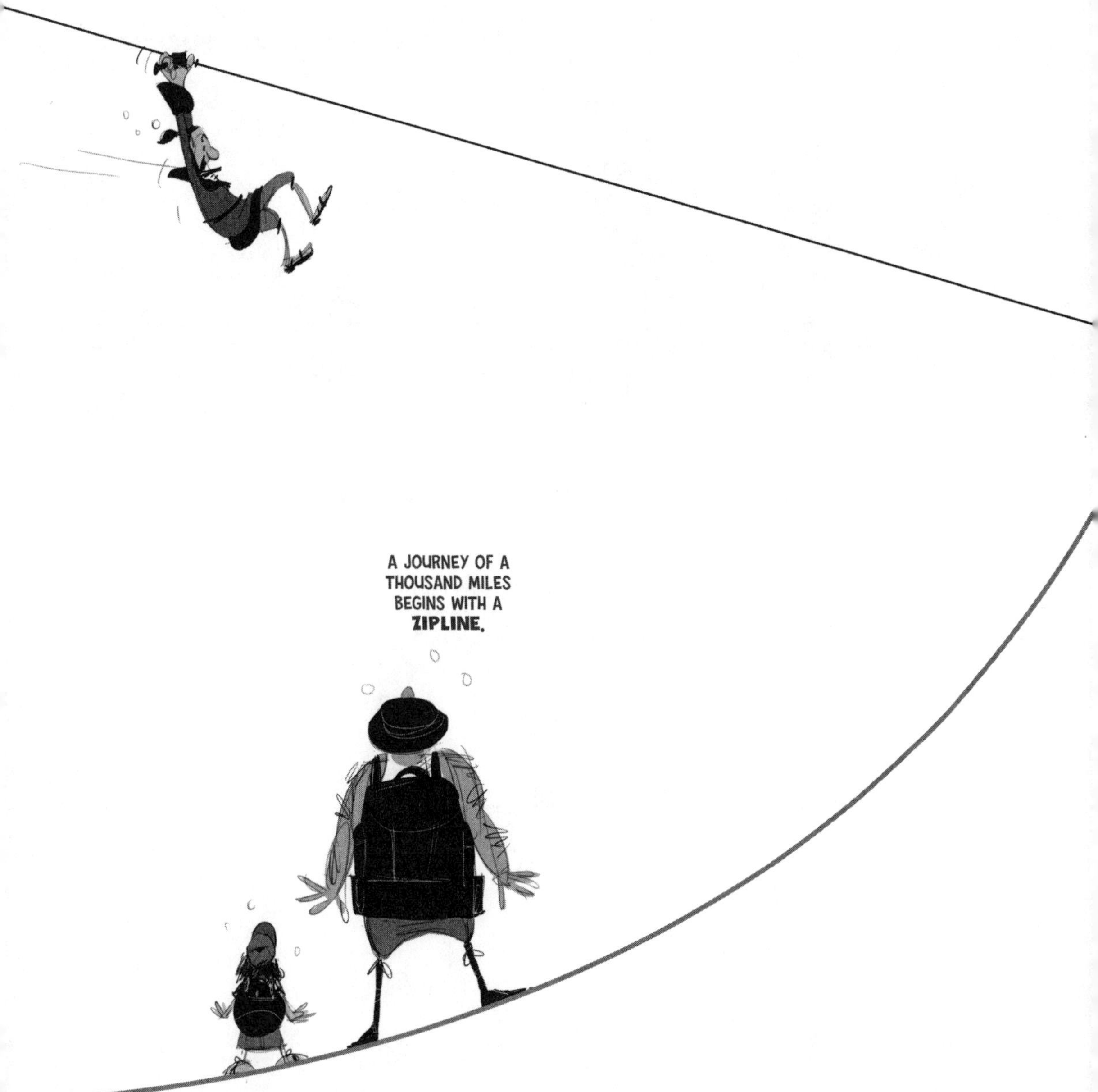
A JOURNEY OF A THOUSAND MILES BEGINS WITH A **ZIPLINE.**

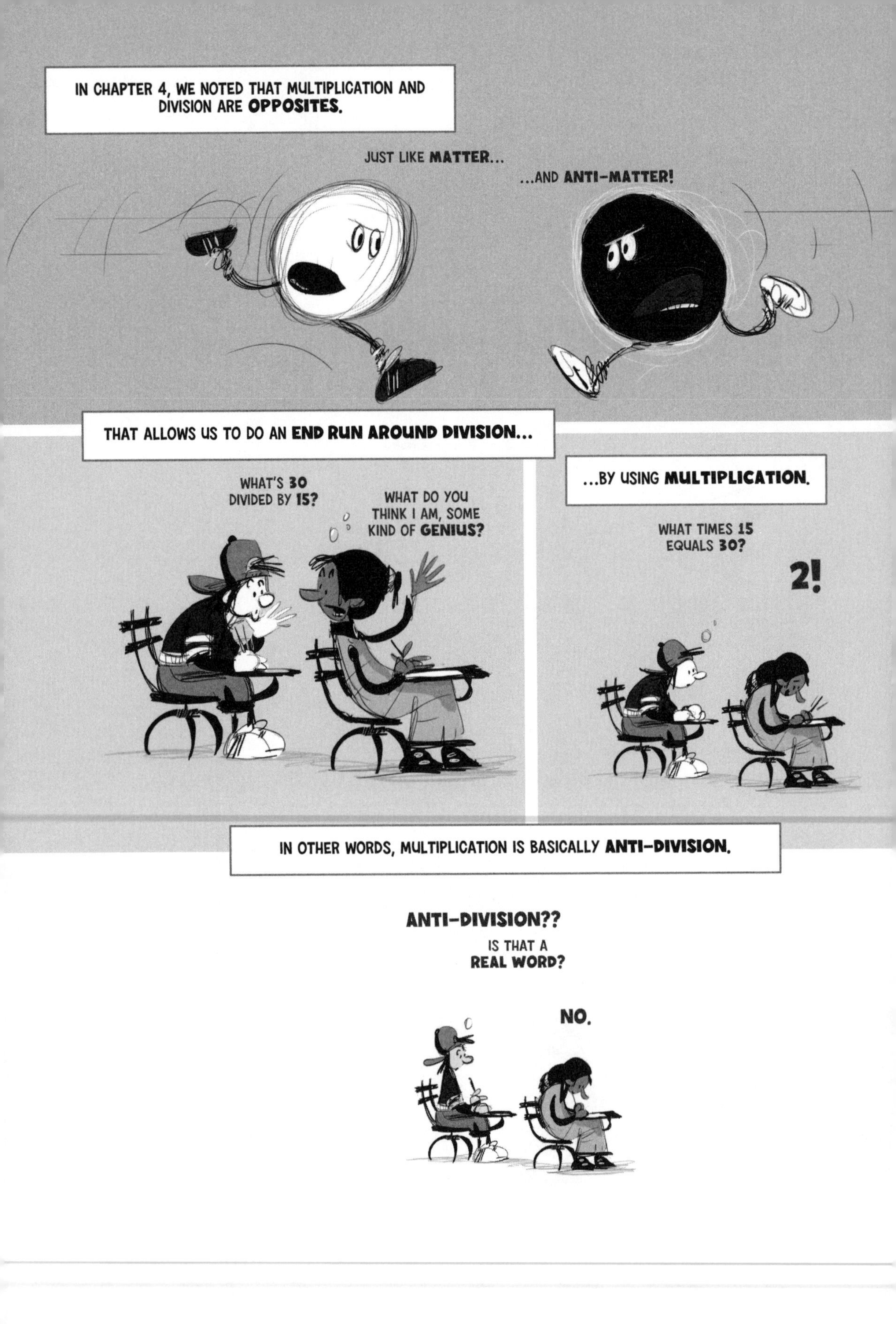
IN CHAPTER 4, WE NOTED THAT MULTIPLICATION AND DIVISION ARE **OPPOSITES.**
JUST LIKE **MATTER...**
...AND **ANTI-MATTER!**
THAT ALLOWS US TO DO AN **END RUN AROUND DIVISION...**
WHAT'S **30** DIVIDED BY **15?**
WHAT DO YOU THINK I AM, SOME KIND OF **GENIUS?**
...BY USING **MULTIPLICATION.**
WHAT TIMES **15** EQUALS **30?**
2!
IN OTHER WORDS, MULTIPLICATION IS BASICALLY **ANTI-DIVISION.**
ANTI-DIVISION??
IS THAT A **REAL WORD?**
NO.

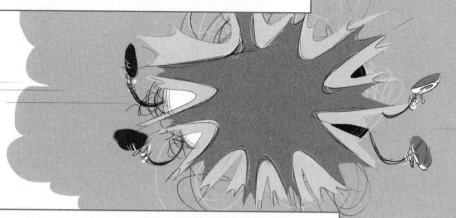

THIS CHAPTER IS BASED ON THE IDEA THAT INTEGRALS AND DERIVATIVES ARE **ALSO OPPOSITES.**

THAT WILL ALLOW US TO DO AN **END RUN AROUND INTEGRALS...**

...BY USING **DERIVATIVES.**

READY TO **CLIMB MT. INTEGRAL?**

NO!

WHEEEEEE!

MORE PRECISELY, WHAT WE NEED ARE **ANTI-DERIVATIVES.**

ANTI-DERIVATIVE??

IS THAT A **REAL WORD?**

YES!

F(x) IS AN **ANTI-DERIVATIVE** OF f(x) IF...

$$\frac{d}{dx} F(x) = f(x)$$

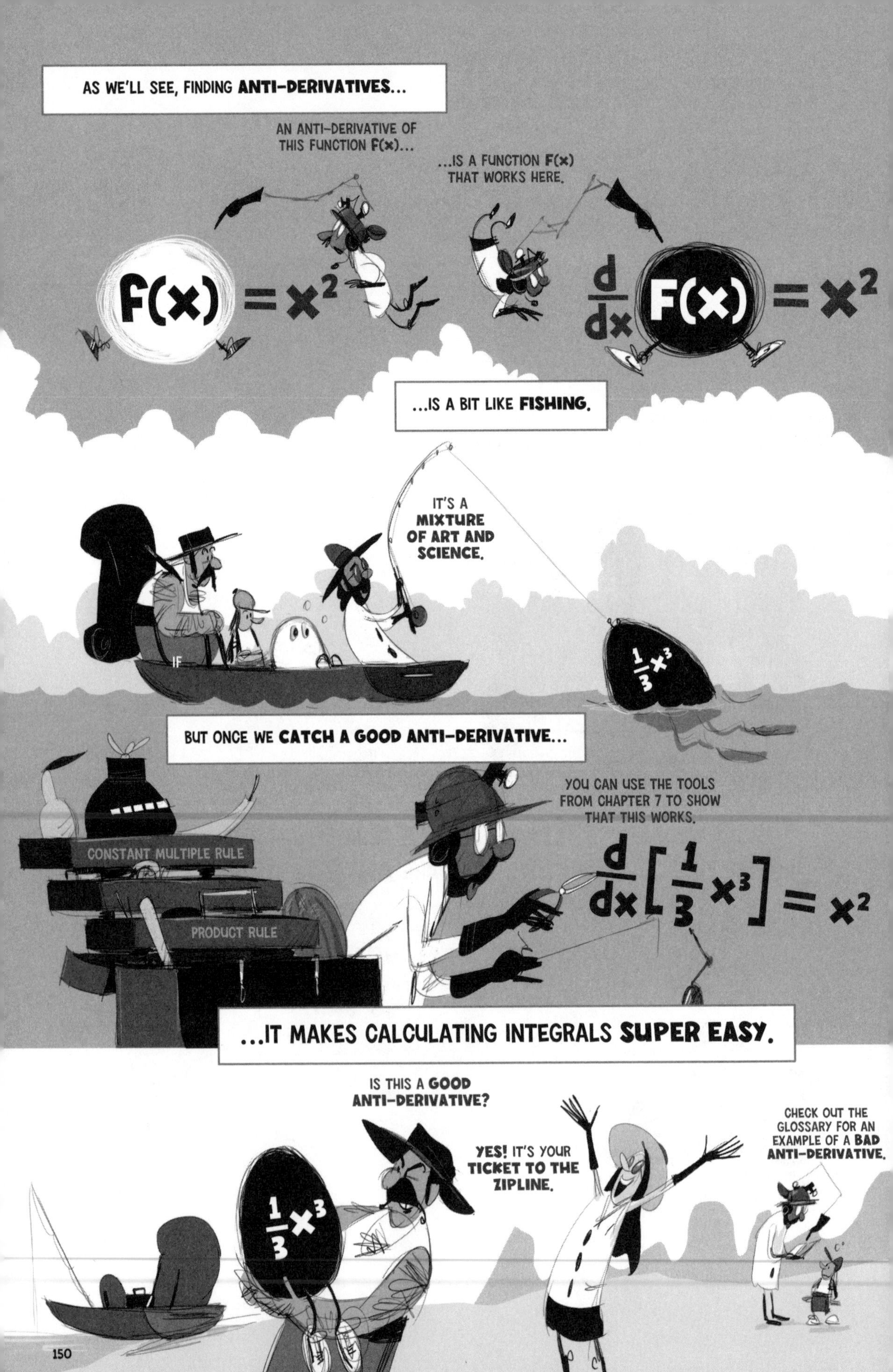
AS WE'LL SEE, FINDING **ANTI-DERIVATIVES**...
AN ANTI-DERIVATIVE OF THIS FUNCTION $f(x)$...
...IS A FUNCTION $F(x)$ THAT WORKS HERE.
$f(x) = x^2$
$\frac{d}{dx} F(x) = x^2$
...IS A BIT LIKE **FISHING**.
IT'S A **MIXTURE OF ART AND SCIENCE.**
$\frac{1}{3}x^3$
BUT ONCE WE **CATCH A GOOD ANTI-DERIVATIVE**...
YOU CAN USE THE TOOLS FROM CHAPTER 7 TO SHOW THAT THIS WORKS.
CONSTANT MULTIPLE RULE
PRODUCT RULE
$\frac{d}{dx}\left[\frac{1}{3}x^3\right] = x^2$
...IT MAKES CALCULATING INTEGRALS **SUPER EASY**.
IS THIS A **GOOD ANTI-DERIVATIVE?**
YES! IT'S YOUR **TICKET TO THE ZIPLINE.**
CHECK OUT THE GLOSSARY FOR AN EXAMPLE OF A **BAD ANTI-DERIVATIVE.**
$\frac{1}{3}x^3$

TO SEE HOW **SUPER EASY** IT IS, LET'S GO BACK TO THE INTEGRAL FROM THE LAST CHAPTER.

CALCULATING THIS THE **HARD** WAY...

$$\int_0^1 x^2 dx$$

...REQUIRED **UPPER AND LOWER RIEMANN SUMS, LIMITS, SANDWICHES...**

...AND LOTS OF **LUCK**.

WITH AN **ANTI-DERIVATIVE** LIKE $F(x) = \frac{1}{3}x^3$, WE JUST HAVE TO DO **TWO EASY CALCULATIONS...**

WE FIND THE VALUE OF $F(x)$ AT $x=0$, THE **STARTING POINT** OF THE INTEGRAL.

AND WE FIND THE VALUE OF $F(x)$ AT $x=1$, THE **ENDPOINT** OF THE INTEGRAL.

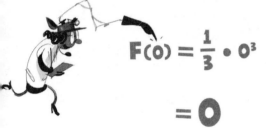

$$F(0) = \frac{1}{3} \cdot 0^3$$

$$= 0$$

$$F(1) = \frac{1}{3} \cdot 1^3$$

$$= \frac{1}{3}$$

...AND THEN **SUBTRACT**...

$$F(1) - F(0) = \frac{1}{3} - 0 = \frac{1}{3}$$

...AND WE'RE **DONE**.

VOILÀ!

$$\int_0^1 x^2 dx = \frac{1}{3}$$

MUCH EASIER THAN ADDING UP TINY RECTANGLES!

MORE GENERALLY, THE **EASY WAY** TO CALCULATE THE INTEGRAL OF **ANY** CONTINUOUS FUNCTION...

WHAT'S THE AREA UNDER THIS FUNCTION...

...BETWEEN x=a AND x=b?

$\int_a^b$

...IS TO FIND A **GOOD ANTI-DERIVATIVE** OF THAT FUNCTION.

$$\frac{d}{dx} \quad = $$

THEN ALL WE HAVE TO DO IS **TWO SIMPLE CALCULATIONS**...

a   b

...AND **SUBTRACT**.

VOILÀ!

$$\int_a^b \quad = \quad b \; - \; a$$

THIS RESULT IS SO IMPORTANT, IT'S PART OF **THE FUNDAMENTAL THEOREM OF CALCULUS.**

The **Fundamental Theorem of Calculus (Part Two)** states that if F(x) is an anti-derivative of a continuous function f(x), then:

$$\int_a^b f(x)dx = F(b) - F(a)$$

In other words:

$$\int_a^b \bigcirc = \mathbf{b} - \mathbf{a}$$

IN THE NEXT CHAPTER, WE'LL EXPLORE **WHY** IT WORKS...

IT LOOKS LIKE **MAGIC.**

BUT WE CAN **PROVE** IT!

...BUT FOR NOW LET'S FOCUS ON **HOW** IT WORKS.

MORE SPECIFICALLY, HOW DO YOU **FIND** A GOOD ANTI-DERIVATIVE?

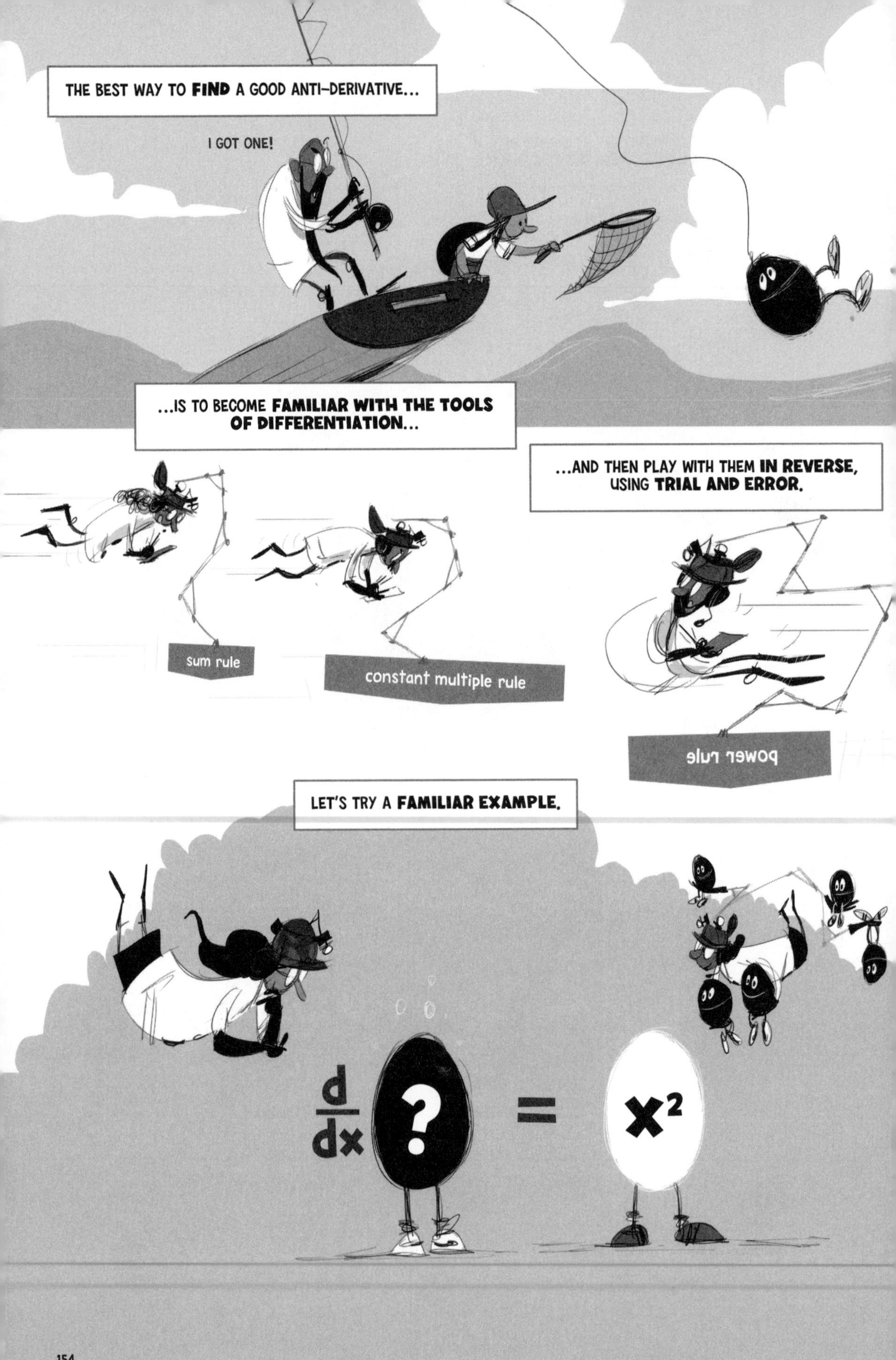
THE BEST WAY TO FIND A GOOD ANTI-DERIVATIVE...
I GOT ONE!
...IS TO BECOME FAMILIAR WITH THE TOOLS OF DIFFERENTIATION...
...AND THEN PLAY WITH THEM IN REVERSE, USING TRIAL AND ERROR.
sum rule
constant multiple rule
power rule
LET'S TRY A FAMILIAR EXAMPLE.
d/dx ? = x²

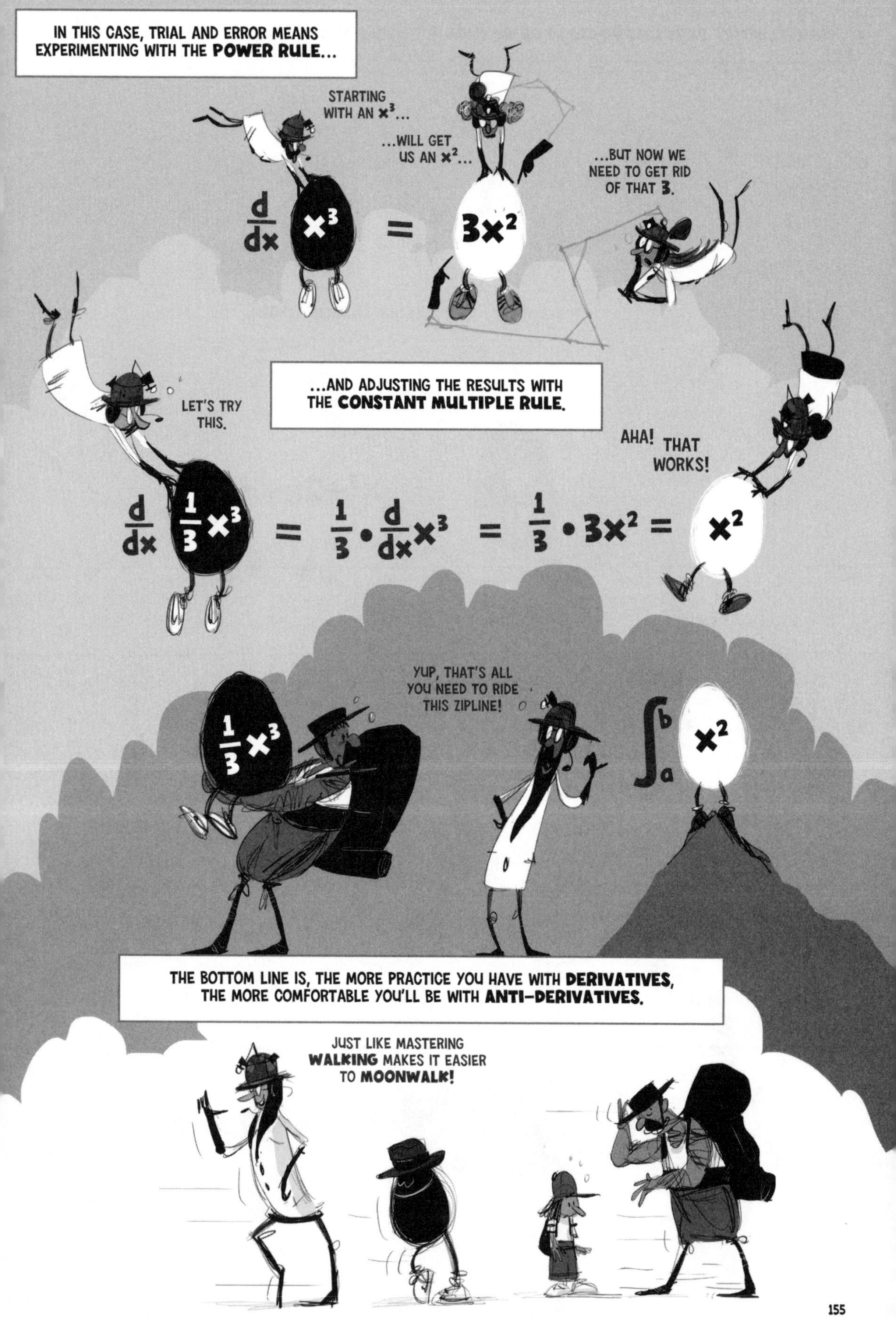
IN THIS CASE, TRIAL AND ERROR MEANS EXPERIMENTING WITH THE **POWER RULE**...
STARTING WITH AN x^3...
...WILL GET US AN x^2...
...BUT NOW WE NEED TO GET RID OF THAT 3.
$\frac{d}{dx} x^3 = 3x^2$
...AND ADJUSTING THE RESULTS WITH THE **CONSTANT MULTIPLE RULE**.
LET'S TRY THIS.
AHA! THAT WORKS!
$\frac{d}{dx} \frac{1}{3}x^3 = \frac{1}{3} \cdot \frac{d}{dx}x^3 = \frac{1}{3} \cdot 3x^2 = x^2$
$\frac{1}{3}x^3$
YUP, THAT'S ALL YOU NEED TO RIDE THIS ZIPLINE!
$\int_a^b x^2$
THE BOTTOM LINE IS, THE MORE PRACTICE YOU HAVE WITH **DERIVATIVES**, THE MORE COMFORTABLE YOU'LL BE WITH **ANTI-DERIVATIVES**.
JUST LIKE MASTERING **WALKING** MAKES IT EASIER TO **MOONWALK**!

UNFORTUNATELY, THERE'S **NO FOOLPROOF METHOD** FOR FINDING ANTI–DERIVATIVES...

EVEN THOUGH I'M PRETTY GOOD AT **WALKING**...

...I STILL CAN'T GET THE HANG OF **MOONWALKING!**

...BUT IT'S OFTEN **EASIER THAN IT LOOKS.**

ZOIKES!

DON'T PANIC, WE CAN HANDLE THIS EXAMPLE.

$$\frac{d}{dx}\ ? = x^2+2x+42$$

CAN YOU FIND AN ANTI–DERIVATIVE FOR $f(x)=x^2$?

YES, WE JUST DID THAT.

$$\frac{d}{dx}\ \tfrac{1}{3}x^3 = x^2$$

CAN YOU FIND AN ANTI–DERIVATIVE FOR $f(x)=2x$?

YES, WITH THE POWER RULE.

$$\frac{d}{dx}\ x^2 = 2x$$

AND CAN YOU FIND AN ANTI–DERIVATIVE FOR $f(x)=42$?

YES, WITH T CONSTAN MULTIPLE R

$$\frac{d}{dx}\ 42x = 42$$

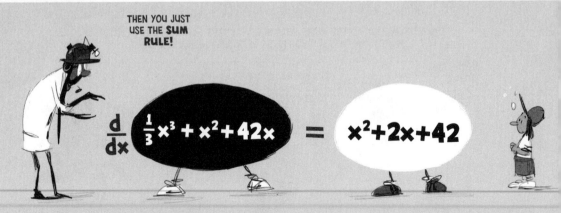

THEN YOU JUST USE THE **SUM RULE!**

$$\frac{d}{dx}\ \tfrac{1}{3}x^3 + x^2 + 42x = x^2+2x+42$$

AND EVERY TIME YOU LEARN SOMETHING NEW ABOUT **CALCULATING DERIVATIVES**...

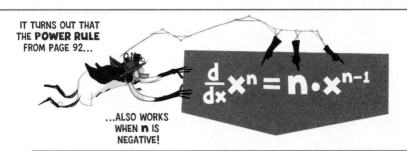

IT TURNS OUT THAT THE **POWER RULE** FROM PAGE 92...

$$\frac{d}{dx}x^n = n \cdot x^{n-1}$$

...ALSO WORKS WHEN **n** IS NEGATIVE!

...YOU'RE ALSO LEARNING MORE ABOUT **SOLVING INTEGRALS THE EASY WAY!**

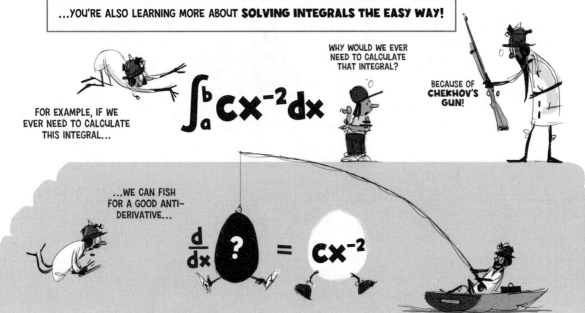

WHY WOULD WE EVER NEED TO CALCULATE THAT INTEGRAL?

BECAUSE OF **CHEKHOV'S GUN!**

FOR EXAMPLE, IF WE EVER NEED TO CALCULATE THIS INTEGRAL...

$$\int_a^b cx^{-2}\,dx$$

...WE CAN FISH FOR A GOOD ANTI-DERIVATIVE...

$$\frac{d}{dx}\ ? = cx^{-2}$$

...BY EXPERIMENTING WITH OUR NEW AND IMPROVED **POWER RULE**...

$$\frac{d}{dx}x^{-1} = -x^{-2}$$

ALMOST!

...AND COMBINING THAT WITH THE **CONSTANT MULTIPLE RULE.**

$$\frac{d}{dx}-cx^{-1} = -c \cdot \frac{d}{dx}x^{-1}$$

THAT WORKS!

$$= cx^{-2}$$

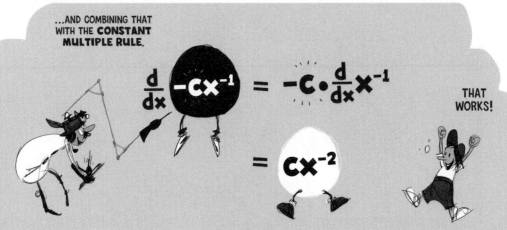

ONE FINAL POINT: ANY TIME YOU CAN FIND **ONE** GOOD ANTI-DERIVATIVE...

...YOU CAN FIND A **WHOLE BUNCH**...

I GOT ONE!

WHAT A GREAT FISHING SPOT!

...BECAUSE ANTI-DERIVATIVES COME IN **FAMILIES.**

IF **F(x)** IS AN ANTI-DERIVATIVE FOR **f(x)**...

...THEN SO IS **F(x) + c.**

YOU CAN PROVE THAT USING THE TOOLS FROM CHAPTER 7.

F(x)   F(x)   F(x)+c

THAT'S WHY WE SAY "**AN**" ANTI-DERIVATIVE INSTEAD OF "**THE**" ANTI-DERIVATIVE.

FIND **ONE** GOOD ONE...

...IT DOESN'T MATTER WHICH ONE...

...AND YOU CAN USE THE **ZIPLINE!**

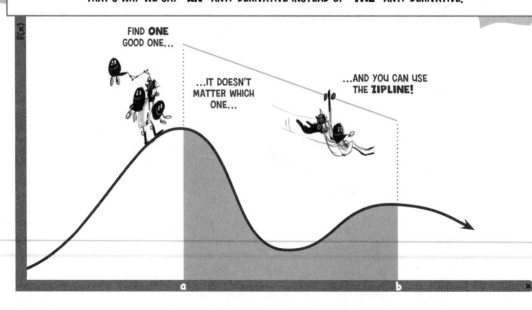

CHAPTER 13
THE FUNDAMENTAL THEOREM, REVISITED

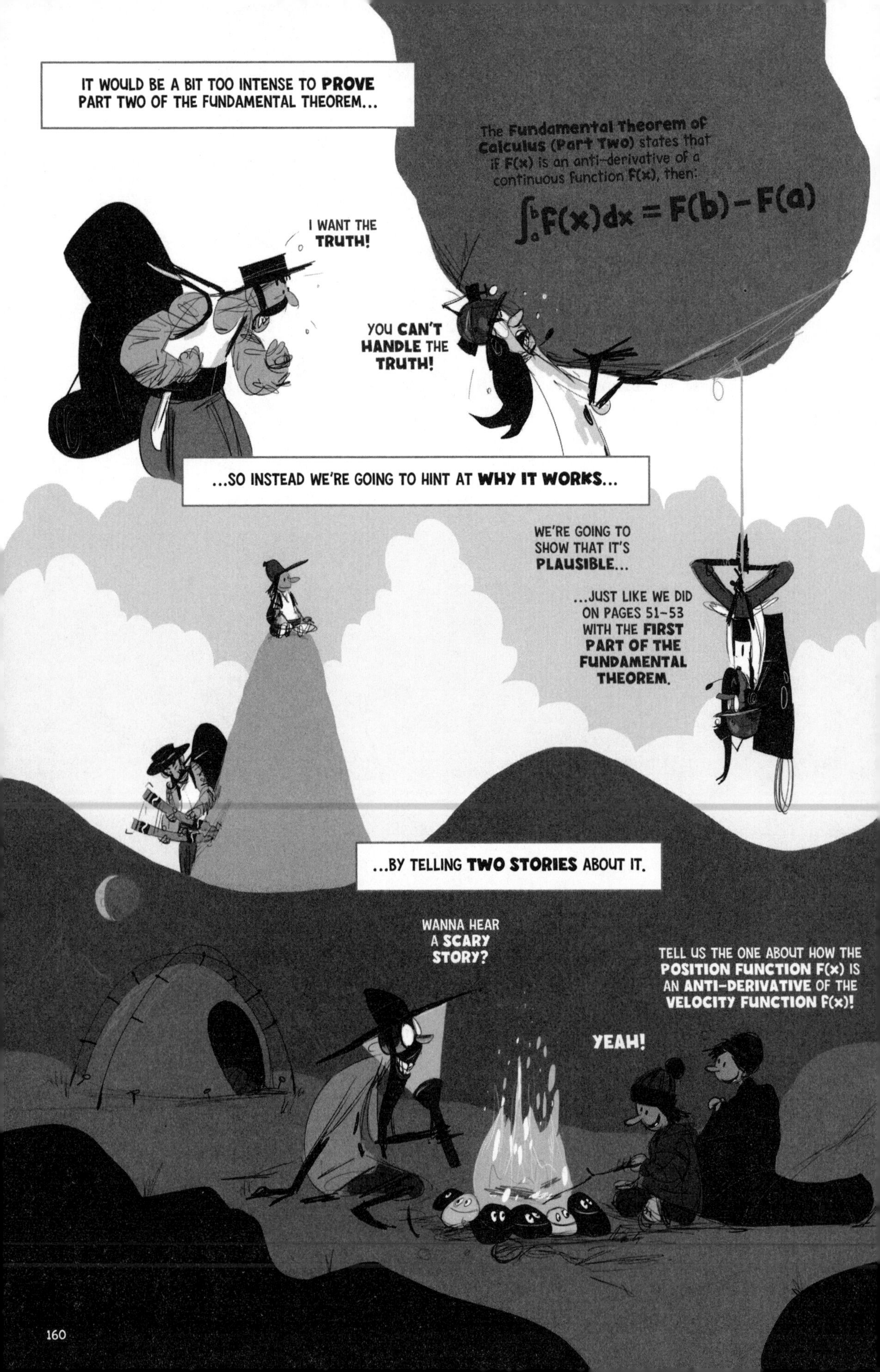
IT WOULD BE A BIT TOO INTENSE TO **PROVE** PART TWO OF THE FUNDAMENTAL THEOREM...
The **Fundamental Theorem of Calculus (Part Two)** states that if **F(x)** is an anti-derivative of a continuous function **f(x)**, then:
$\int_a^b f(x)dx = F(b) - F(a)$
I WANT THE **TRUTH!**
YOU **CAN'T HANDLE** THE **TRUTH!**
...SO INSTEAD WE'RE GOING TO HINT AT **WHY IT WORKS**...
WE'RE GOING TO SHOW THAT IT'S **PLAUSIBLE**...
...JUST LIKE WE DID ON PAGES 51–53 WITH THE **FIRST PART OF THE FUNDAMENTAL THEOREM**.
...BY TELLING **TWO STORIES** ABOUT IT.
WANNA HEAR A **SCARY STORY?**
TELL US THE ONE ABOUT HOW THE **POSITION FUNCTION F(x)** IS AN **ANTI-DERIVATIVE** OF THE **VELOCITY FUNCTION f(x)!**
YEAH!

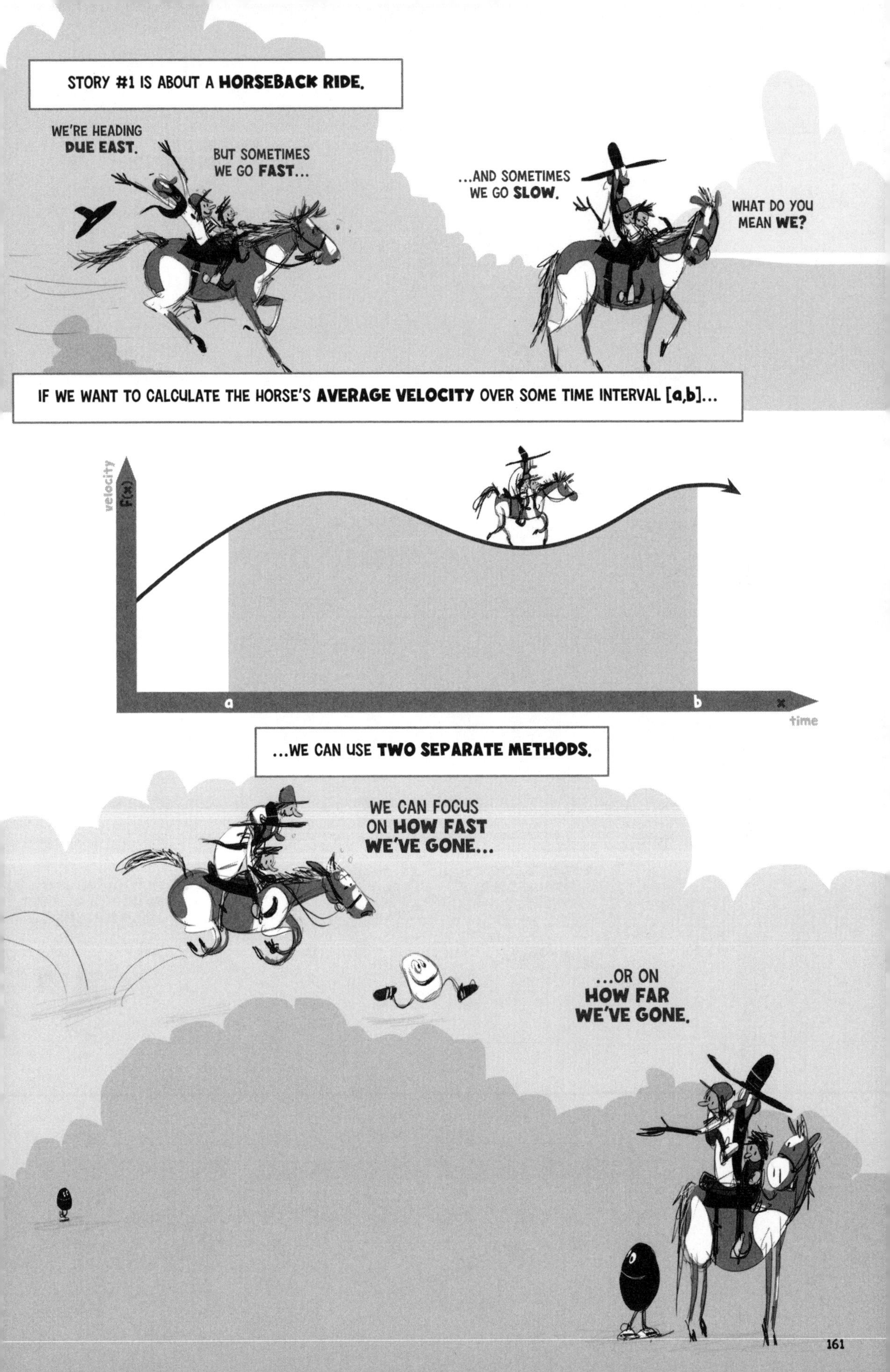
STORY #1 IS ABOUT A **HORSEBACK RIDE.**
WE'RE HEADING **DUE EAST.**
BUT SOMETIMES WE GO **FAST...**
...AND SOMETIMES WE GO **SLOW.**
WHAT DO YOU MEAN **WE?**
IF WE WANT TO CALCULATE THE HORSE'S **AVERAGE VELOCITY** OVER SOME TIME INTERVAL [a,b]...
velocity
f(x)
a
b
x
time
...WE CAN USE **TWO SEPARATE METHODS.**
WE CAN FOCUS ON **HOW FAST WE'VE GONE...**
...OR ON **HOW FAR WE'VE GONE.**

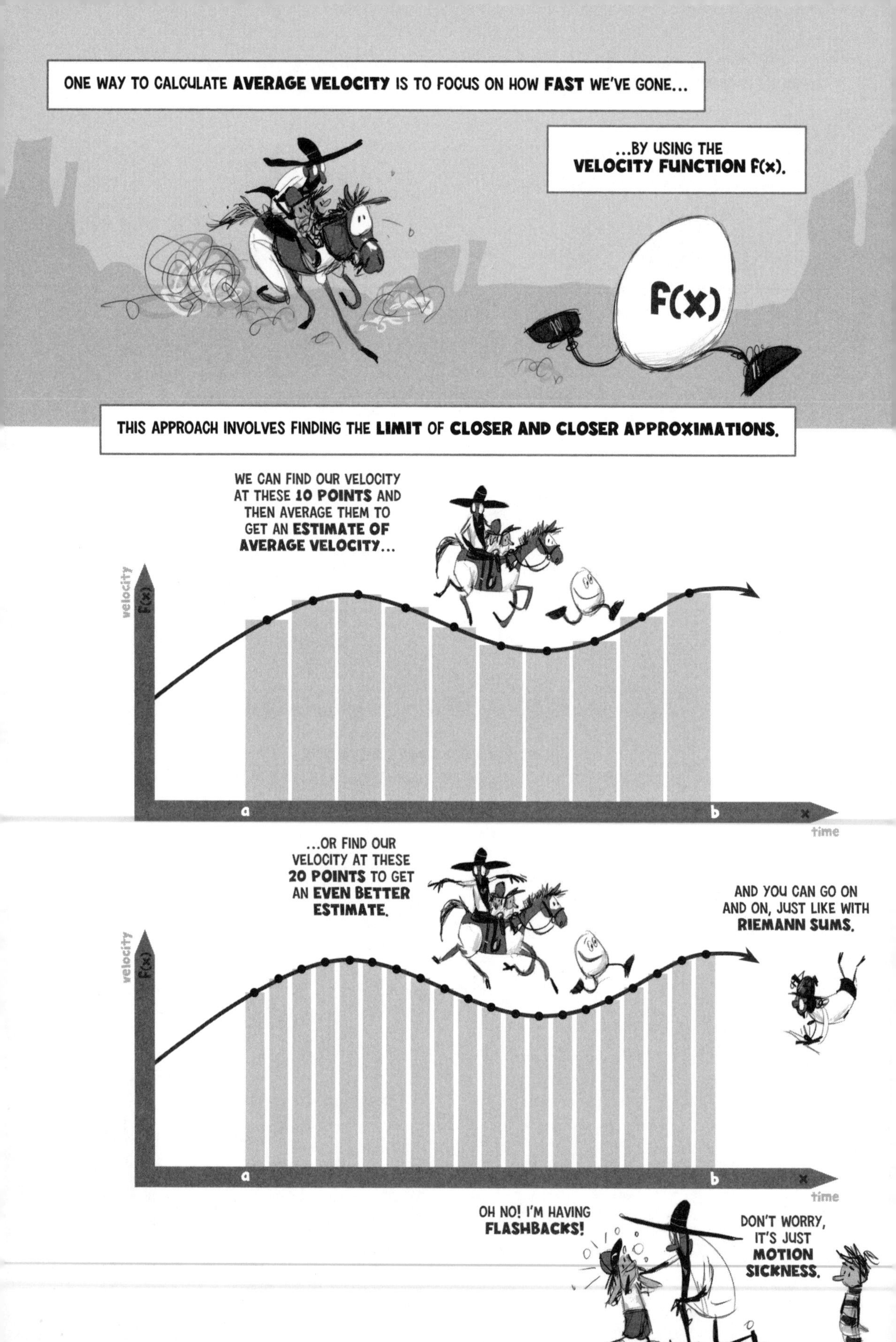
ONE WAY TO CALCULATE AVERAGE VELOCITY IS TO FOCUS ON HOW FAST WE'VE GONE...
...BY USING THE VELOCITY FUNCTION f(x).
f(x)
THIS APPROACH INVOLVES FINDING THE LIMIT OF CLOSER AND CLOSER APPROXIMATIONS.
WE CAN FIND OUR VELOCITY AT THESE 10 POINTS AND THEN AVERAGE THEM TO GET AN ESTIMATE OF AVERAGE VELOCITY...
velocity
f(x)
a
b
x
time
...OR FIND OUR VELOCITY AT THESE 20 POINTS TO GET AN EVEN BETTER ESTIMATE.
AND YOU CAN GO ON AND ON, JUST LIKE WITH RIEMANN SUMS.
velocity
f(x)
a
b
x
time
OH NO! I'M HAVING FLASHBACKS!
DON'T WORRY, IT'S JUST MOTION SICKNESS.

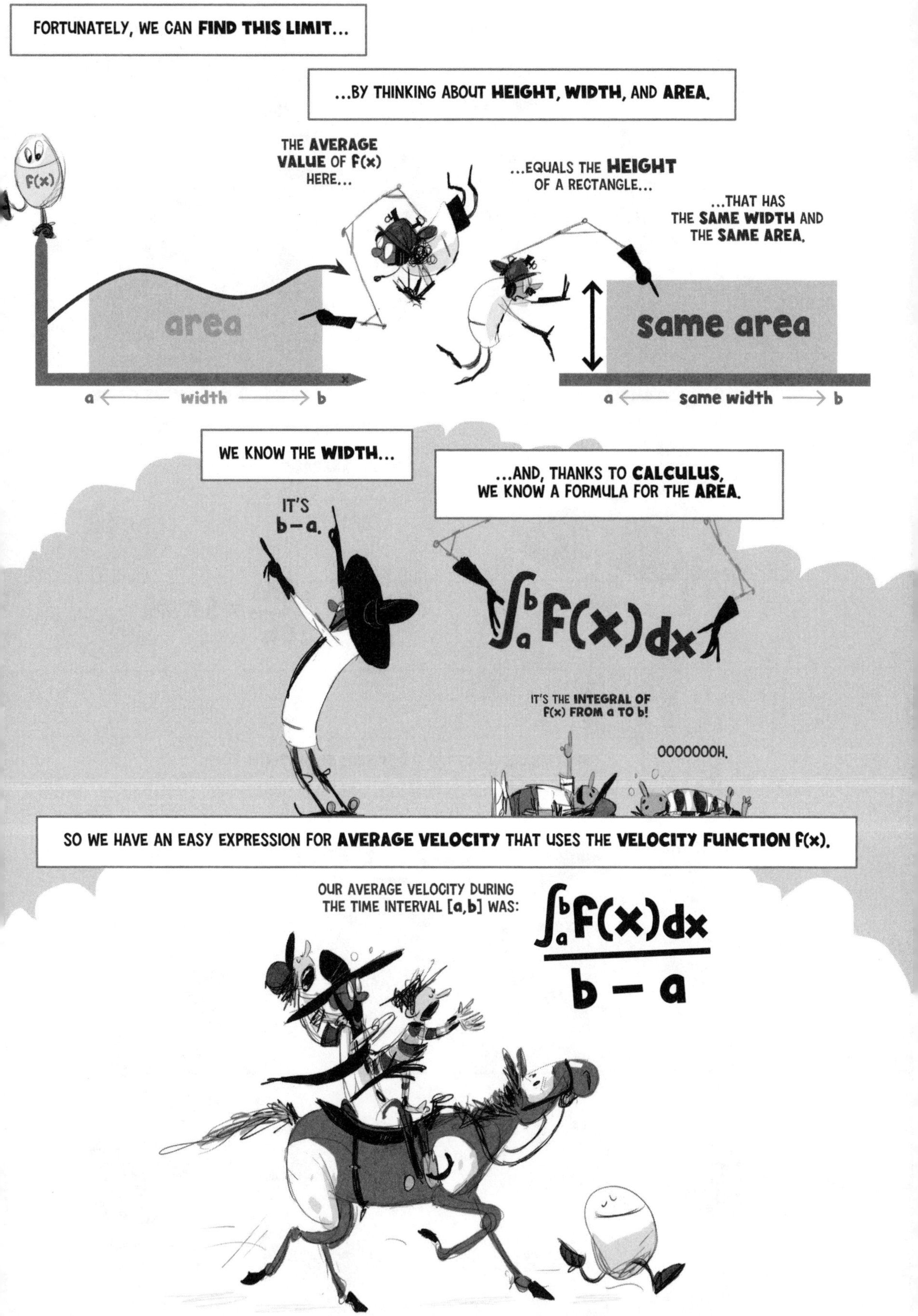
FORTUNATELY, WE CAN FIND THIS LIMIT...
...BY THINKING ABOUT HEIGHT, WIDTH, AND AREA.
F(x)
THE AVERAGE VALUE OF f(x) HERE...
...EQUALS THE HEIGHT OF A RECTANGLE...
...THAT HAS THE SAME WIDTH AND THE SAME AREA.
area
a
width
b
same area
a
same width
b
WE KNOW THE WIDTH...
IT'S b – a.
...AND, THANKS TO CALCULUS, WE KNOW A FORMULA FOR THE AREA.
$\int_a^b f(x)dx$
IT'S THE INTEGRAL OF f(x) FROM a TO b!
OOOOOOOH.
SO WE HAVE AN EASY EXPRESSION FOR AVERAGE VELOCITY THAT USES THE VELOCITY FUNCTION f(x).
OUR AVERAGE VELOCITY DURING THE TIME INTERVAL [a,b] WAS:
$\dfrac{\int_a^b f(x)dx}{b-a}$

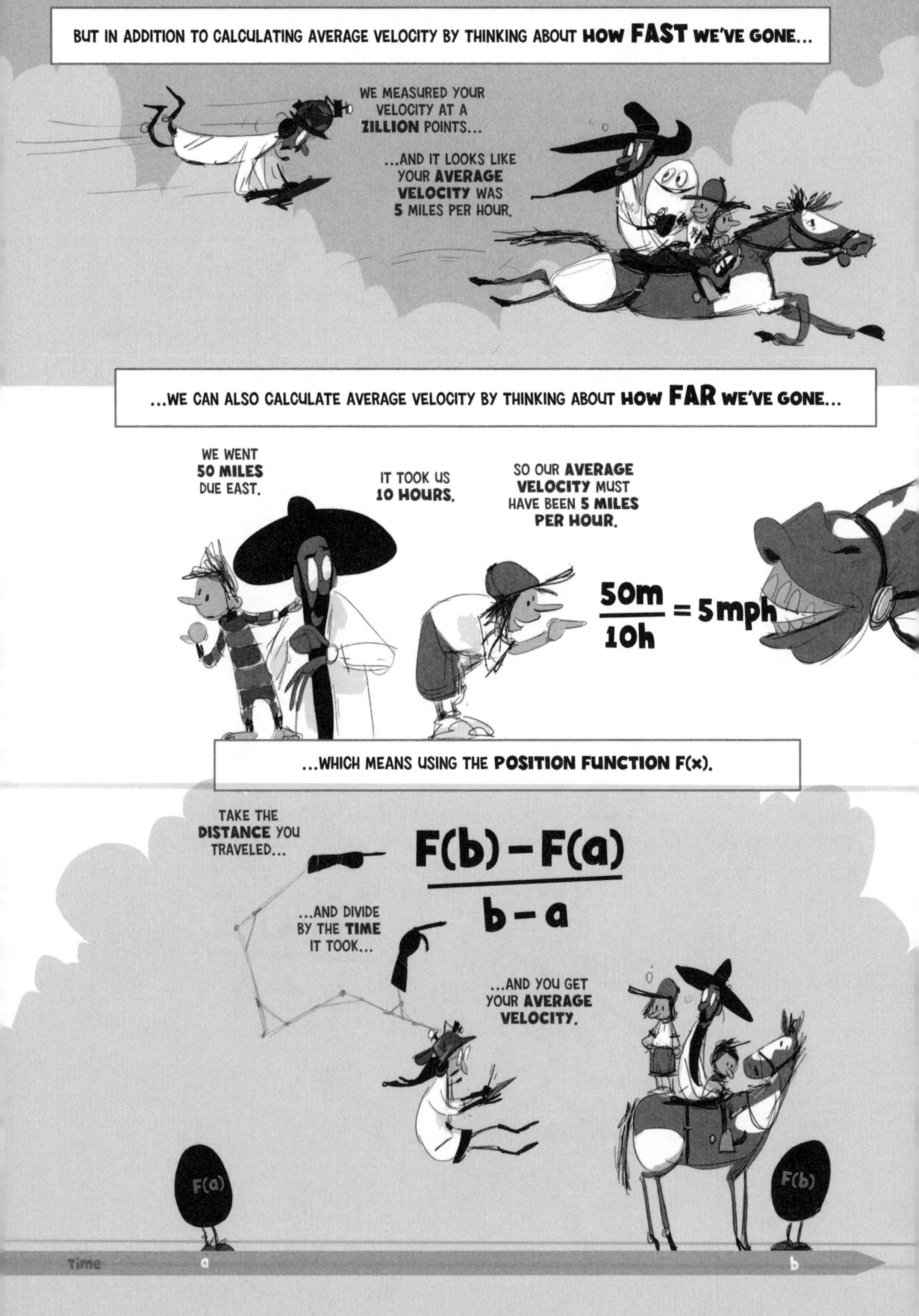
BUT IN ADDITION TO CALCULATING AVERAGE VELOCITY BY THINKING ABOUT HOW FAST WE'VE GONE...
WE MEASURED YOUR VELOCITY AT A ZILLION POINTS...
...AND IT LOOKS LIKE YOUR AVERAGE VELOCITY WAS 5 MILES PER HOUR.
...WE CAN ALSO CALCULATE AVERAGE VELOCITY BY THINKING ABOUT HOW FAR WE'VE GONE...
WE WENT 50 MILES DUE EAST.
IT TOOK US 10 HOURS.
SO OUR AVERAGE VELOCITY MUST HAVE BEEN 5 MILES PER HOUR.
50m / 10h = 5mph
...WHICH MEANS USING THE POSITION FUNCTION F(x).
TAKE THE DISTANCE YOU TRAVELED...
F(b) – F(a) / b – a
...AND DIVIDE BY THE TIME IT TOOK...
...AND YOU GET YOUR AVERAGE VELOCITY.
F(a)
F(b)
Time
a
b

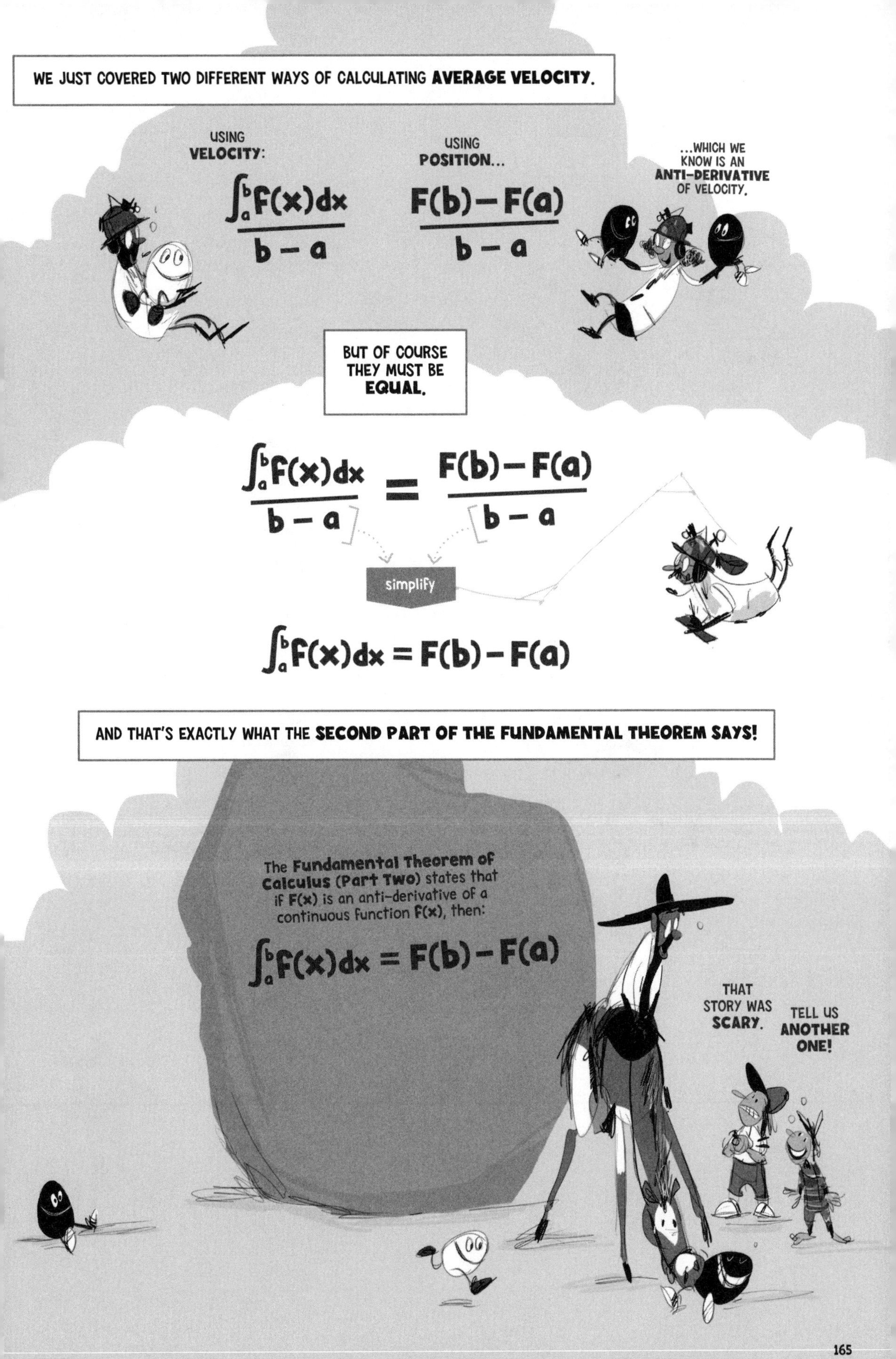
WE JUST COVERED TWO DIFFERENT WAYS OF CALCULATING **AVERAGE VELOCITY.**
USING **VELOCITY:**
$\frac{\int_a^b f(x)dx}{b-a}$
USING **POSITION...**
$\frac{F(b)-F(a)}{b-a}$
...WHICH WE KNOW IS AN **ANTI-DERIVATIVE** OF VELOCITY.
BUT OF COURSE THEY MUST BE **EQUAL.**
$\frac{\int_a^b f(x)dx}{b-a} = \frac{F(b)-F(a)}{b-a}$
simplify
$\int_a^b f(x)dx = F(b)-F(a)$
AND THAT'S EXACTLY WHAT THE **SECOND PART OF THE FUNDAMENTAL THEOREM SAYS!**
The **Fundamental Theorem of Calculus (Part Two)** states that if **F(x)** is an anti-derivative of a continuous function **f(x)**, then:
$\int_a^b f(x)dx = F(b)-F(a)$
THAT STORY WAS **SCARY.**
TELL US **ANOTHER ONE!**

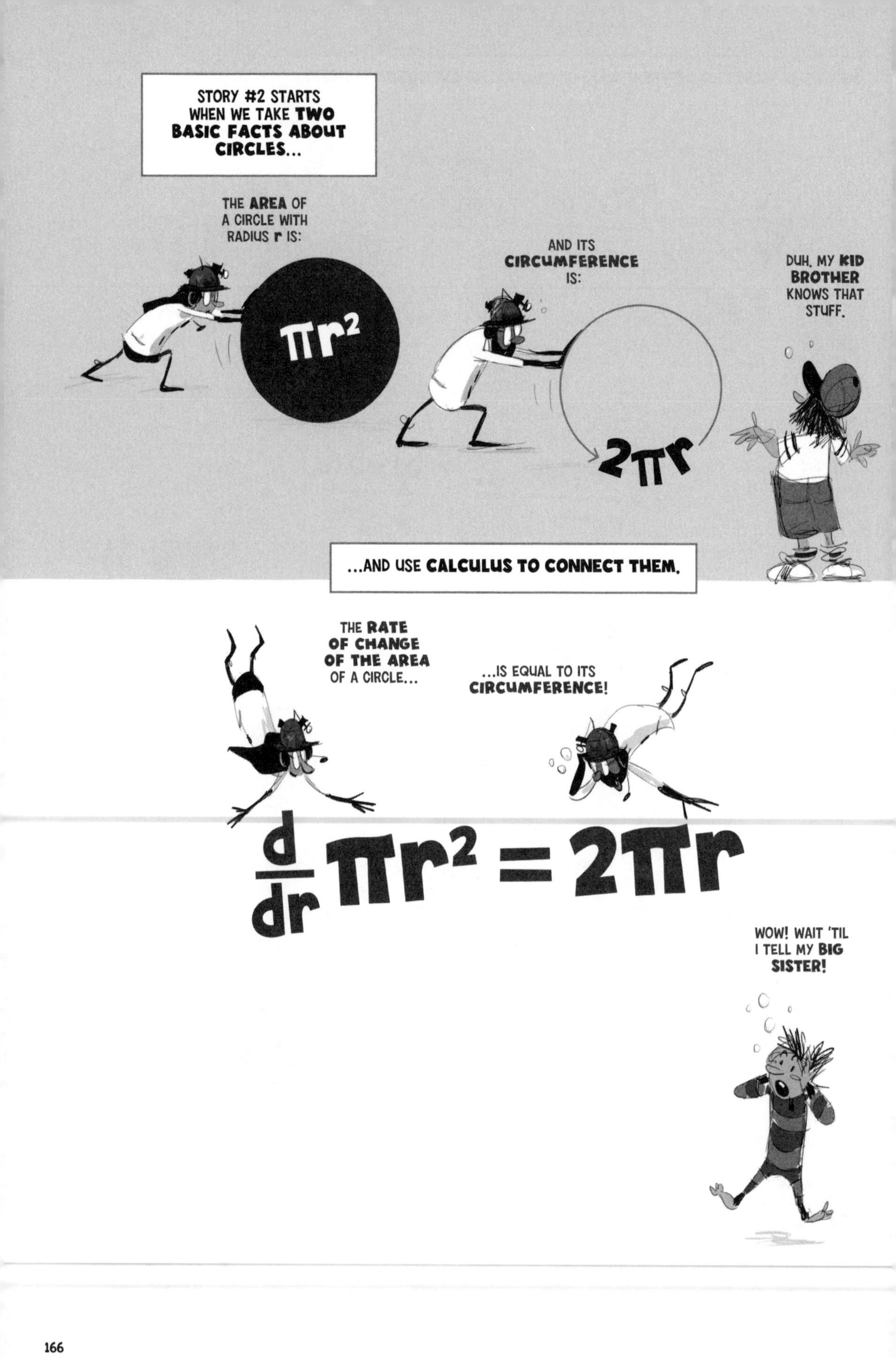

STORY #2 STARTS WHEN WE TAKE **TWO BASIC FACTS ABOUT CIRCLES**...
THE **AREA** OF A CIRCLE WITH RADIUS **r** IS:
πr^2
AND ITS **CIRCUMFERENCE** IS:
$2\pi r$
DUH, MY **KID BROTHER** KNOWS THAT STUFF.
...AND USE **CALCULUS TO CONNECT THEM.**
THE **RATE OF CHANGE OF THE AREA** OF A CIRCLE...
...IS EQUAL TO ITS **CIRCUMFERENCE**!
$\frac{d}{dr}\pi r^2 = 2\pi r$
WOW! WAIT 'TIL I TELL MY **BIG SISTER**!

WE CAN USE THE TOOLS FROM CHAPTER 7 TO **PROVE THIS CONNECTION...**

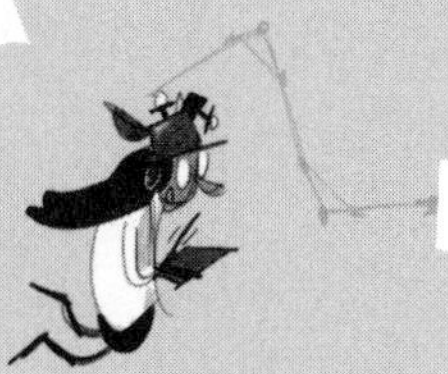

$$\frac{d}{dr}\pi r^2$$

constant multiple rule

$$= \pi \cdot \frac{d}{dr} r^2$$

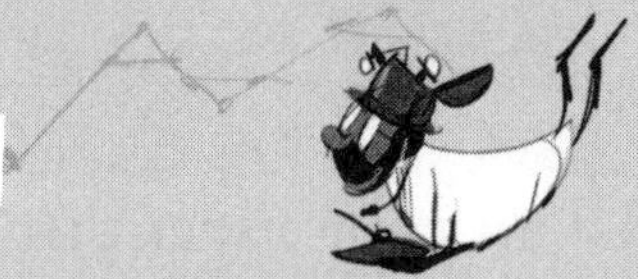

power rule

$$= 2\pi r$$

...AND WE CAN USE THE LANGUAGE OF CHAPTER 12 TO **DESCRIBE IT.**

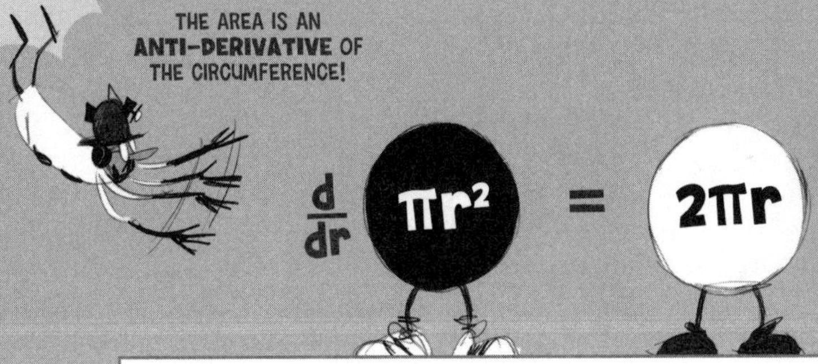
THE AREA IS AN **ANTI-DERIVATIVE** OF THE CIRCUMFERENCE!

$$\frac{d}{dr}\pi r^2 = 2\pi r$$

AND, IN A MINUTE, WE'LL SEE HOW IT RELATES TO THE **FUNDAMENTAL THEOREM.**

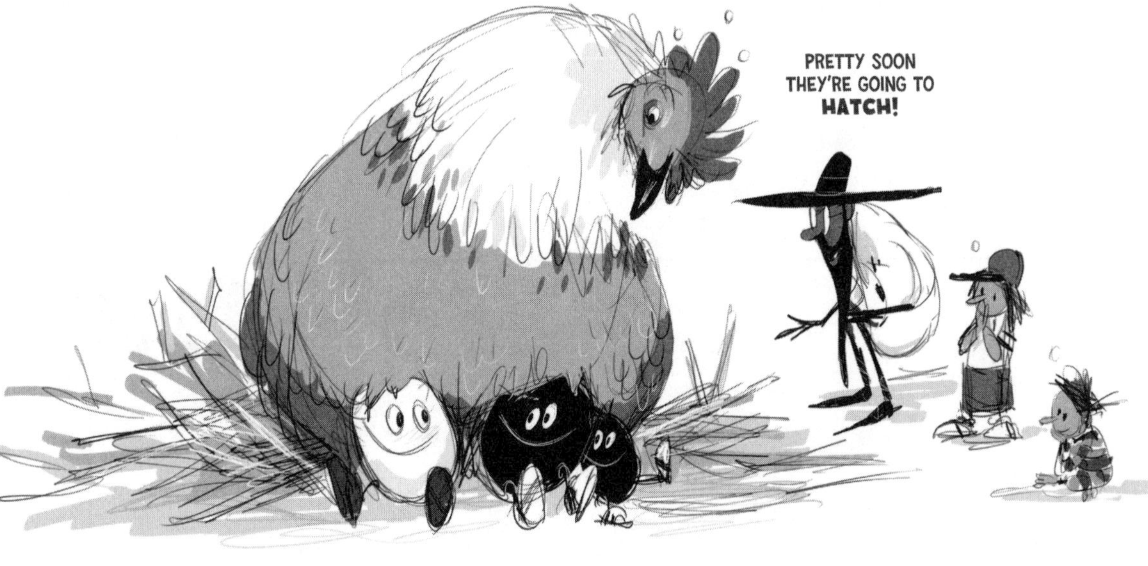
PRETTY SOON THEY'RE GOING TO **HATCH!**

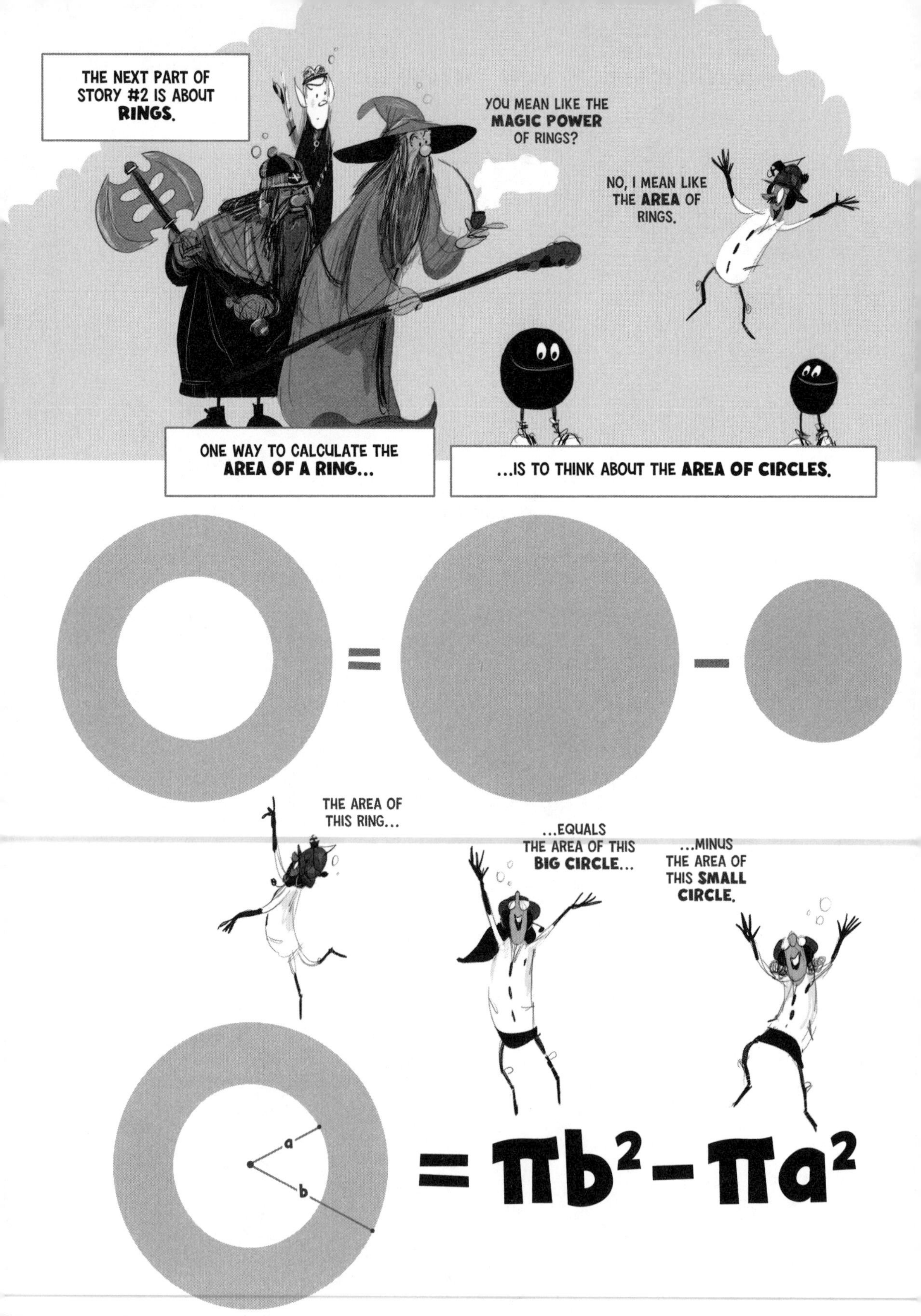
THE NEXT PART OF STORY #2 IS ABOUT **RINGS.**
YOU MEAN LIKE THE **MAGIC POWER** OF RINGS?
NO, I MEAN LIKE THE **AREA** OF RINGS.
ONE WAY TO CALCULATE THE **AREA OF A RING...**
...IS TO THINK ABOUT THE **AREA OF CIRCLES.**
=
−
THE AREA OF THIS RING...
...EQUALS THE AREA OF THIS **BIG CIRCLE...**
...MINUS THE AREA OF THIS **SMALL CIRCLE.**
a
b
$= \pi b^2 - \pi a^2$

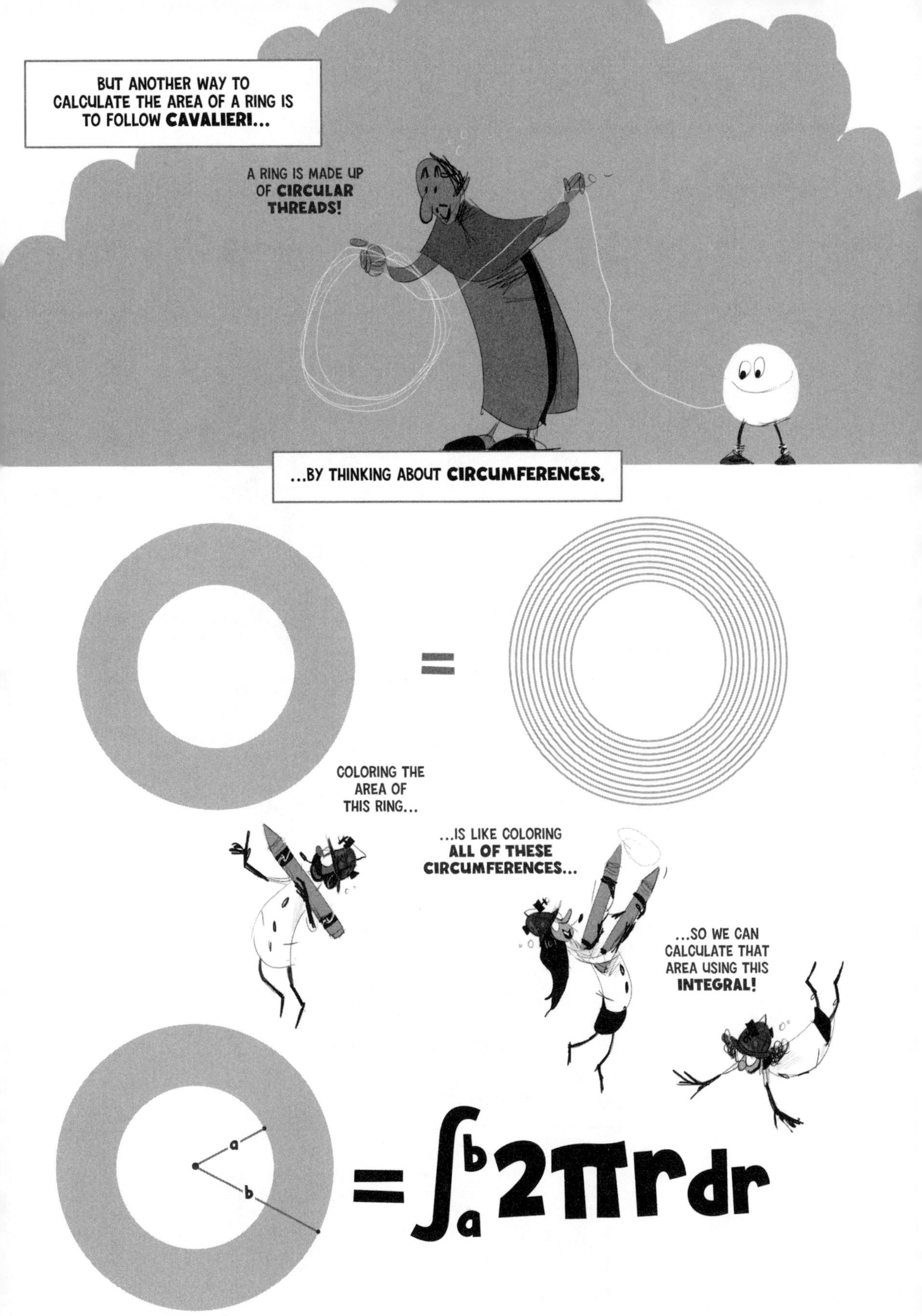
BUT ANOTHER WAY TO CALCULATE THE AREA OF A RING IS TO FOLLOW **CAVALIERI...**
A RING IS MADE UP OF **CIRCULAR THREADS!**
...BY THINKING ABOUT **CIRCUMFERENCES.**
=
COLORING THE AREA OF THIS RING...
...IS LIKE COLORING **ALL OF THESE CIRCUMFERENCES...**
...SO WE CAN CALCULATE THAT AREA USING THIS **INTEGRAL!**
a
b
$= \int_a^b 2\pi r\,dr$

OF COURSE, THESE TWO WAYS OF CALCULATING THE AREA OF A RING **MUST GIVE US THE SAME ANSWER...**

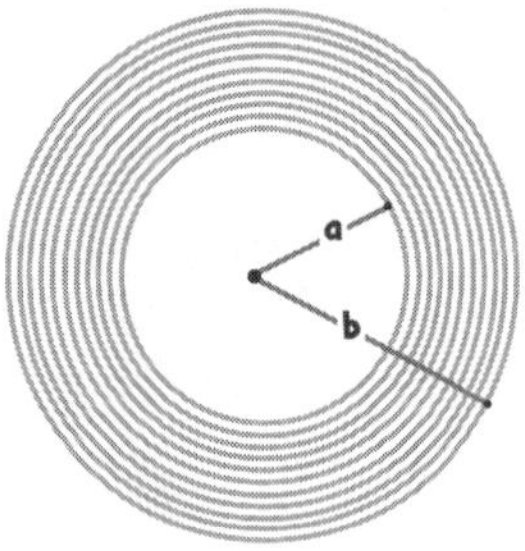
a
b

$$\int_a^b 2\pi r\,dr = \pi b^2 - \pi a^2$$

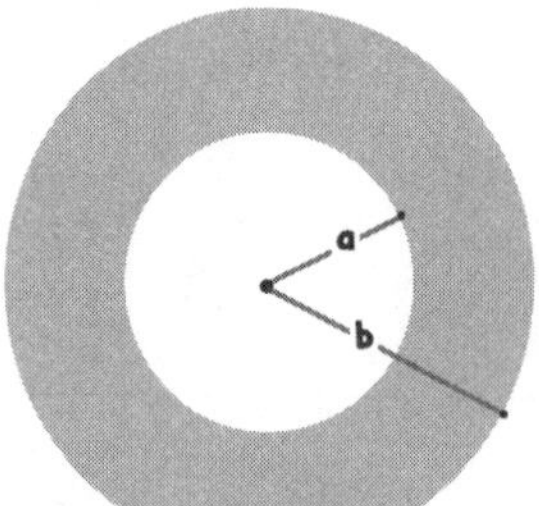
a
b

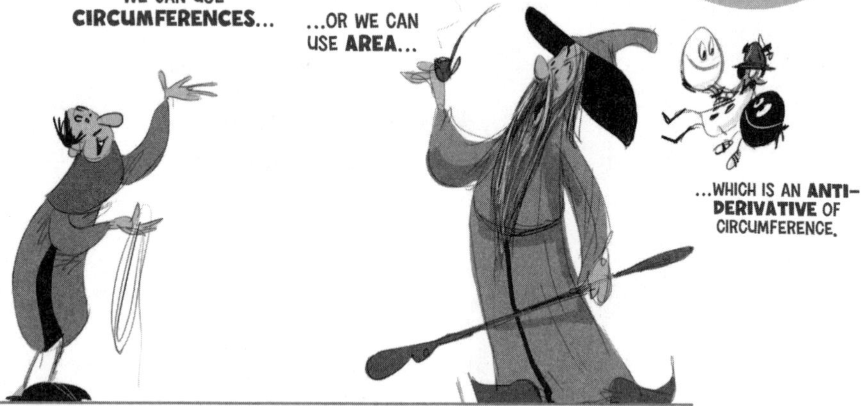
WE CAN USE **CIRCUMFERENCES...**
...OR WE CAN USE **AREA...**
...WHICH IS AN **ANTI-DERIVATIVE** OF CIRCUMFERENCE.

...AND THAT'S EXACTLY WHAT THE **SECOND PART OF THE FUNDAMENTAL THEOREM SAYS!**

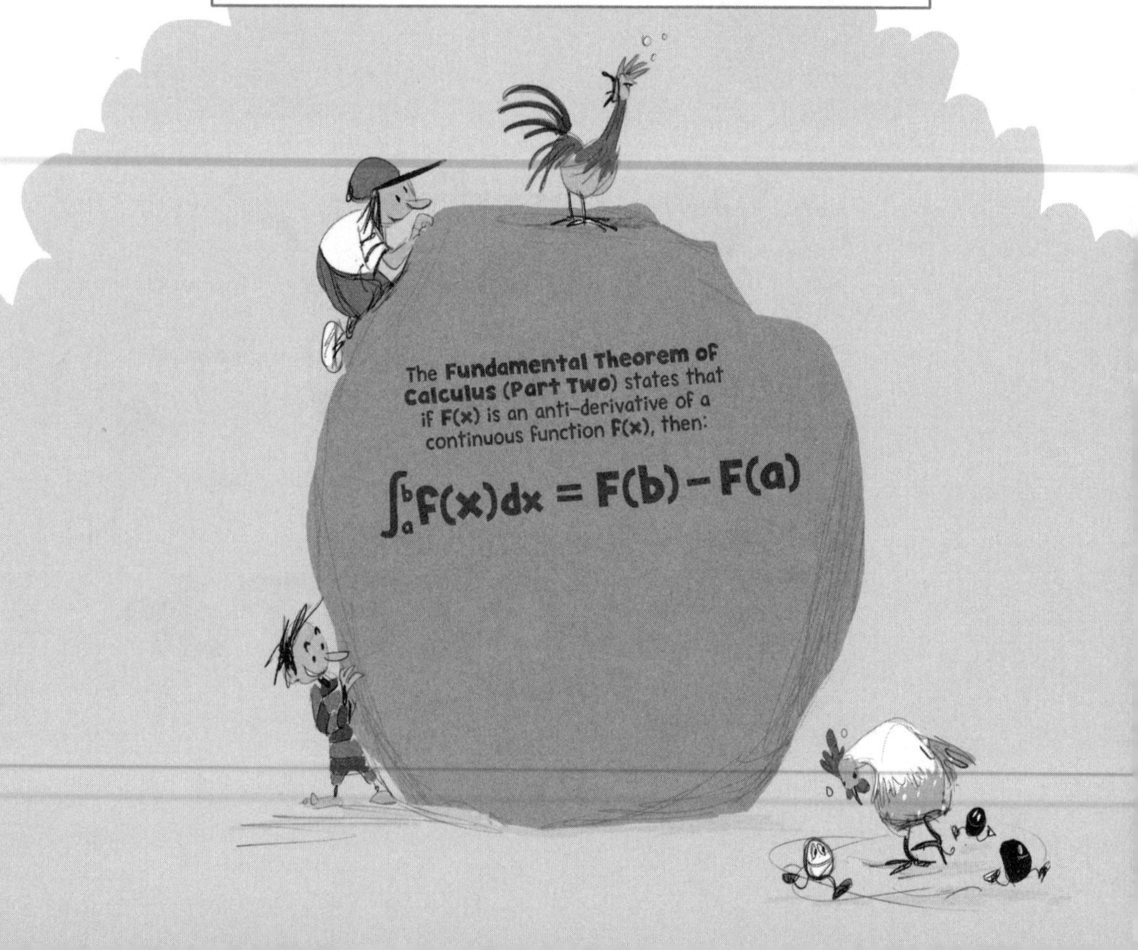
The **Fundamental Theorem of Calculus (Part Two)** states that if **F(x)** is an anti-derivative of a continuous function **f(x)**, then:
$\int_a^b f(x)dx = F(b) - F(a)$

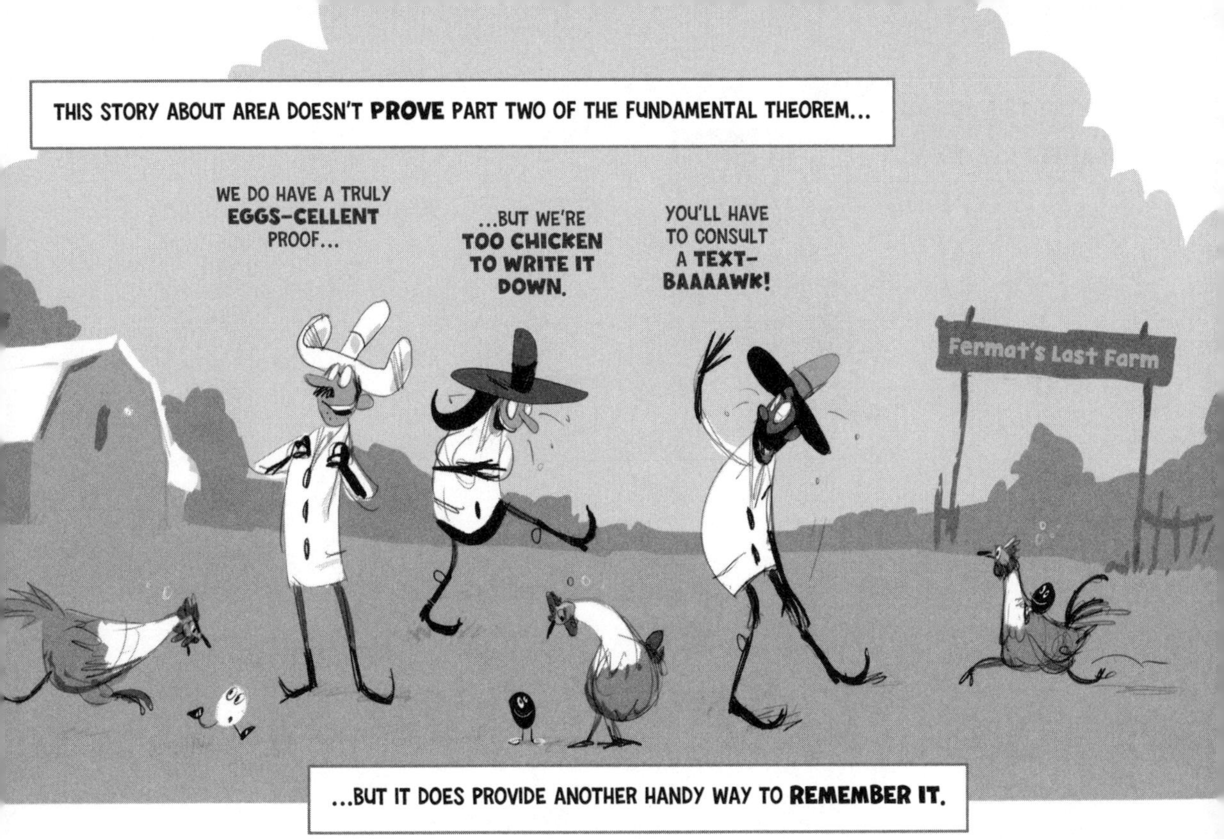
THIS STORY ABOUT AREA DOESN'T **PROVE** PART TWO OF THE FUNDAMENTAL THEOREM...
WE DO HAVE A TRULY **EGGS-CELLENT** PROOF...
...BUT WE'RE **TOO CHICKEN TO WRITE IT DOWN.**
YOU'LL HAVE TO CONSULT A **TEXT-BAAAAWK!**
Fermat's Last Farm
...BUT IT DOES PROVIDE ANOTHER HANDY WAY TO **REMEMBER IT.**

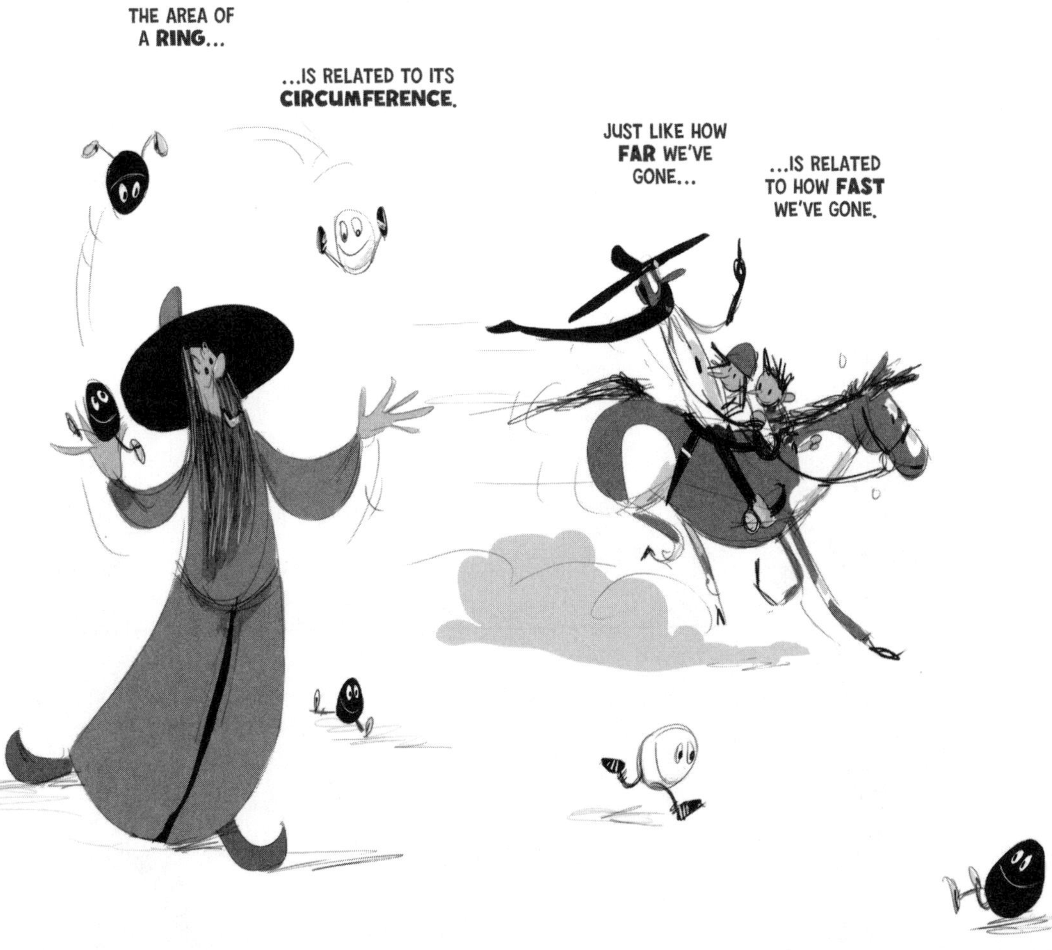
THE AREA OF A **RING**...
...IS RELATED TO ITS **CIRCUMFERENCE.**
JUST LIKE HOW **FAR** WE'VE GONE...
...IS RELATED TO HOW **FAST** WE'VE GONE.

AND THERE'S A **BONUS:** THINK ABOUT ADDING **LAYERS** TO A RING...

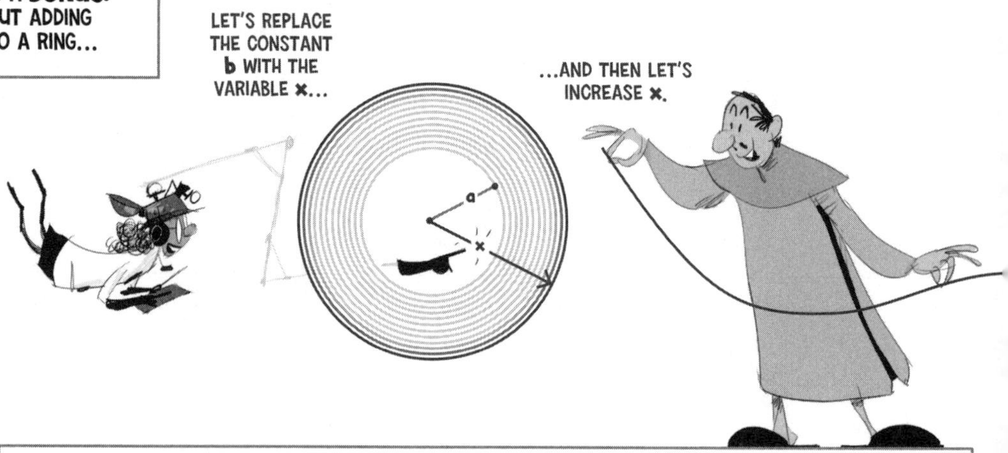
LET'S REPLACE THE CONSTANT b WITH THE VARIABLE x...
...AND THEN LET'S INCREASE x.
a
x

...AND YOU GET A RING VERSION OF **PART ONE OF THE FUNDAMENTAL THEOREM.**

$$\frac{d}{dx}\left[\int_a^x 2\pi r\,dr\right] = 2\pi x$$

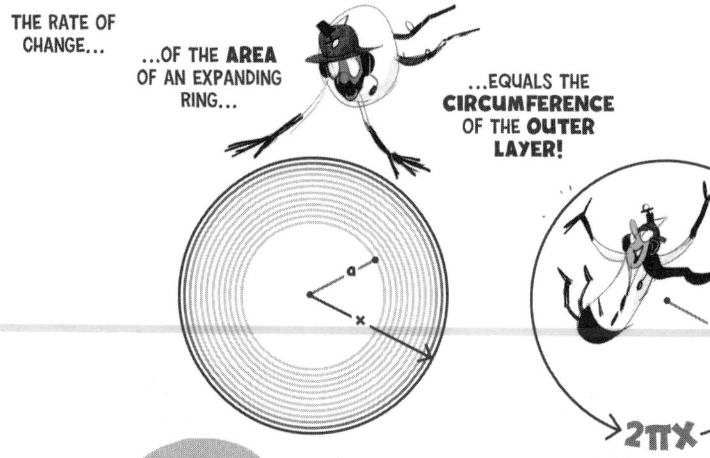
THE RATE OF CHANGE...
...OF THE AREA OF AN EXPANDING RING...
...EQUALS THE CIRCUMFERENCE OF THE OUTER LAYER!
a
x
x
2πx

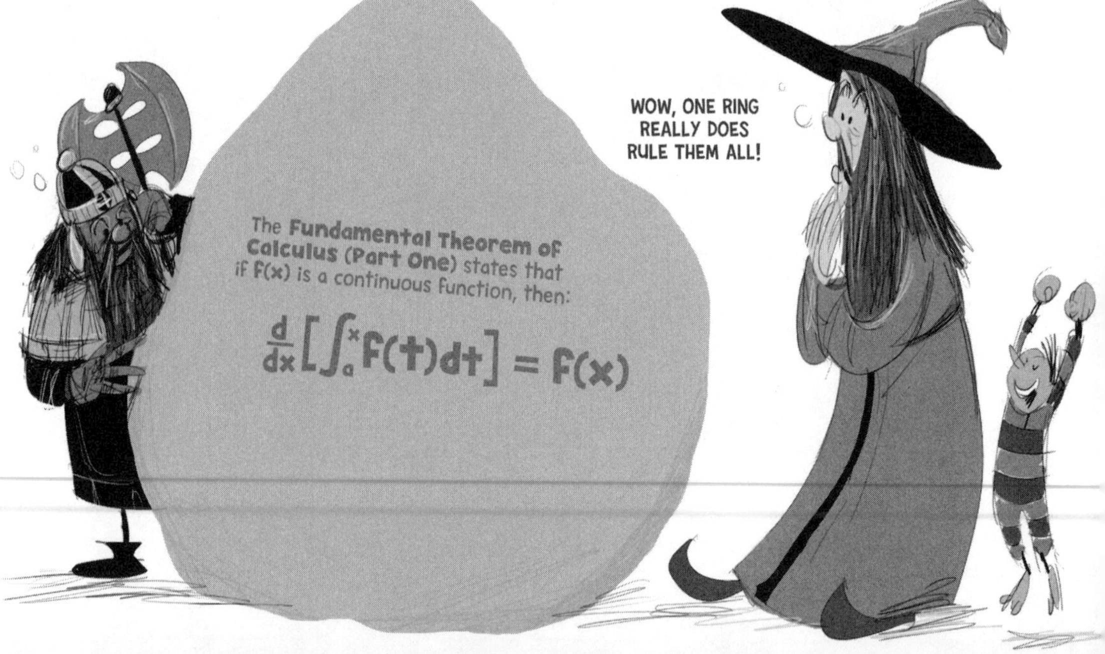
WOW, ONE RING REALLY DOES RULE THEM ALL!
The Fundamental Theorem of Calculus (Part One) states that if F(x) is a continuous function, then:
$\frac{d}{dx}\left[\int_a^x F(t)\,dt\right] = F(x)$

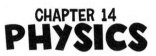

CHAPTER 14
PHYSICS

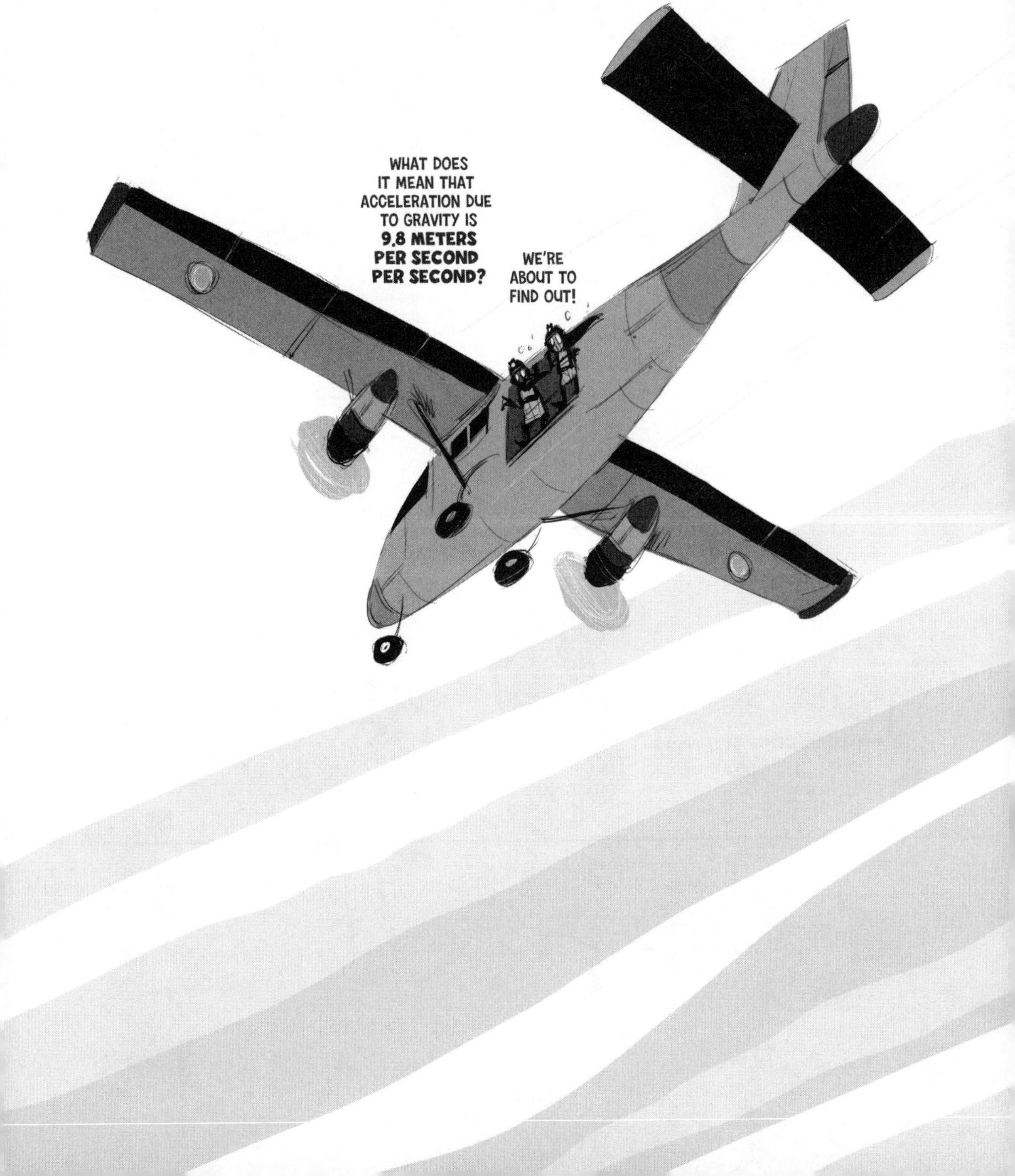
WHAT DOES IT MEAN THAT ACCELERATION DUE TO GRAVITY IS **9.8 METERS PER SECOND PER SECOND?**
WE'RE ABOUT TO FIND OUT!

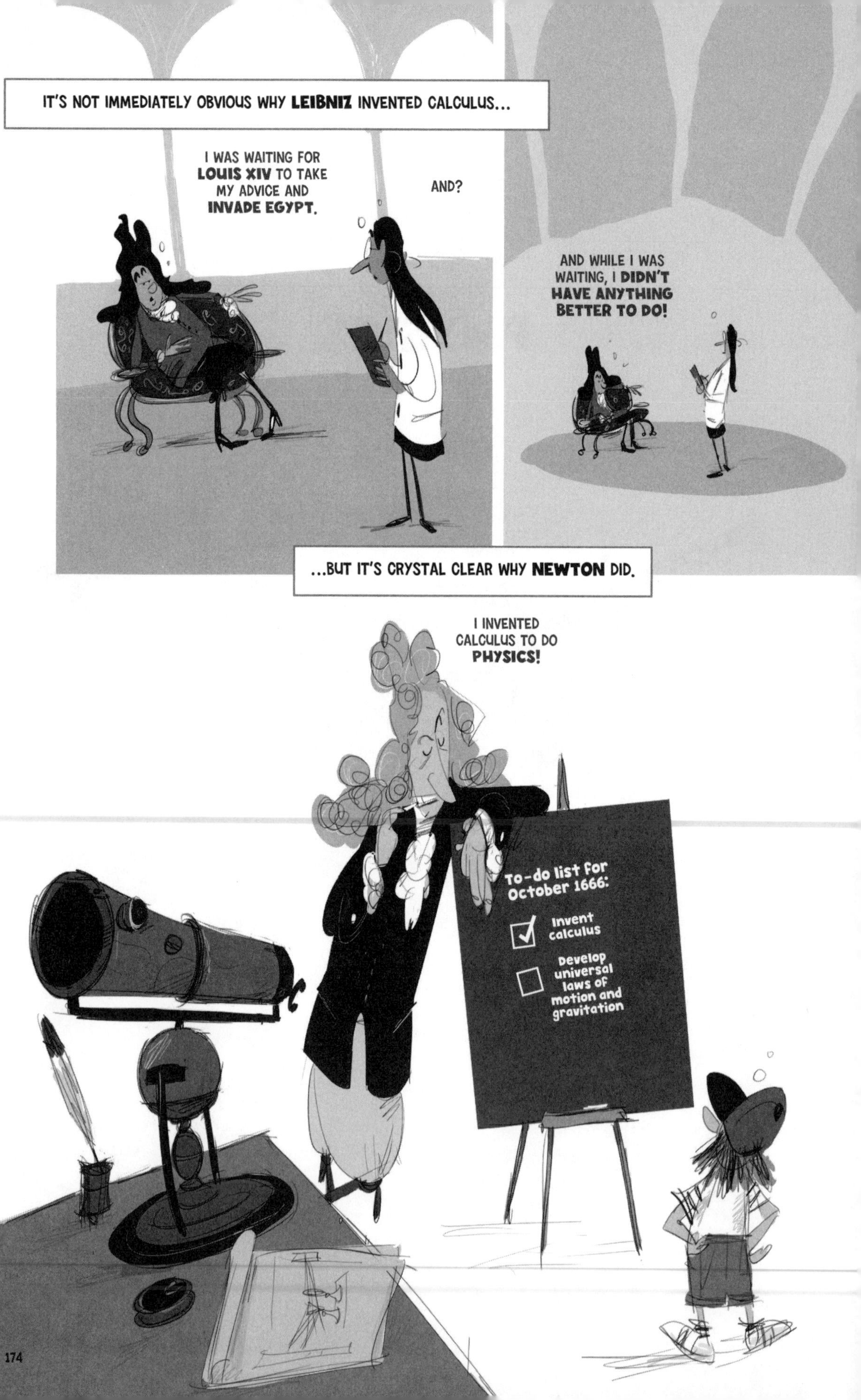

IT'S NOT IMMEDIATELY OBVIOUS WHY LEIBNIZ INVENTED CALCULUS...
I WAS WAITING FOR LOUIS XIV TO TAKE MY ADVICE AND INVADE EGYPT.
AND?
AND WHILE I WAS WAITING, I DIDN'T HAVE ANYTHING BETTER TO DO!
...BUT IT'S CRYSTAL CLEAR WHY NEWTON DID.
I INVENTED CALCULUS TO DO PHYSICS!
To-do list for October 1666:
Invent calculus
Develop universal laws of motion and gravitation

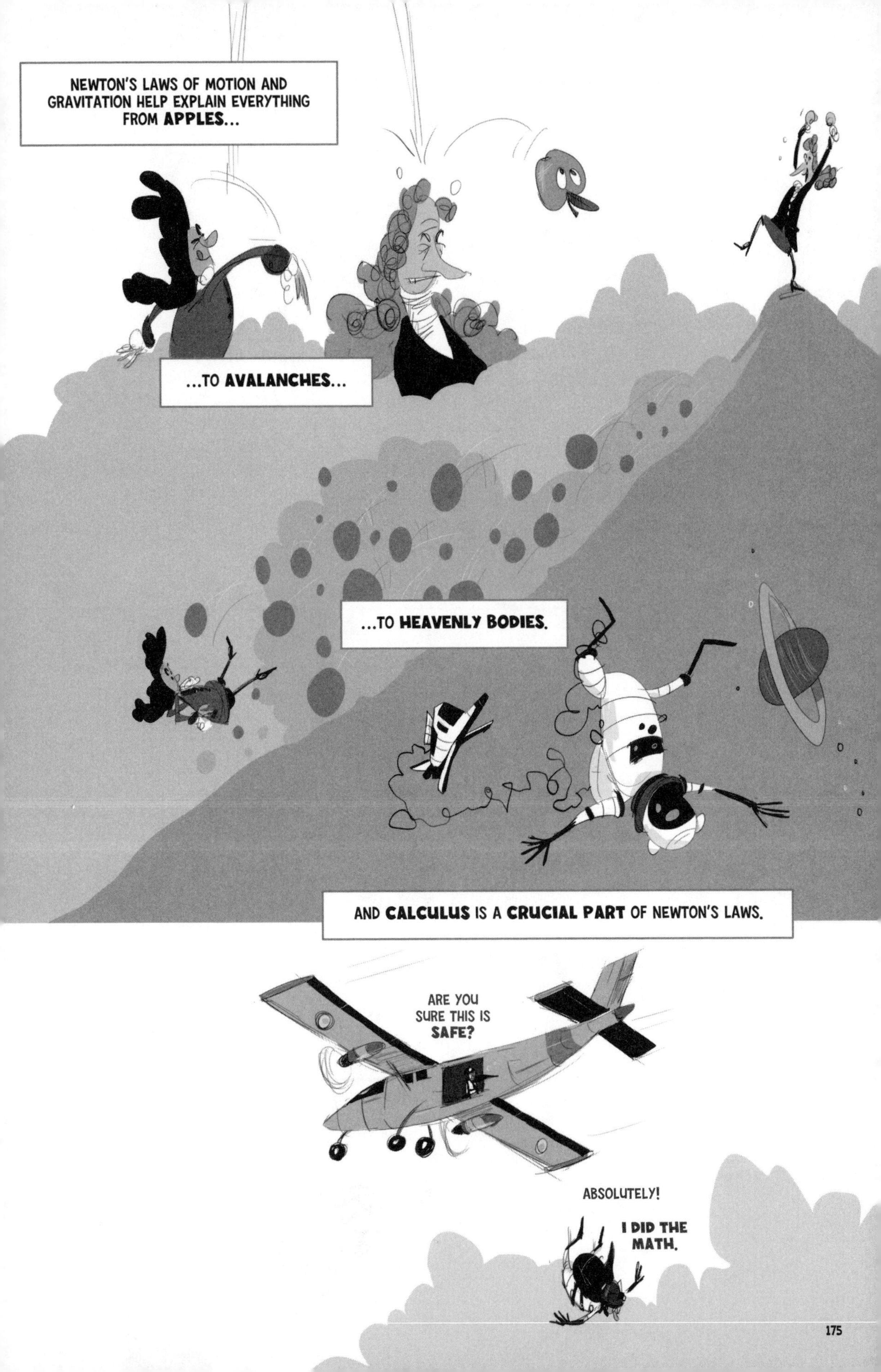
NEWTON'S LAWS OF MOTION AND GRAVITATION HELP EXPLAIN EVERYTHING FROM **APPLES**...
...TO **AVALANCHES**...
...TO **HEAVENLY BODIES**.
AND **CALCULUS** IS A **CRUCIAL PART** OF NEWTON'S LAWS.
ARE YOU SURE THIS IS **SAFE?**
ABSOLUTELY!
I DID THE MATH.

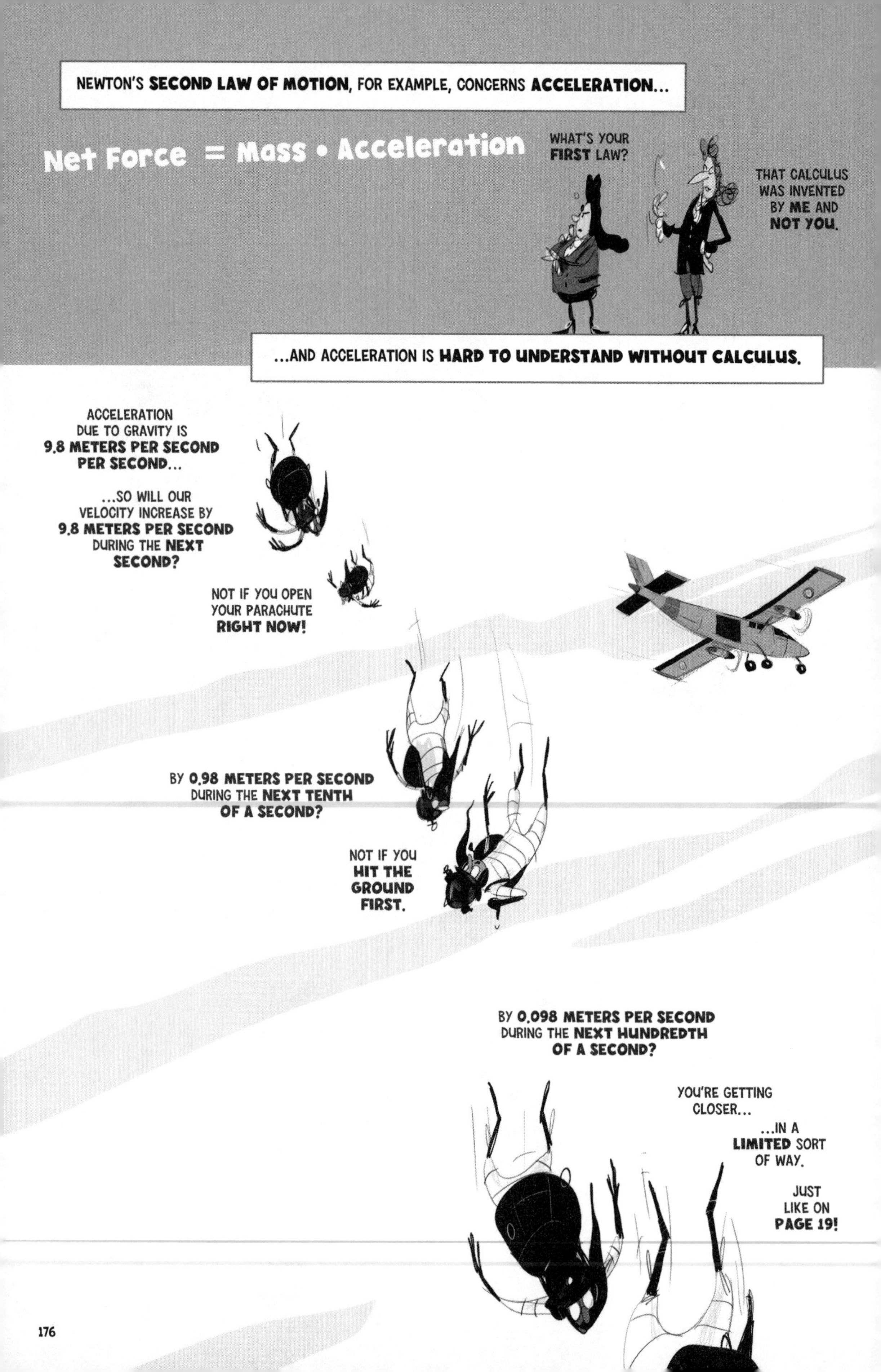

NEWTON'S **SECOND LAW OF MOTION**, FOR EXAMPLE, CONCERNS **ACCELERATION**...

Net Force = Mass • Acceleration

WHAT'S YOUR **FIRST** LAW?

THAT CALCULUS WAS INVENTED BY **ME** AND **NOT YOU.**

...AND ACCELERATION IS **HARD TO UNDERSTAND WITHOUT CALCULUS.**

ACCELERATION DUE TO GRAVITY IS **9.8 METERS PER SECOND PER SECOND**...

...SO WILL OUR VELOCITY INCREASE BY **9.8 METERS PER SECOND** DURING THE **NEXT SECOND?**

NOT IF YOU OPEN YOUR PARACHUTE **RIGHT NOW!**

BY **0.98 METERS PER SECOND** DURING THE **NEXT TENTH OF A SECOND?**

NOT IF YOU **HIT THE GROUND FIRST.**

BY **0.098 METERS PER SECOND** DURING THE **NEXT HUNDREDTH OF A SECOND?**

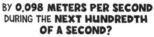

YOU'RE GETTING CLOSER...

...IN A **LIMITED** SORT OF WAY.

JUST LIKE ON **PAGE 19!**

THAT'S BECAUSE WE NEED **LIMITS** TO DEFINE ACCELERATION.

ACCELERATION MEASURES SOMETHING **INSTANTANEOUS**...

...JUST LIKE **VELOCITY**!

IN EXACTLY THE SAME WAY THAT **VELOCITY**...

...MEASURES THE **INSTANTANEOUS RATE OF CHANGE OF POSITION**...

$$v(t) = \frac{d}{dt}f(t) = \lim_{h \to 0} \frac{f(t+h) - f(t)}{h}$$

...**ACCELERATION**...

...MEASURES THE **INSTANTANEOUS RATE OF CHANGE OF VELOCITY**.

$$a(t) = \frac{d}{dt}v(t) = \lim_{h \to 0} \frac{v(t+h) - v(t)}{h}$$

IN THE LANGUAGE OF MATHEMATICS, ACCELERATION IS A **SECOND DERIVATIVE.**

YOU TAKE THE DERIVATIVE OF **POSITION** TO GET **VELOCITY...**

...AND YOU TAKE THE DERIVATIVE OF **VELOCITY** TO GET **ACCELERATION.**

SO A SECOND DERIVATIVE IS **A DERIVATIVE OF A DERIVATIVE!**

$$\frac{d}{dx}\left[\frac{d}{dx}f(x)\right]$$

WE CAN MAKE THINGS A BIT **LESS AWKWARD** WITH SOME **NEW NOTATION...**

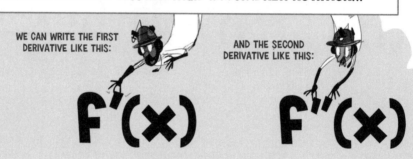

WE CAN WRITE THE FIRST DERIVATIVE LIKE THIS:

$$f'(x)$$

AND THE SECOND DERIVATIVE LIKE THIS:

$$f''(x)$$

...BUT THE TRUTH IS THAT WRAPPING YOUR HEAD AROUND SECOND DERIVATIVES CAN BE **TOUGH.**

THE SECOND DERIVATIVE OF f(x) MEASURES **CURVATURE.**

IT'S THE INSTANTANEOUS RATE OF CHANGE...

...OF THE INSTANTANEOUS RATE OF CHANGE OF f(x).

MY BRAIN IS INSTANTANEOUSLY **CHANGING INTO MUSH!**

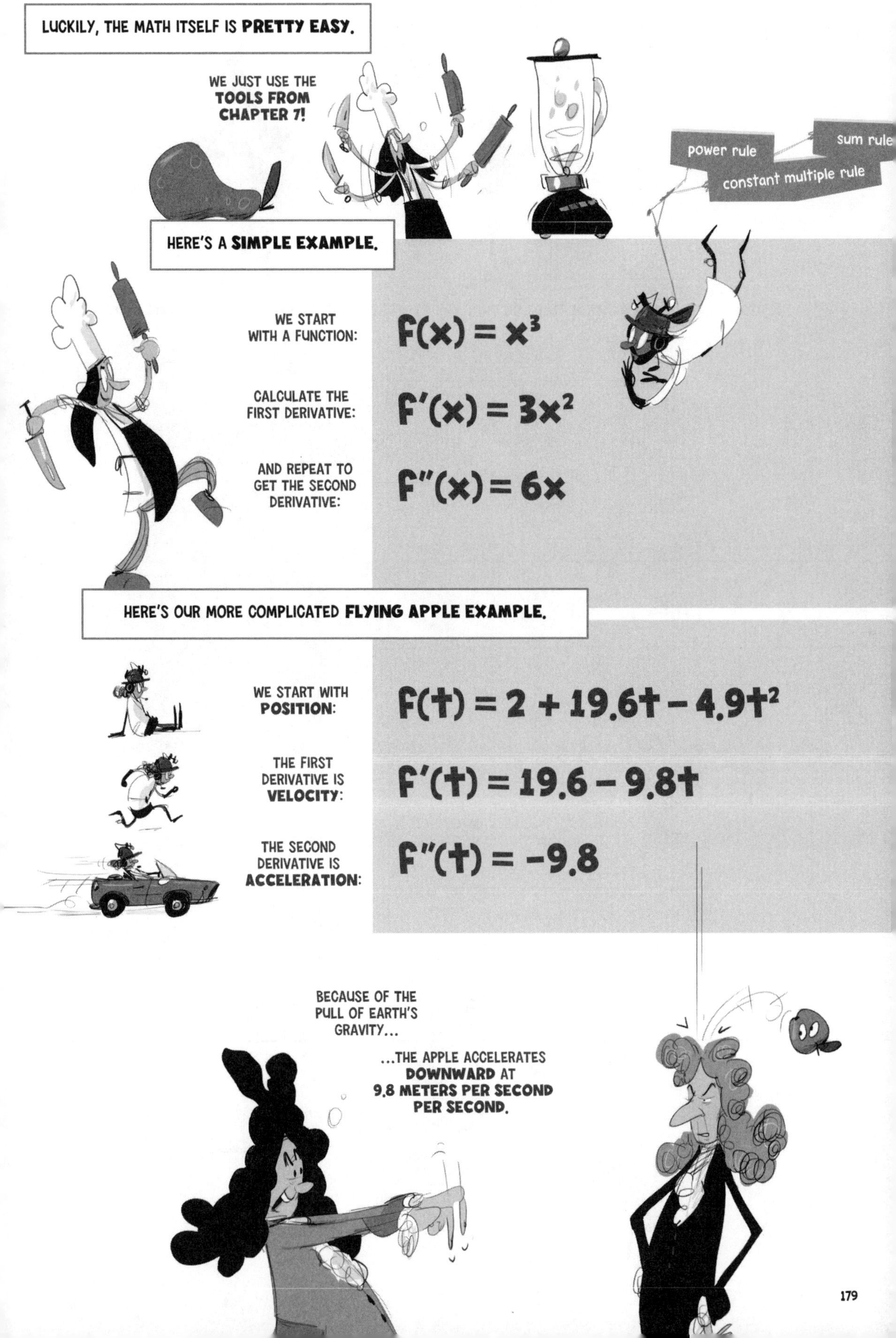
LUCKILY, THE MATH ITSELF IS PRETTY EASY.
WE JUST USE THE TOOLS FROM CHAPTER 7!
power rule
sum rule
constant multiple rule
HERE'S A SIMPLE EXAMPLE.
WE START WITH A FUNCTION:
$f(x) = x^3$
CALCULATE THE FIRST DERIVATIVE:
$f'(x) = 3x^2$
AND REPEAT TO GET THE SECOND DERIVATIVE:
$f''(x) = 6x$
HERE'S OUR MORE COMPLICATED FLYING APPLE EXAMPLE.
WE START WITH POSITION:
$f(t) = 2 + 19.6t - 4.9t^2$
THE FIRST DERIVATIVE IS VELOCITY:
$f'(t) = 19.6 - 9.8t$
THE SECOND DERIVATIVE IS ACCELERATION:
$f''(t) = -9.8$
BECAUSE OF THE PULL OF EARTH'S GRAVITY...
...THE APPLE ACCELERATES DOWNWARD AT 9.8 METERS PER SECOND PER SECOND.

OF COURSE, THE PULL OF EARTH'S GRAVITY **ISN'T CONSTANT.**

OF COURSE, IT ONLY SAYS "OF COURSE" BECAUSE OF **MY THEORIES.**

NEWTON'S LAW OF GRAVITATION TELLS US THAT THE STRENGTH OF GRAVITY **CHANGES WITH ALTITUDE.**

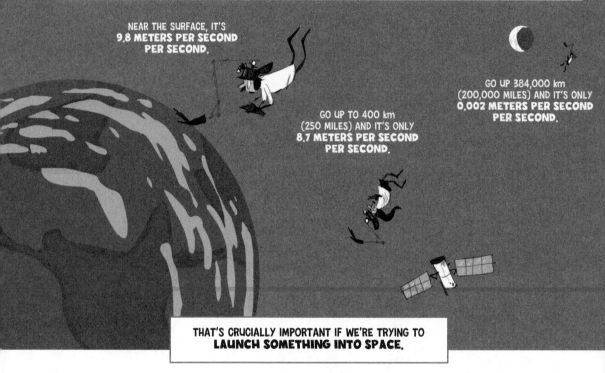

NEAR THE SURFACE, IT'S **9.8 METERS PER SECOND PER SECOND.**

GO UP TO 400 km (250 MILES) AND IT'S ONLY **8.7 METERS PER SECOND PER SECOND.**

GO UP 384,000 km (200,000 MILES) AND IT'S ONLY **0.002 METERS PER SECOND PER SECOND.**

THAT'S CRUCIALLY IMPORTANT IF WE'RE TRYING TO **LAUNCH SOMETHING INTO SPACE.**

HELP! HELP!

WHAT'S WRONG, MAJOR TOM?

I'M **HUNGRY,** THROW ME AN **APPLE!**

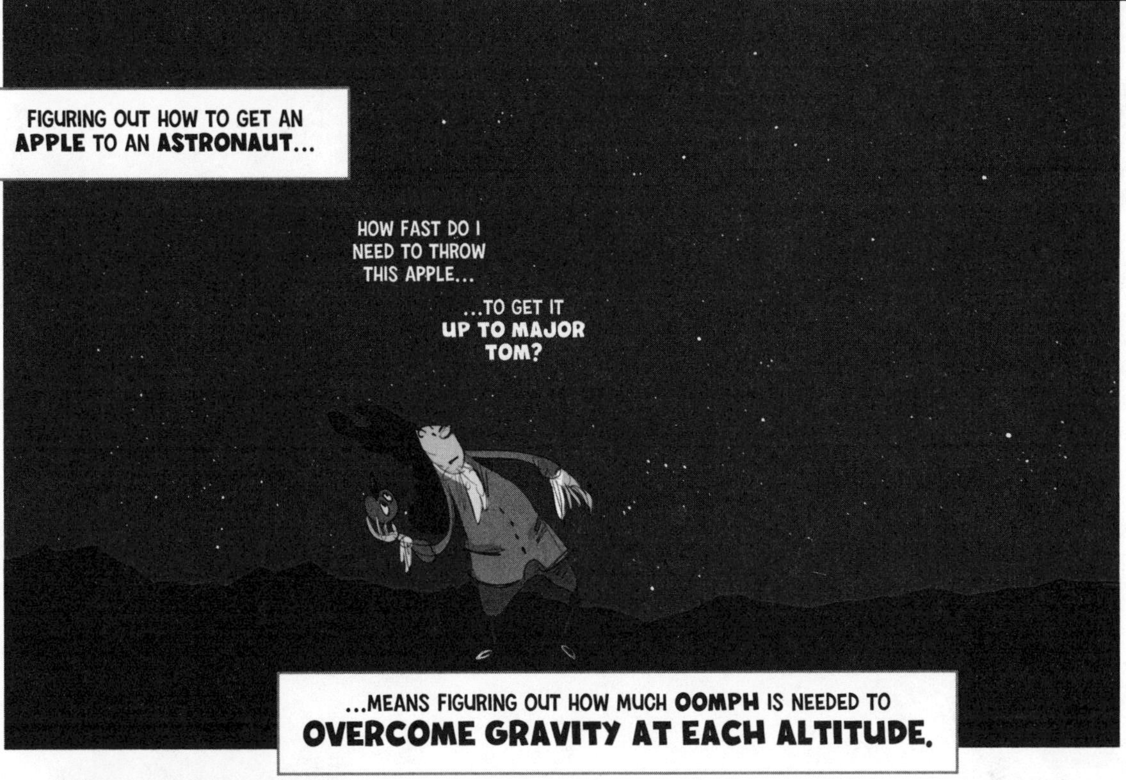
FIGURING OUT HOW TO GET AN **APPLE** TO AN **ASTRONAUT**...
HOW FAST DO I NEED TO THROW THIS APPLE...
...TO GET IT **UP TO MAJOR TOM?**
...MEANS FIGURING OUT HOW MUCH **OOMPH** IS NEEDED TO
OVERCOME GRAVITY AT EACH ALTITUDE.

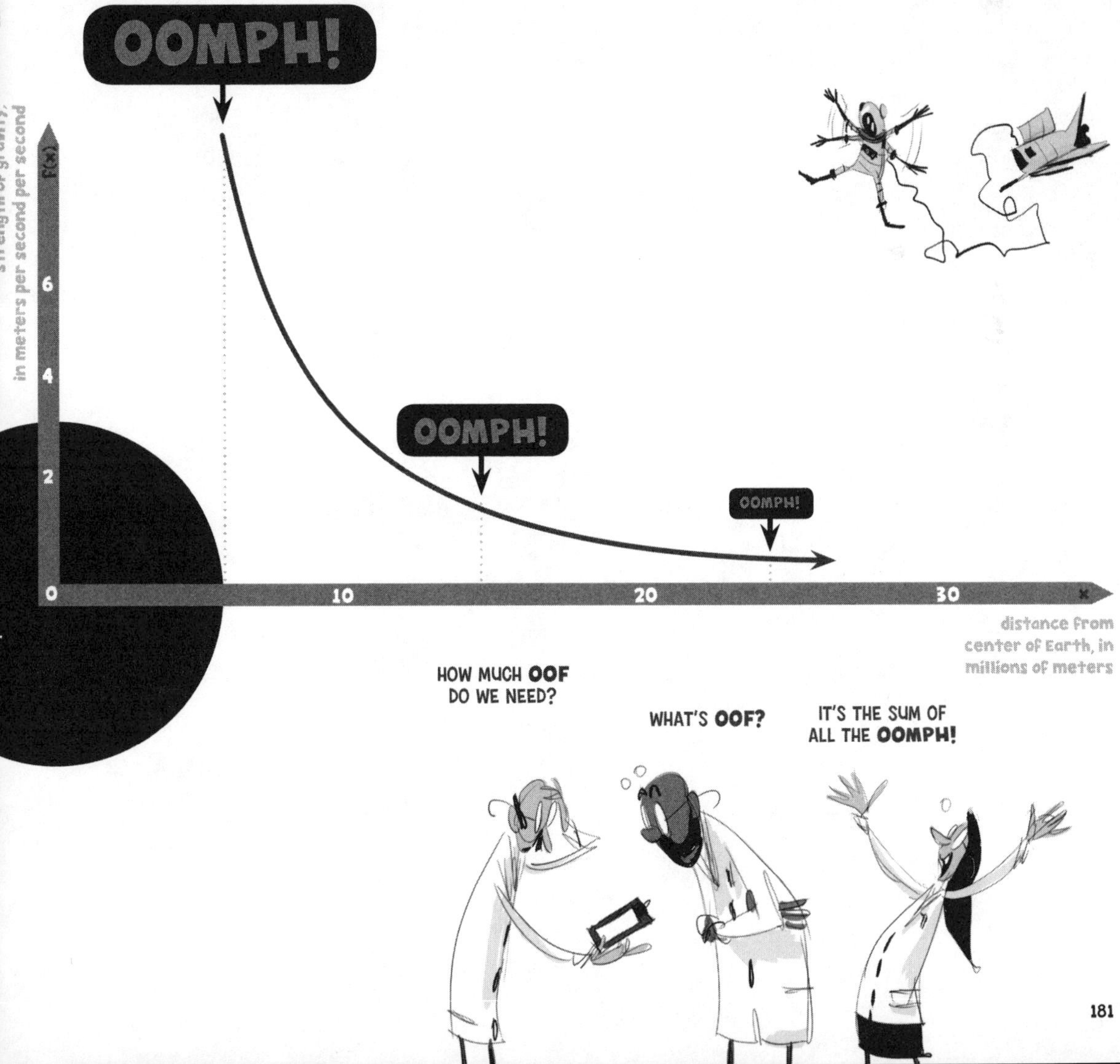
OOMPH!
f(x)
in meters per second per second
6
4
2
0
OOMPH!
OOMPH!
10
20
30
x
distance from center of Earth, in millions of meters
HOW MUCH **OOF** DO WE NEED?
WHAT'S **OOF?**
IT'S THE SUM OF ALL THE **OOMPH!**

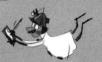

THIS IS A **TOUGH PHYSICS PROBLEM...**

a = RADIUS OF EARTH [$\approx 6.4 \cdot 10^6$ meters]

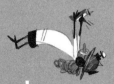

b = DISTANCE FROM EARTH'S CENTER TO ASTRONAUT [LET'S SAY $2.5 \cdot 10^7$ meters]

M = MASS OF EARTH [$\approx 6.0 \cdot 10^{24}$ kg]

G = GRAVITATIONAL CONSTANT [$\approx 6.7 \cdot 10^{-11}$ meters^3/(kg$\cdot$sec^2)]

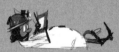

m = MASS OF APPLE [LET'S SAY 0.2 kg]

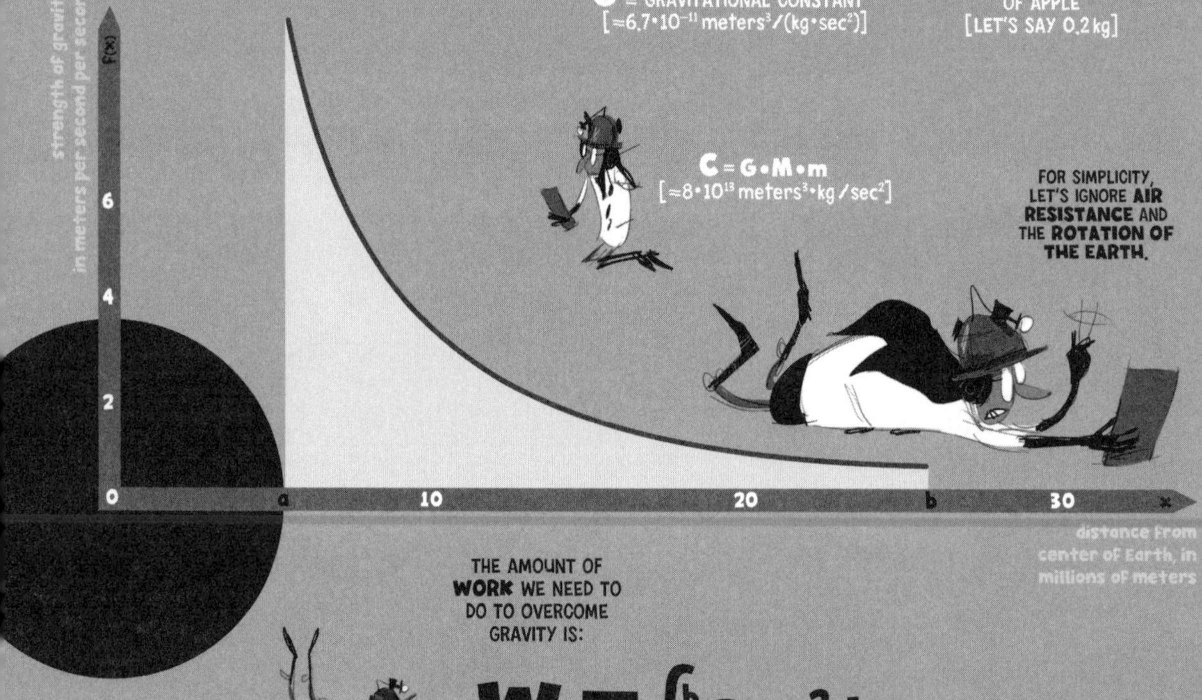

strength of gravity, in meters per second per second
f(x)
6
4
2
0
a
10
20
b
30
x
distance from center of Earth, in millions of meters

C = **G** $\cdot$ **M** $\cdot$ **m** [$\approx 8 \cdot 10^{13}$ meters$^3 \cdot$kg/sec^2]

FOR SIMPLICITY, LET'S IGNORE **AIR RESISTANCE** AND THE **ROTATION OF THE EARTH.**

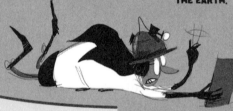

THE AMOUNT OF **WORK** WE NEED TO DO TO OVERCOME GRAVITY IS:

$$W = \int_a^b cx^{-2}dx$$

TO TRANSLATE THAT INTO THE **LAUNCH VELOCITY...**

...WE NEED TO PLUG WORK INTO THIS EQUATION AND THEN SOLVE FOR **v**.

$$W = \frac{1}{2}mv^2$$

WORK IS MEASURED IN **JOULES.** VELOCITY IS MEASURED IN **METERS PER SECOND.** HUNGER IS MEASURED IN **GRUMBLIES.**

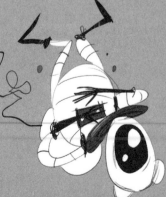

...BUT THE **MATH** BOILS DOWN TO **THIS QUESTION:**

WHAT'S THE VALUE OF THIS INTEGRAL?
$\int_a^b cx^{-2}dx$
f(x)
0
a
b
x

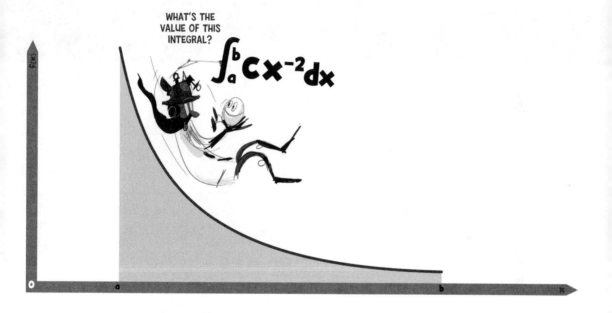

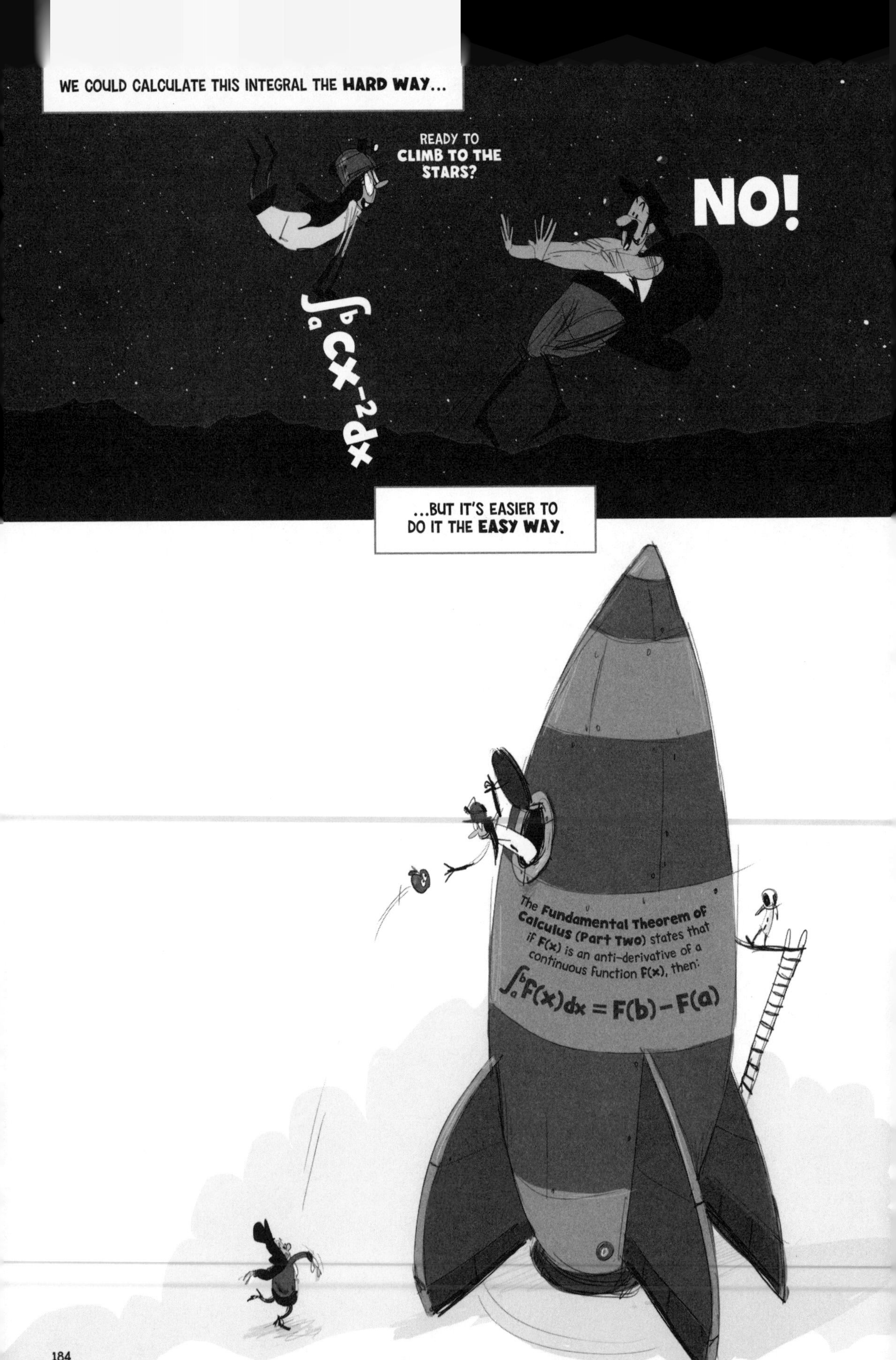

WE COULD CALCULATE THIS INTEGRAL THE **HARD WAY**...
READY TO **CLIMB TO THE STARS?**
$\int_a^b cx^{-2}dx$
NO!
...BUT IT'S EASIER TO DO IT THE **EASY WAY**.
The **Fundamental Theorem of Calculus (Part Two)** states that if F(x) is an anti-derivative of a continuous function f(x), then:
$\int_a^b f(x)dx = F(b) - F(a)$

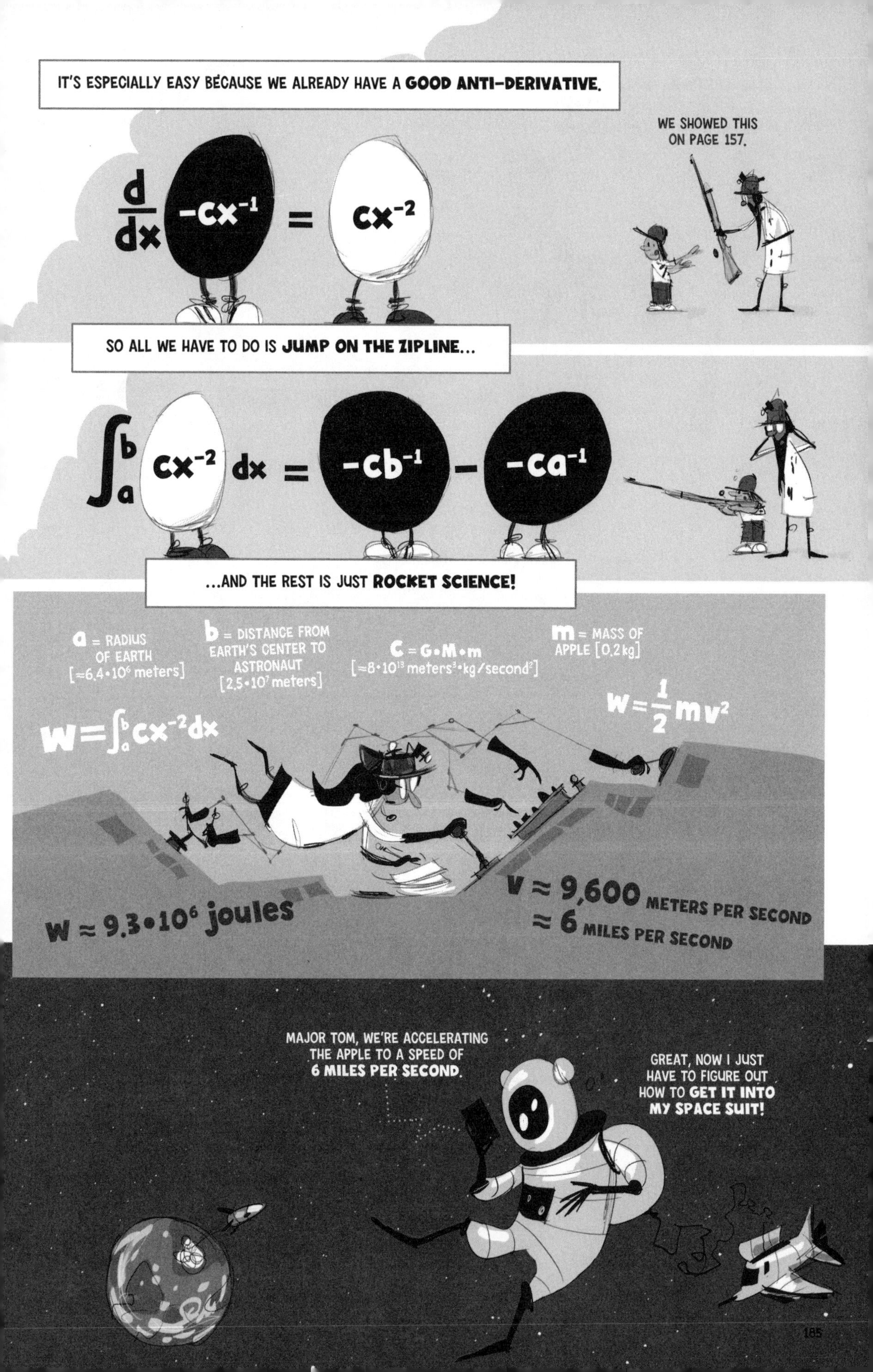

IT'S ESPECIALLY EASY BECAUSE WE ALREADY HAVE A **GOOD ANTI-DERIVATIVE.**

WE SHOWED THIS ON PAGE 157.

$$\frac{d}{dx} \; -cx^{-1} = cx^{-2}$$

SO ALL WE HAVE TO DO IS **JUMP ON THE ZIPLINE**...

$$\int_a^b cx^{-2}\,dx = -cb^{-1} - -ca^{-1}$$

...AND THE REST IS JUST **ROCKET SCIENCE!**

a = RADIUS OF EARTH [≈6.4·10⁶ meters]

b = DISTANCE FROM EARTH'S CENTER TO ASTRONAUT [2.5·10⁷ meters]

C = G·M·m [≈8·10¹³ meters³·kg/second²]

m = MASS OF APPLE [0.2 kg]

$$W = \int_a^b cx^{-2}\,dx$$

$$W = \frac{1}{2}mv^2$$

$$W \approx 9.3 \cdot 10^6 \; \text{joules}$$

$$v \approx 9,600 \; \text{METERS PER SECOND}$$
$$\approx 6 \; \text{MILES PER SECOND}$$

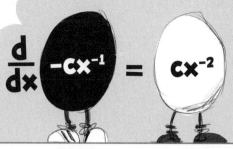

MAJOR TOM, WE'RE ACCELERATING THE APPLE TO A SPEED OF **6 MILES PER SECOND.**

GREAT, NOW I JUST HAVE TO FIGURE OUT HOW TO **GET IT INTO MY SPACE SUIT!**

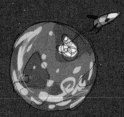

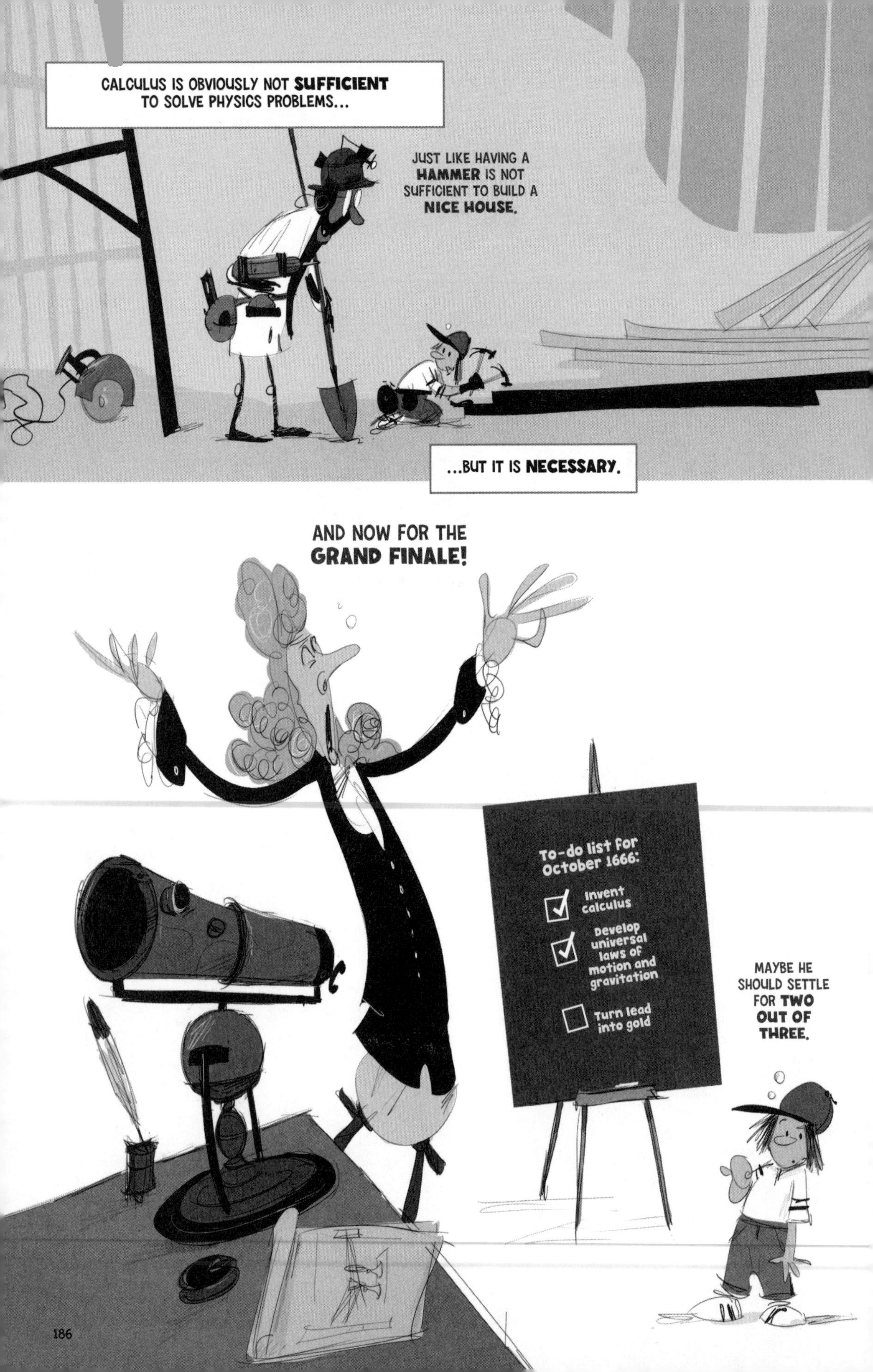
CALCULUS IS OBVIOUSLY NOT SUFFICIENT TO SOLVE PHYSICS PROBLEMS...
JUST LIKE HAVING A HAMMER IS NOT SUFFICIENT TO BUILD A NICE HOUSE.
...BUT IT IS NECESSARY.
AND NOW FOR THE GRAND FINALE!
To-do list for October 1666:
Invent calculus
Develop universal laws of motion and gravitation
Turn lead into gold
MAYBE HE SHOULD SETTLE FOR TWO OUT OF THREE.

CHAPTER 15

LIMITS BEYOND LIMITS

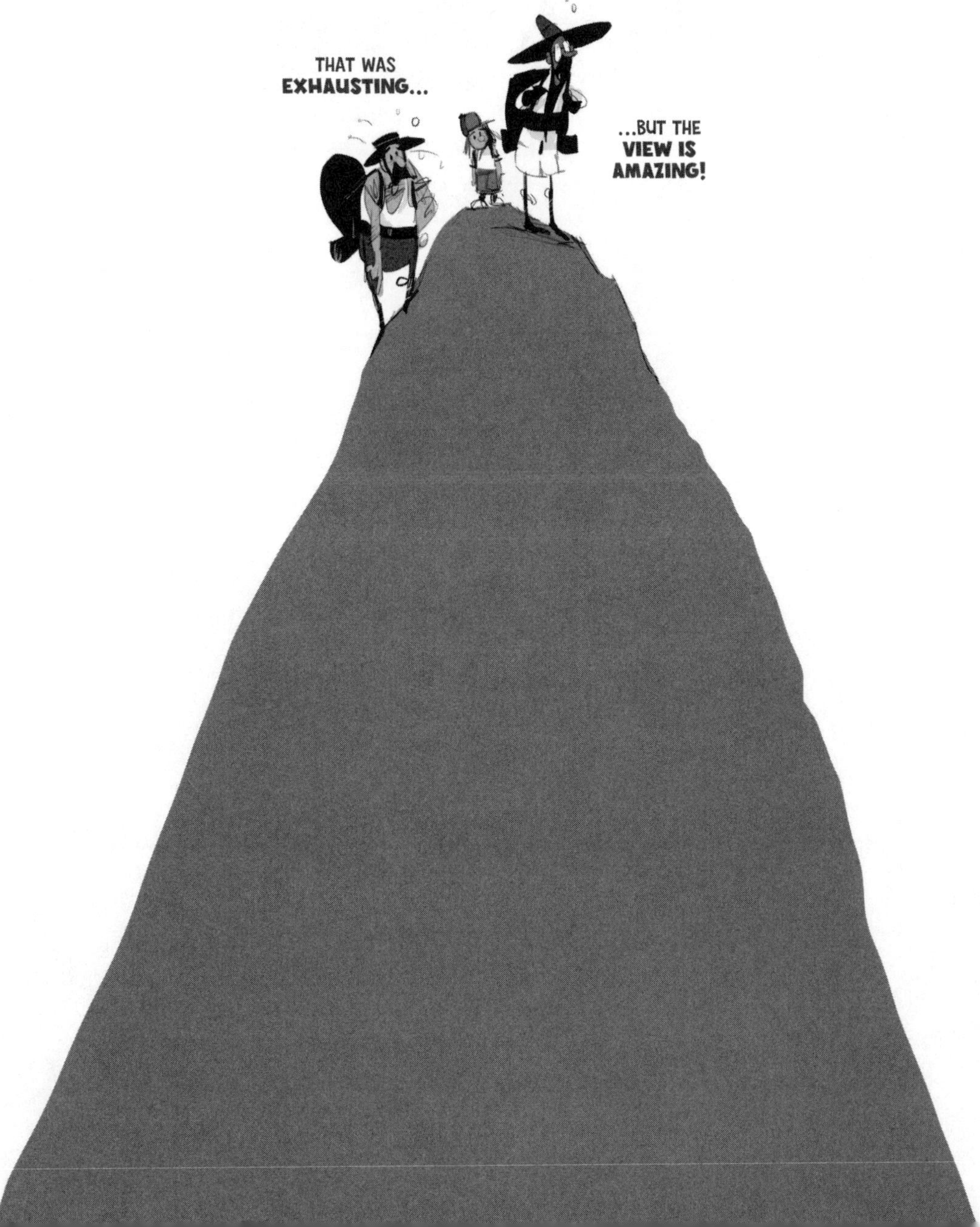
THAT WAS EXHAUSTING...
...BUT THE VIEW IS AMAZING!

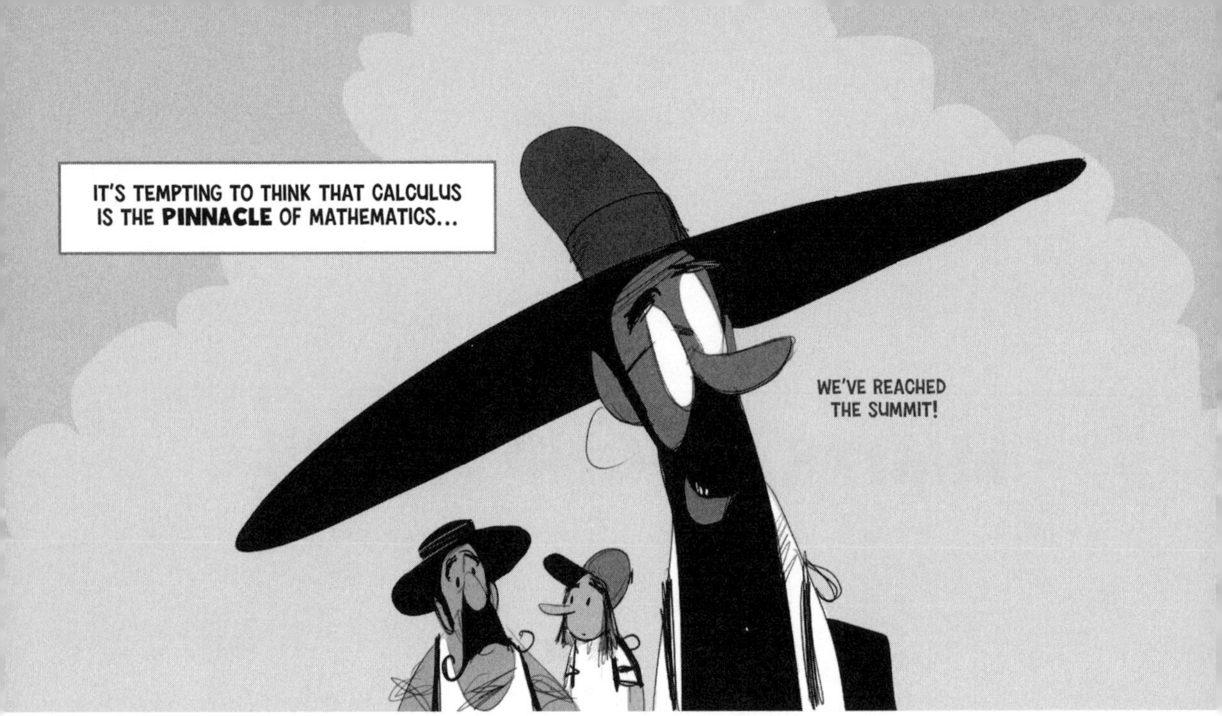
IT'S TEMPTING TO THINK THAT CALCULUS IS THE **PINNACLE** OF MATHEMATICS...
WE'VE REACHED THE SUMMIT!

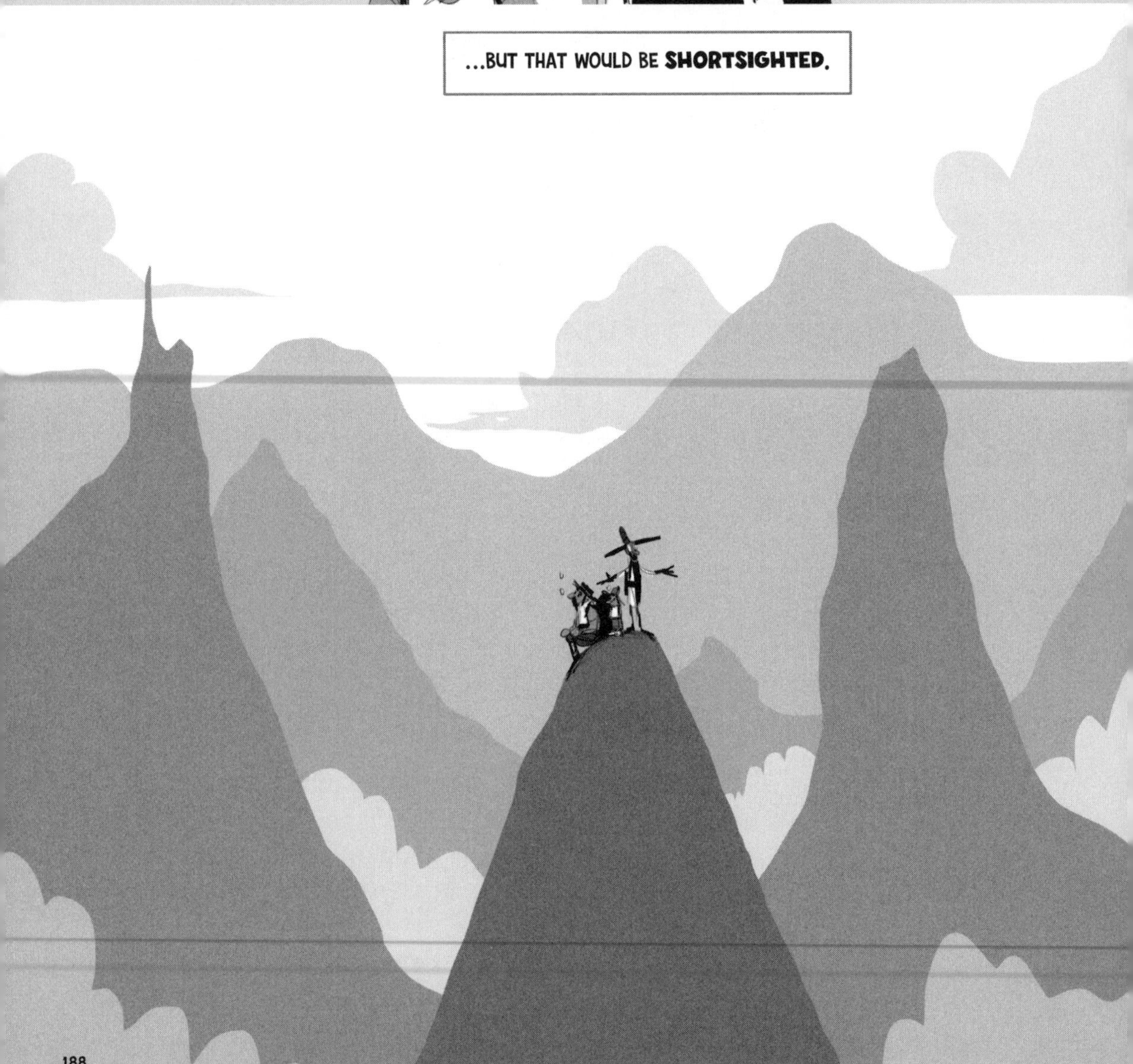
...BUT THAT WOULD BE **SHORTSIGHTED.**

THAT'S BECAUSE MATHEMATICS IS ABOUT FINDING **PATTERNS**...

IN **ANYTHING**.

IN **NUMBERS**.

IN **KNOTS**.

IN **SHELLS**.

...AND THEN WORKING TO FORMALIZE THOSE PATTERNS WITH **PROOFS**.

THIS BOOK
INCLUDES PROOF BY
INDUCTION...

...PROOF BY
CONTRADICTION...

...AND PROOF BY
SANDWICH.

THE GOAL OF THIS CHAPTER IS TO OPEN YOUR EYES
TO THE WORLD OF MATHEMATICS...

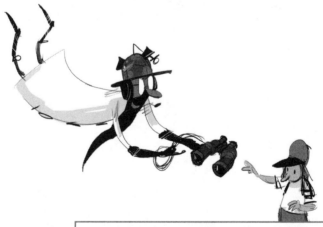

...BY FURTHER EXPLORING ONE OF THIS BOOK'S **KEY IDEAS**.

LIMITS

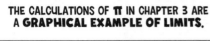

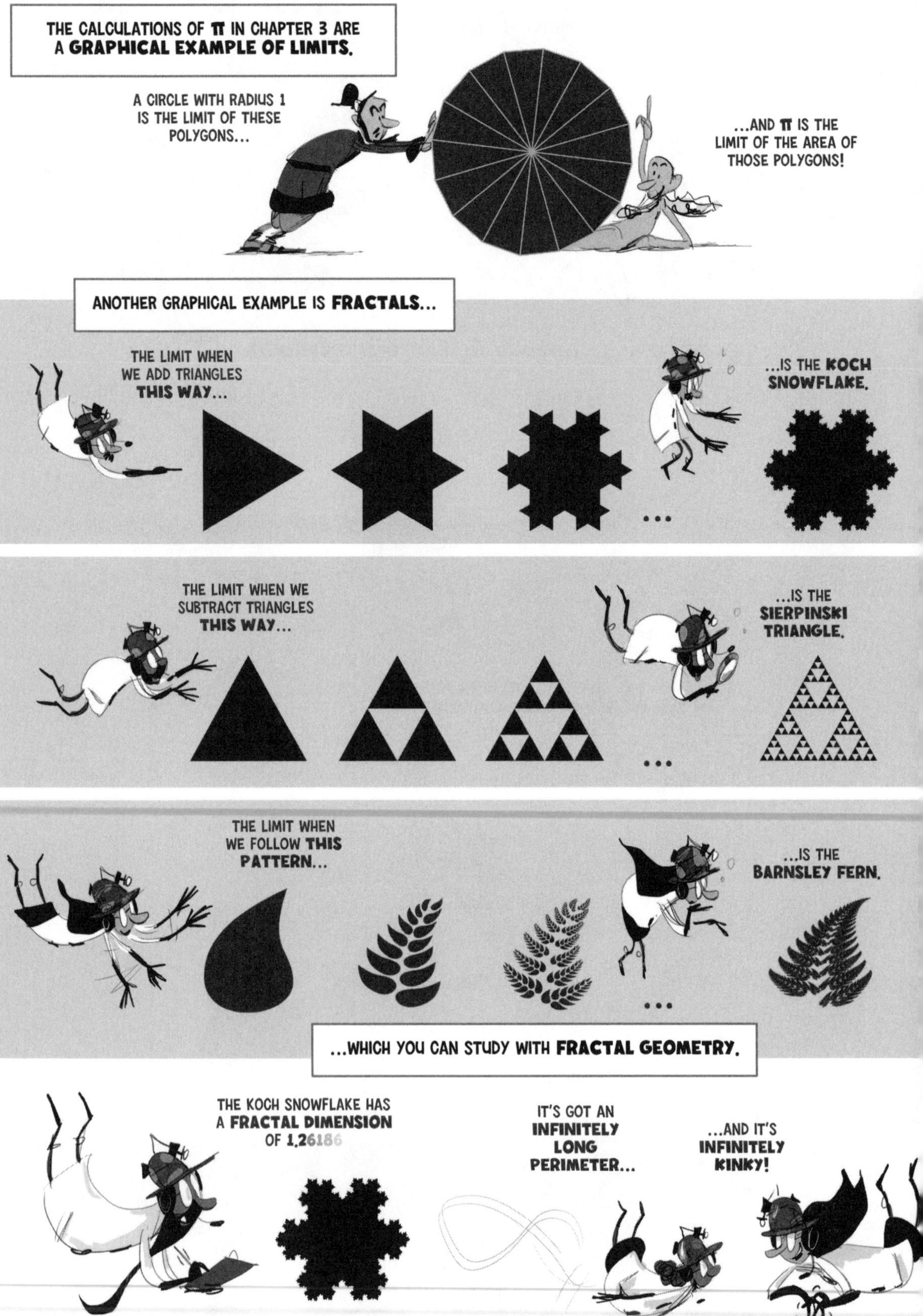

THE CALCULATIONS OF π IN CHAPTER 3 ARE A **GRAPHICAL EXAMPLE OF LIMITS.**

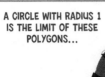

A CIRCLE WITH RADIUS 1 IS THE LIMIT OF THESE POLYGONS...

...AND π IS THE LIMIT OF THE AREA OF THOSE POLYGONS!

ANOTHER GRAPHICAL EXAMPLE IS **FRACTALS**...

THE LIMIT WHEN WE ADD TRIANGLES **THIS WAY**...

...IS THE **KOCH SNOWFLAKE.**

THE LIMIT WHEN WE SUBTRACT TRIANGLES **THIS WAY**...

...IS THE **SIERPINSKI TRIANGLE.**

THE LIMIT WHEN WE FOLLOW **THIS PATTERN**...

...IS THE **BARNSLEY FERN.**

...WHICH YOU CAN STUDY WITH **FRACTAL GEOMETRY.**

THE KOCH SNOWFLAKE HAS A **FRACTAL DIMENSION** OF 1.26186

IT'S GOT AN **INFINITELY LONG PERIMETER**...

...AND IT'S **INFINITELY KINKY!**

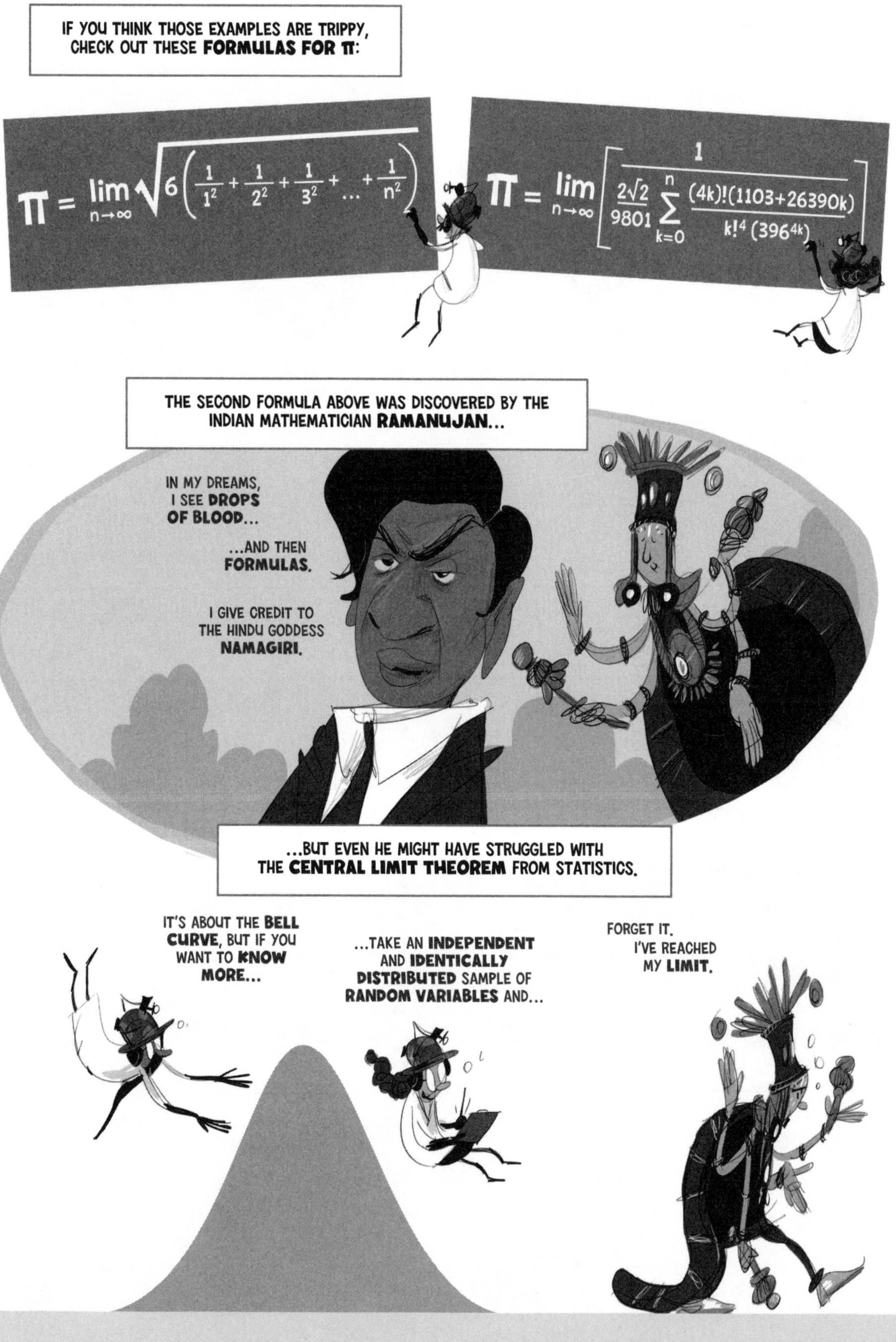
IF YOU THINK THOSE EXAMPLES ARE TRIPPY, CHECK OUT THESE **FORMULAS FOR π**:
$\pi = \lim_{n\to\infty} \sqrt{6\left(\frac{1}{1^2}+\frac{1}{2^2}+\frac{1}{3^2}+\dots+\frac{1}{n^2}\right)}$
$\pi = \lim_{n\to\infty} \left[\frac{1}{\frac{2\sqrt{2}}{9801}\sum_{k=0}^{n}\frac{(4k)!(1103+26390k)}{k!^4\,(396^{4k})}}\right]$
THE SECOND FORMULA ABOVE WAS DISCOVERED BY THE INDIAN MATHEMATICIAN **RAMANUJAN**...
IN MY DREAMS, I SEE **DROPS OF BLOOD**...
...AND THEN **FORMULAS**.
I GIVE CREDIT TO THE HINDU GODDESS **NAMAGIRI**.
...BUT EVEN HE MIGHT HAVE STRUGGLED WITH THE **CENTRAL LIMIT THEOREM** FROM STATISTICS.
IT'S ABOUT THE **BELL CURVE**, BUT IF YOU WANT TO **KNOW MORE**...
...TAKE AN **INDEPENDENT** AND **IDENTICALLY DISTRIBUTED** SAMPLE OF **RANDOM VARIABLES** AND...
FORGET IT. I'VE REACHED MY **LIMIT**.

FOR ANOTHER EXAMPLE OF LIMITS, CONSIDER **PRIME NUMBERS.**

PRIME NUMBERS, LIKE THESE...

2, 3, 5, 7, 11, 13, 17, 19, 23, 29, 31

...ARE NUMBERS THAT ARE **DIVISIBLE ONLY BY 1 AND THEMSELVES.**

TELL ME MORE, **GENIUS.**

THE ANCIENT GREEK MATHEMATICIAN **EUCLID** USED **PROOF BY CONTRADICTION** TO SHOW THAT THERE ARE **INFINITELY MANY PRIME NUMBERS.**

IF THERE WERE **ONLY A FINITE NUMBER OF PRIMES**, THEN THERE WOULD BE A **LARGEST ONE**...

...BUT IF YOU GIVE ME ANY PRIME NUMBER, I CAN PROVE THAT THERE **MUST BE A LARGER ONE.**

THAT WOULD BE A **CONTRADICTION**, SO THERE **MUST BE INFINITELY MANY PRIME NUMBERS.**

BUT WHAT IF WE COUNT HOW MANY PRIME NUMBERS THERE ARE THAT ARE **LESS THAN 10?**

OR **LESS THAN 100?**

I COUNT **4.**

25.

OR LESS THAN **1,000?**

OR **LESS THAN 10,000?**

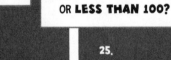

168.

192

...ARE **APPROXIMATED** BY THIS FUNCTION.

WE CAN SEE THIS BY LOOKING AT THE **RATIO** BETWEEN THEM.

x	# of primes < x	$f(x)= x/\ln(x)$	$\dfrac{\text{# of primes} < x}{x/\ln(x)}$ = Ratio
10	4	$f(10) = $ **4.3429**	$4/4.3429 = $ **0.9210**
100	25	$f(100) = $ **21.7147**	$25/21.7147 = $ **1.1513**
1000	168	$f(1000) = $ **144.7648**	$168/144.7648 = $ **1.1605**
10000	1229	$f(10000) = $ **1085.7362**	$1229/1085.7362 = $ **1.1320**
100000	9592	$f(100000) = $ **8685.8896**	$9592/8685.8896 = $ **1.1043**

BUT WE NEED **LIMITS** IN ORDER TO **PRECISELY DEFINE THE RELATIONSHIP.**

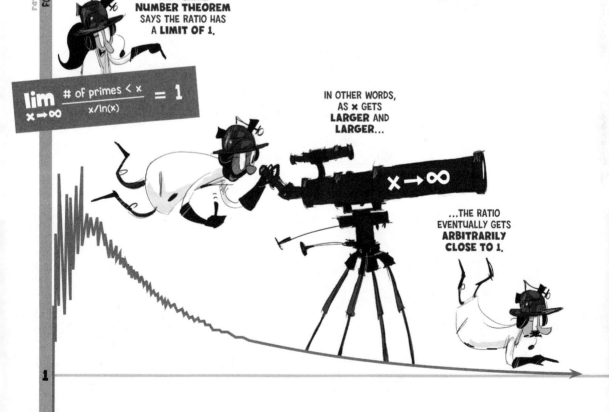

THE **PRIME NUMBER THEOREM** SAYS THE RATIO HAS A **LIMIT OF 1.**

$$\lim_{x \to \infty} \frac{\text{# of primes} < x}{x/\ln(x)} = 1$$

IN OTHER WORDS, AS x GETS **LARGER** AND **LARGER**...

$x \to \infty$

...THE RATIO EVENTUALLY GETS **ARBITRARILY CLOSE TO 1.**

ratio $\frac{f(x)}{}$

1

10^{12} 10^{24} 10^{36} x

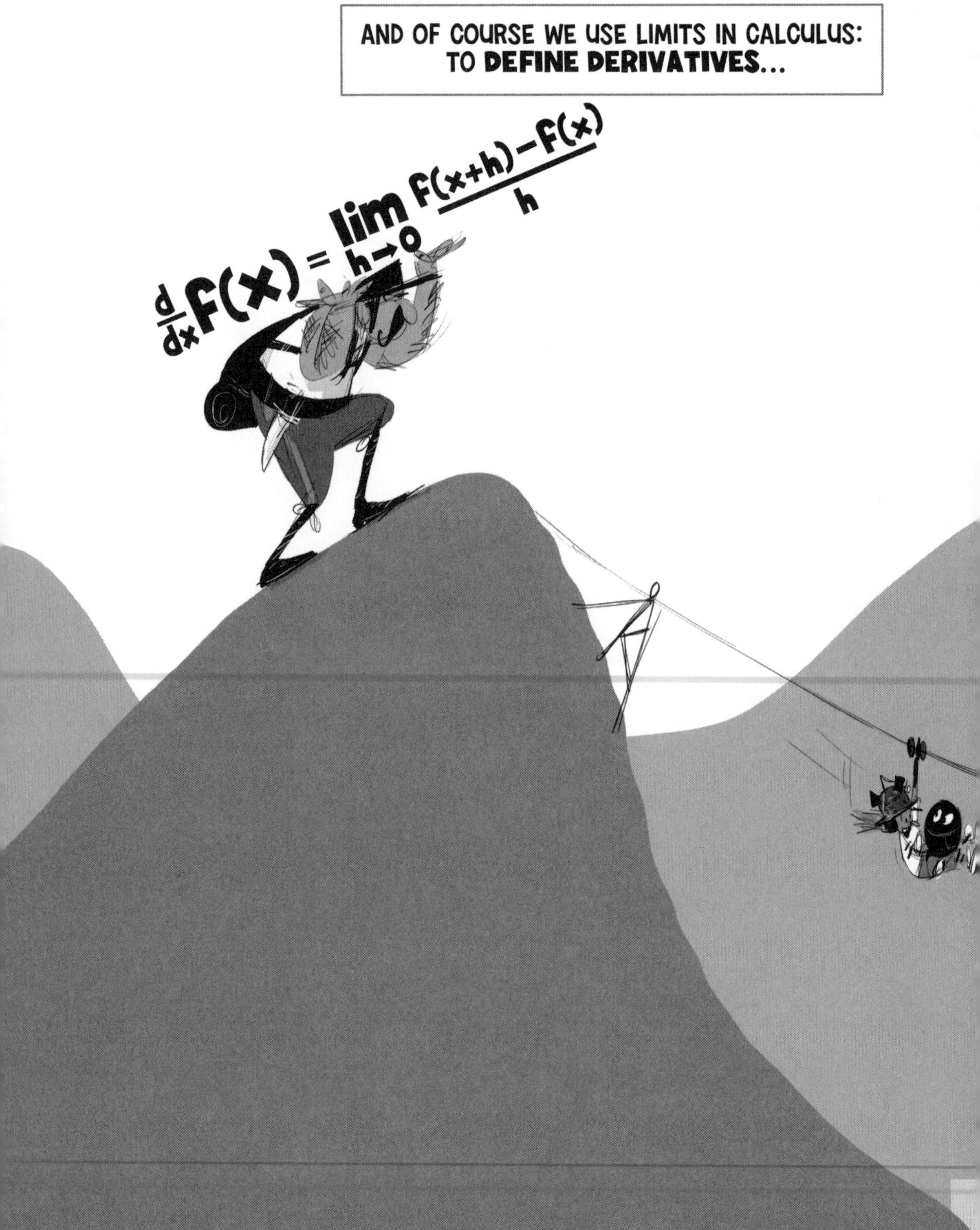
AND OF COURSE WE USE LIMITS IN CALCULUS:
TO **DEFINE DERIVATIVES**...
$\frac{d}{dx}f(x) = \lim_{h \to 0} \frac{f(x+h)-f(x)}{h}$

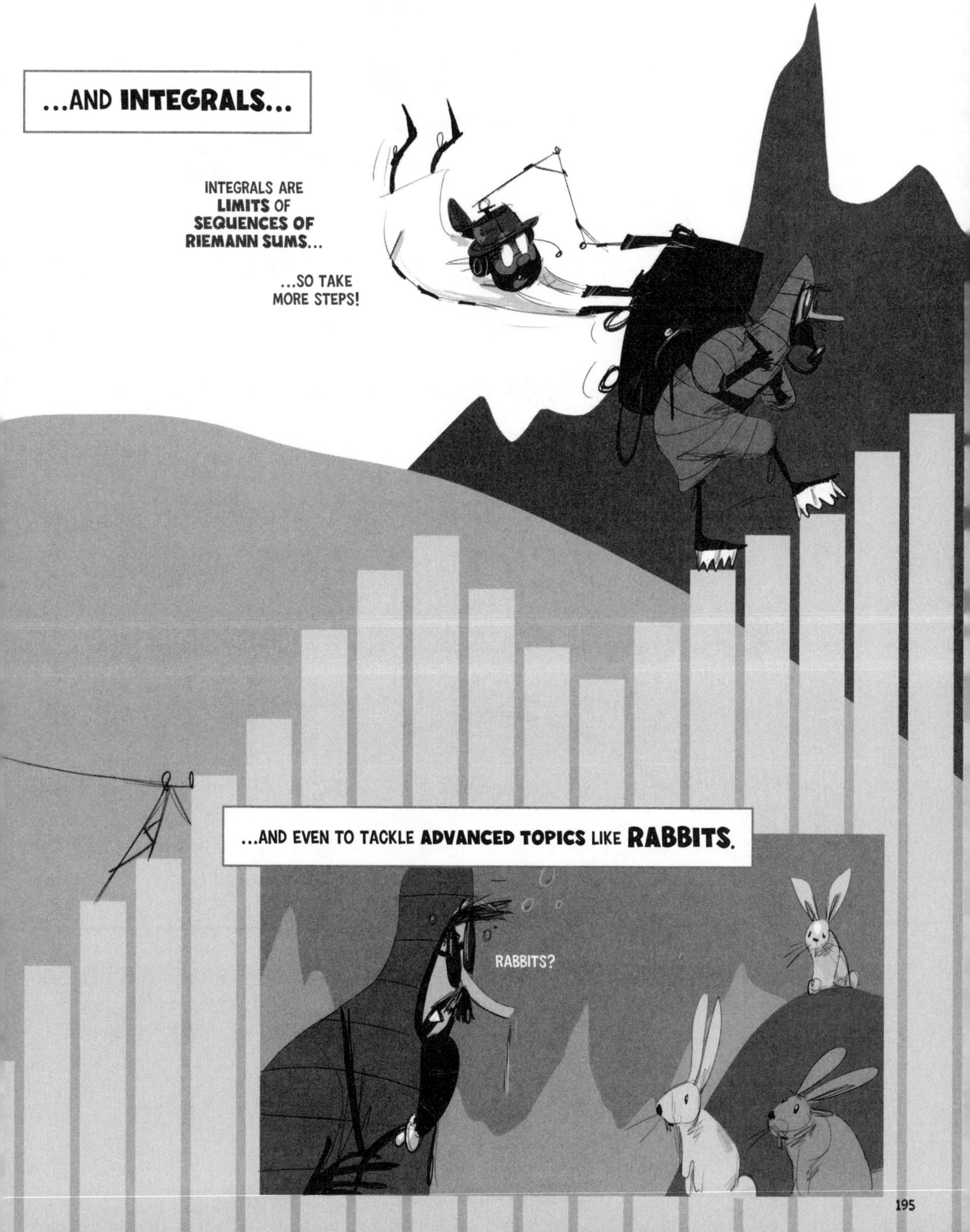
...AND INTEGRALS...
INTEGRALS ARE LIMITS OF SEQUENCES OF RIEMANN SUMS...
...SO TAKE MORE STEPS!
...AND EVEN TO TACKLE ADVANCED TOPICS LIKE RABBITS.
RABBITS?

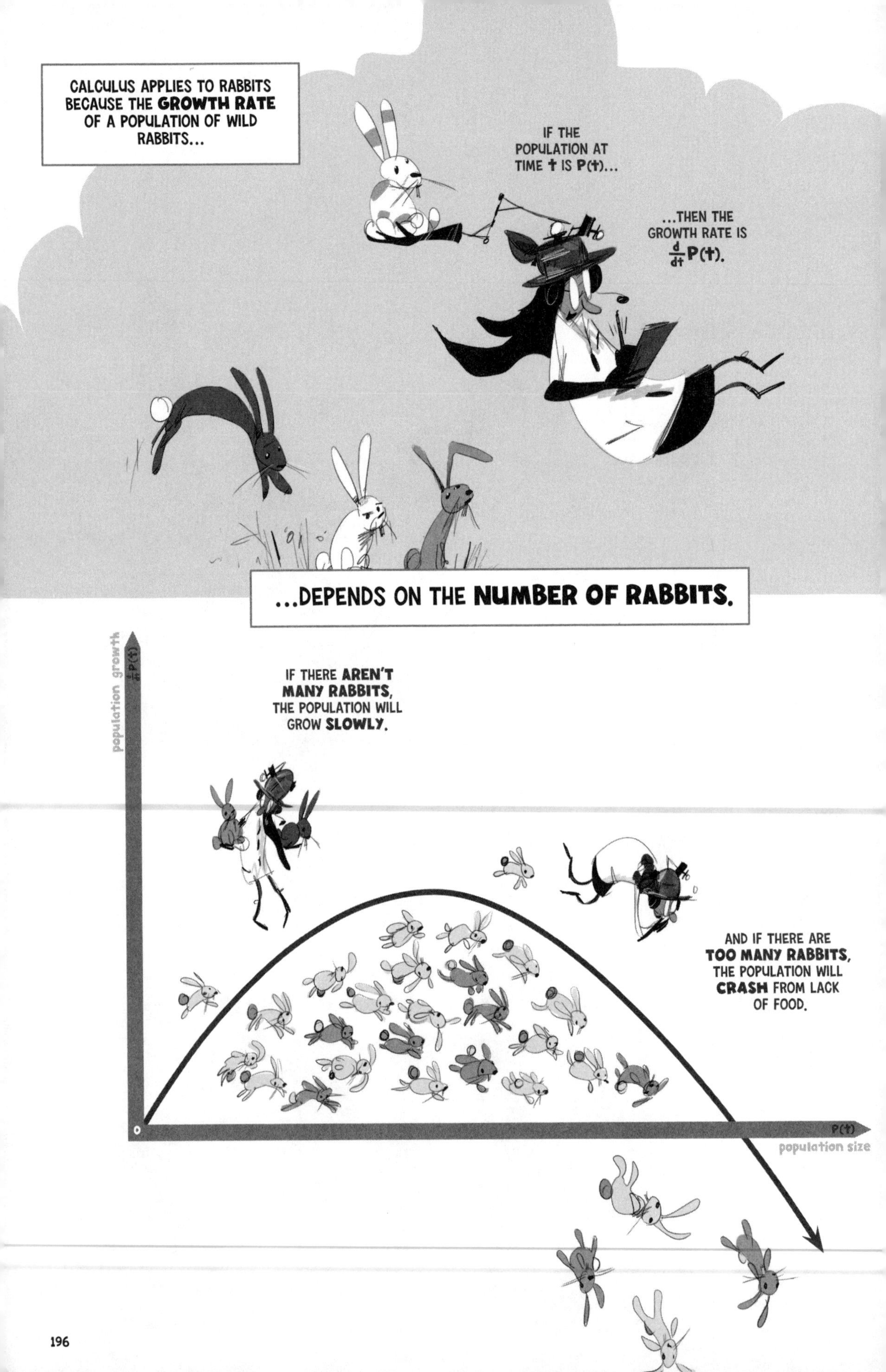
CALCULUS APPLIES TO RABBITS BECAUSE THE GROWTH RATE OF A POPULATION OF WILD RABBITS...
IF THE POPULATION AT TIME t IS P(t)...
...THEN THE GROWTH RATE IS $\frac{d}{dt}P(t)$.
...DEPENDS ON THE NUMBER OF RABBITS.
population growth
$\frac{d}{dt}P(t)$
IF THERE AREN'T MANY RABBITS, THE POPULATION WILL GROW SLOWLY.
AND IF THERE ARE TOO MANY RABBITS, THE POPULATION WILL CRASH FROM LACK OF FOOD.
0
P(t)
population size

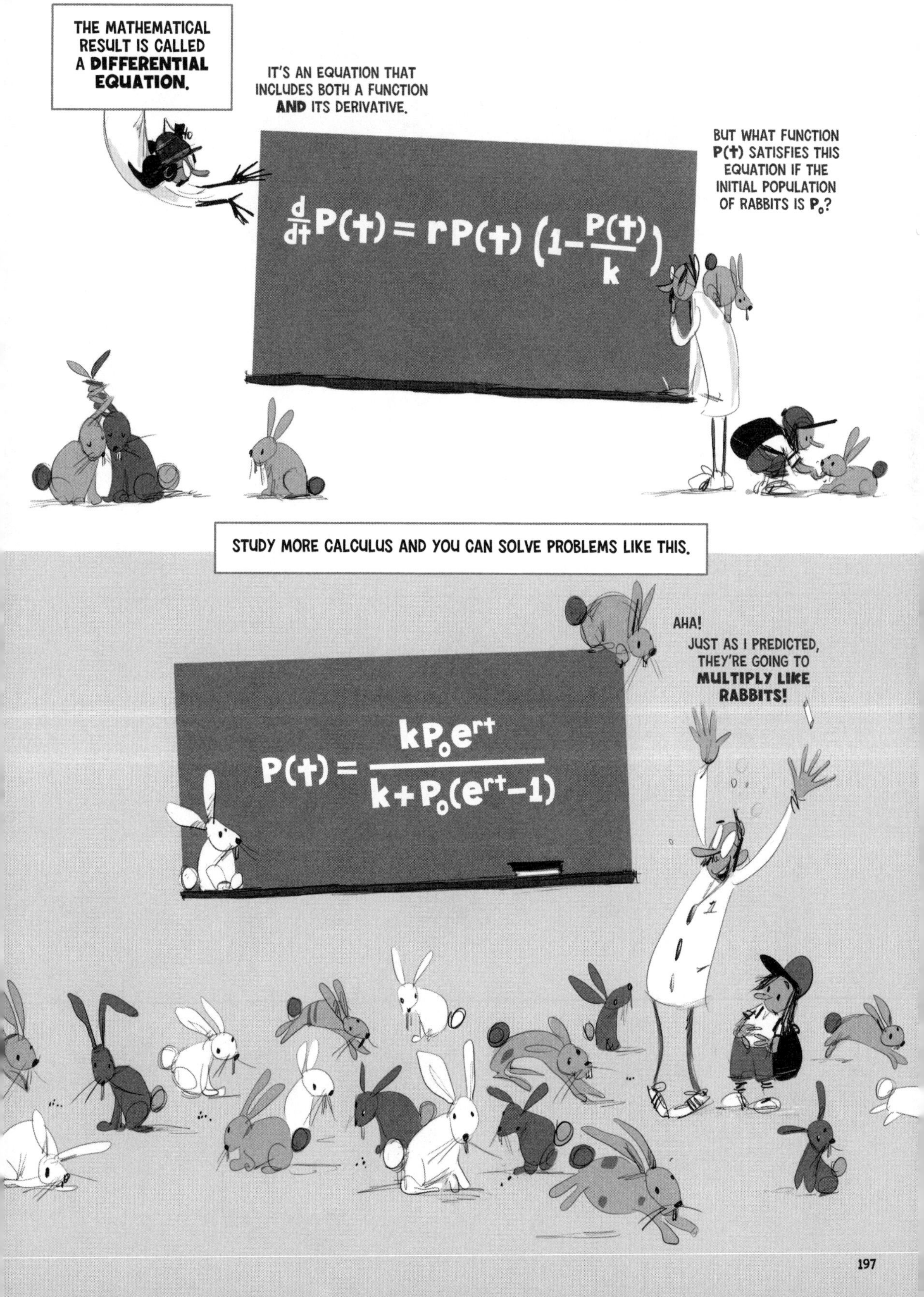
THE MATHEMATICAL RESULT IS CALLED A **DIFFERENTIAL EQUATION.**
IT'S AN EQUATION THAT INCLUDES BOTH A FUNCTION **AND** ITS DERIVATIVE.
$\frac{d}{dt}P(t) = rP(t)\left(1 - \frac{P(t)}{k}\right)$
BUT WHAT FUNCTION **P(t)** SATISFIES THIS EQUATION IF THE INITIAL POPULATION OF RABBITS IS P_0?
STUDY MORE CALCULUS AND YOU CAN SOLVE PROBLEMS LIKE THIS.
AHA!
JUST AS I PREDICTED, THEY'RE GOING TO **MULTIPLY LIKE RABBITS!**
$P(t) = \frac{kP_0e^{rt}}{k + P_0(e^{rt} - 1)}$

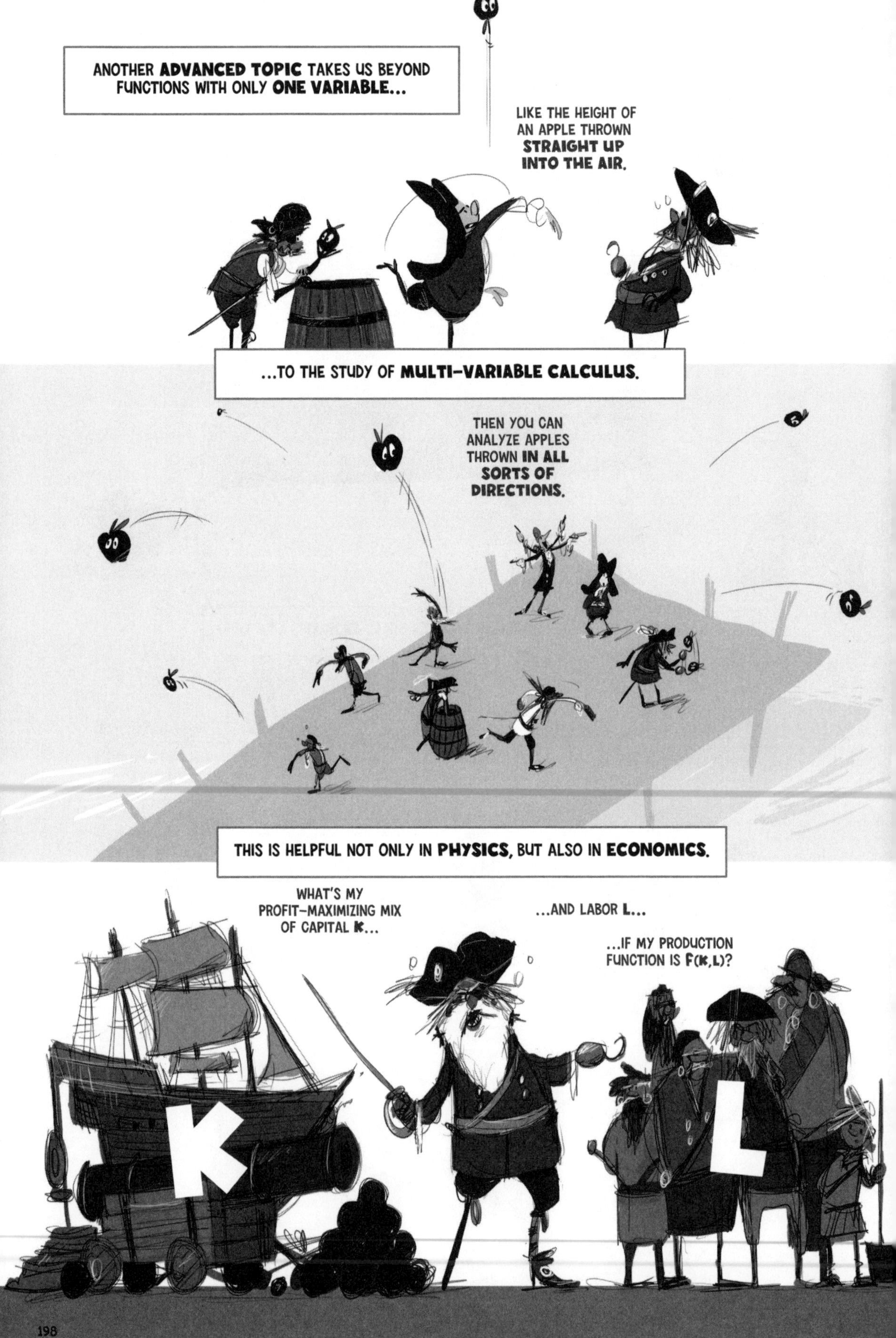

ANOTHER ADVANCED TOPIC TAKES US BEYOND FUNCTIONS WITH ONLY ONE VARIABLE...
LIKE THE HEIGHT OF AN APPLE THROWN STRAIGHT UP INTO THE AIR.
...TO THE STUDY OF MULTI-VARIABLE CALCULUS.
THEN YOU CAN ANALYZE APPLES THROWN IN ALL SORTS OF DIRECTIONS.
THIS IS HELPFUL NOT ONLY IN PHYSICS, BUT ALSO IN ECONOMICS.
WHAT'S MY PROFIT-MAXIMIZING MIX OF CAPITAL K...
...AND LABOR L...
...IF MY PRODUCTION FUNCTION IS F(K,L)?
K
L

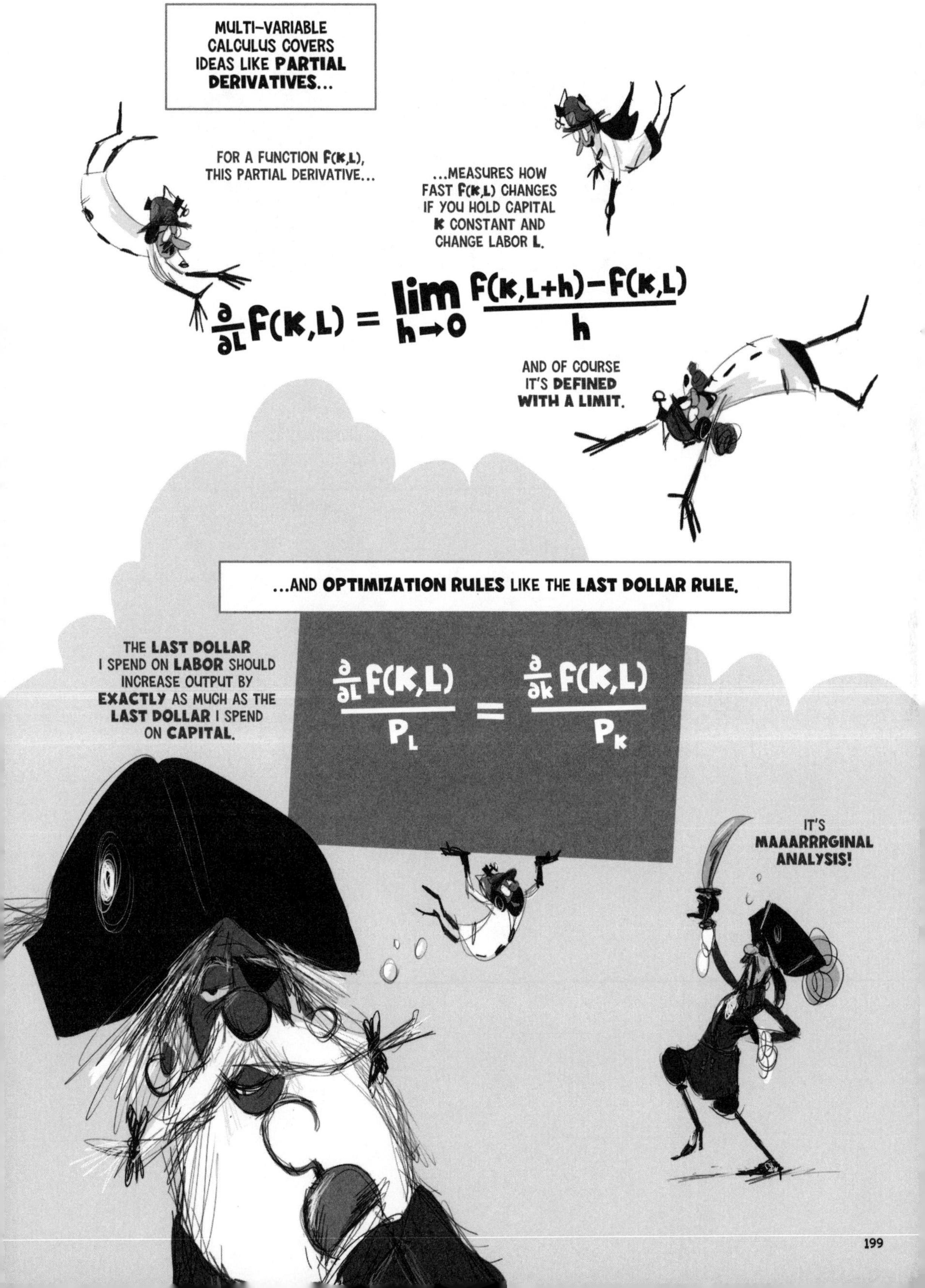
MULTI-VARIABLE CALCULUS COVERS IDEAS LIKE **PARTIAL DERIVATIVES...**
FOR A FUNCTION **F(K,L)**, THIS PARTIAL DERIVATIVE...
...MEASURES HOW FAST **F(K,L)** CHANGES IF YOU HOLD CAPITAL **K** CONSTANT AND CHANGE LABOR **L**.
$\frac{\partial}{\partial L} f(K,L) = \lim_{h \to 0} \frac{f(K,L+h) - f(K,L)}{h}$
AND OF COURSE IT'S **DEFINED WITH A LIMIT.**
...AND **OPTIMIZATION RULES** LIKE THE **LAST DOLLAR RULE.**
THE **LAST DOLLAR** I SPEND ON **LABOR** SHOULD INCREASE OUTPUT BY **EXACTLY** AS MUCH AS THE **LAST DOLLAR** I SPEND ON **CAPITAL.**
$\frac{\frac{\partial}{\partial L} f(K,L)}{P_L} = \frac{\frac{\partial}{\partial K} f(K,L)}{P_K}$
IT'S **MAAARRRGINAL ANALYSIS!**

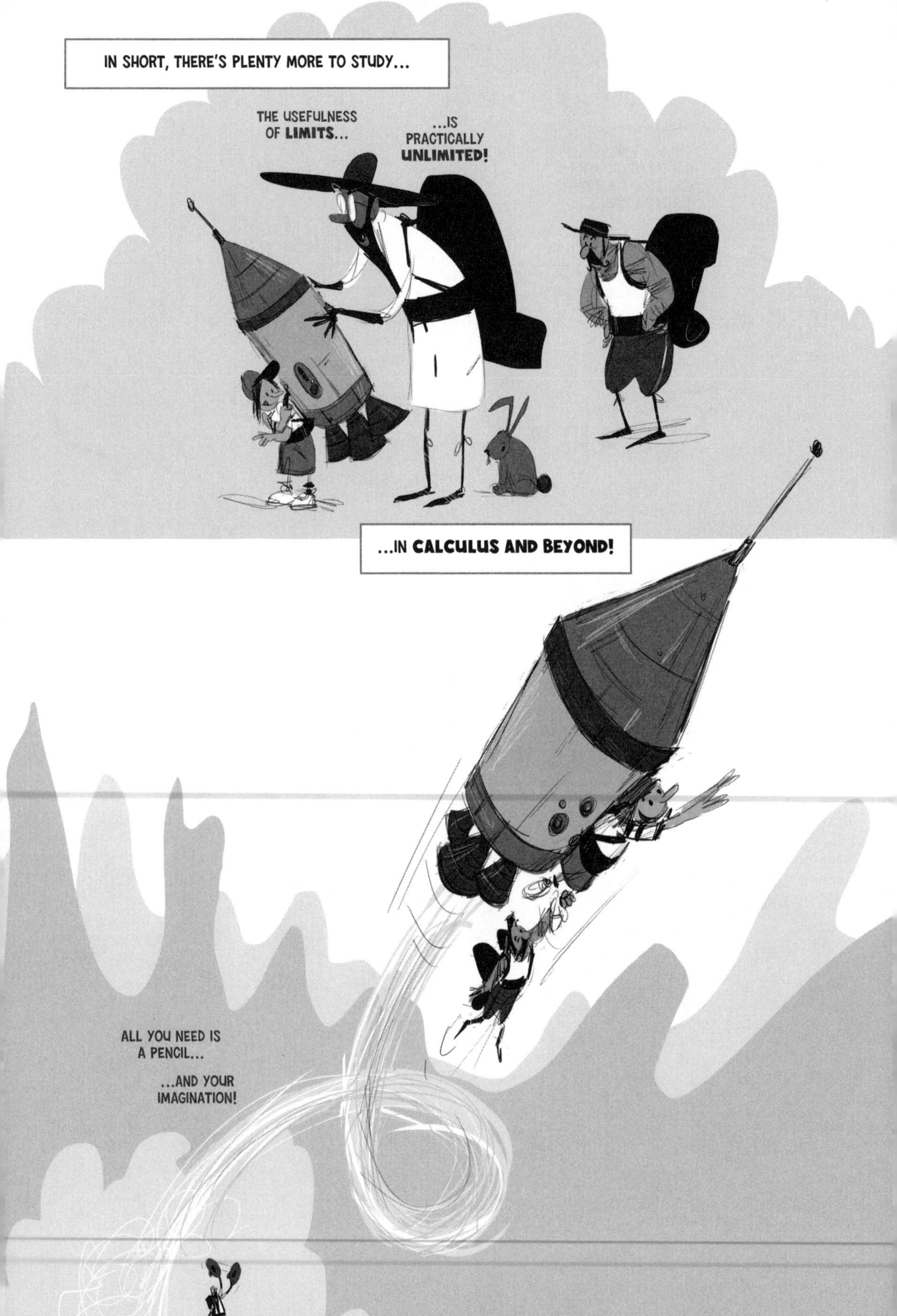
IN SHORT, THERE'S PLENTY MORE TO STUDY...
THE USEFULNESS OF **LIMITS**...
...IS PRACTICALLY **UNLIMITED!**
...IN **CALCULUS AND BEYOND!**
ALL YOU NEED IS A PENCIL...
...AND YOUR IMAGINATION!

GLOSSARY

SYMBOLS

$\frac{d}{dx}$ SEE **DERIVATIVE**.

ε-δ THE GREEK LETTERS **EPSILON** AND **DELTA**; TOGETHER THESE REFER TO A TYPE OF PROOF OF CONTINUITY. 63–64

∞ **INFINITY**, USED IN PARTICULAR IN EXPRESSIONS LIKE $\lim_{x \to \infty} f(x) = a$, MEANING THAT AS **x** GETS LARGER AND LARGER, **f(x)** EVENTUALLY GETS ARBITRARILY CLOSE TO **a**. 63, 193

$\int$ SEE **INTEGRAL**.

lim   SEE **LIMIT**.

π THE GREEK LETTER **PI**, $\pi \approx 3.1415926...$, IS EQUAL TO THE AREA OF A CIRCLE WITH RADIUS 1, OR THE RATIO OF THE CIRCUMFERENCE OF A CIRCLE TO ITS DIAMETER. 34–40, 166, 191

TOOLS

CONSTANT MULTIPLE RULE 88–89
THE DERIVATIVE OF $c \bullet f(x)$ IS $\frac{d}{dx}[c \bullet f(x)] = c \bullet \frac{d}{dx} f(x)$.

POWER RULE 92–93
FOR ANY INTEGER **n** (AND IN FACT FOR ANY REAL NUMBER **n**), THE DERIVATIVE OF $f(x) = x^n$ IS $\frac{d}{dx} x^n = nx^{n-1}$.

PRODUCT RULE 90–91
FOR DIFFERENTIABLE FUNCTIONS **f(x)** AND **g(x)**, THE DERIVATIVE OF THEIR PRODUCT IS $\frac{d}{dx}[f(x) \bullet g(x)] = f(x) \bullet \frac{d}{dx} g(x) + g(x) \bullet \frac{d}{dx} f(x)$.

SUM RULE 86–87
FOR DIFFERENTIABLE FUNCTIONS **f(x)** AND **g(x)**, THE DERIVATIVE OF THEIR SUM IS $\frac{d}{dx}[f(x) + g(x)] = \frac{d}{dx} f(x) + \frac{d}{dx} g(x)$.

WORDS AND PHRASES

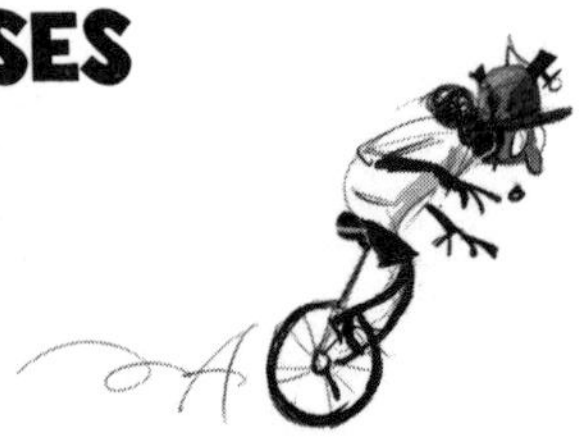

ACCELERATION 176–78

THE INSTANTANEOUS RATE OF CHANGE OF VELOCITY.

ANTI-DERIVATIVE 149–52

A FUNCTION $F(x)$ IS AN ANTI-DERIVATIVE OF $f(x)$ IF THE DERIVATIVE OF $F(x)$ IS $f(x)$.

A **GOOD ANTI-DERIVATIVE** IS ONE THAT HELPS US CALCULATE AN INTEGRAL USING THE FUNDAMENTAL THEOREM OF CALCULUS; A **BAD ANTI-DERIVATIVE** IS ONE THAT DOESN'T. WITH $\int_0^1 3x^2 dx$, FOR EXAMPLE, THE FUNDAMENTAL THEOREM SAYS WE NEED TO FIND AN $F(x)$ THAT IS AN ANTI-DERIVATIVE OF $f(x) = 3x^2$ AND THEN CALCULATE $\int_0^1 3x^2 dx = F(1) - F(0)$.
IT FOLLOWS THAT $F(x) = x^3$ IS A GOOD ANTI-DERIVATIVE BECAUSE IT'S EASY TO CALCULATE $F(1) - F(0) = 1^3 - 0^3 = 1$; SIMILARLY, $F(x) = \int_0^x 3r^2 dr$ IS A BAD ANTI-DERIVATIVE BECAUSE $F(1) - F(0) = \int_0^1 3r^2 dr - \int_0^0 3r^2 dr = \int_0^1 3r^2 dr$, WHICH DOESN'T HELP US AT ALL.

ARCHIMEDES 8, 34–42

A GREEK MATHEMATICIAN AND SCIENTIST, ARCHIMEDES (≈287–212 B.C.E.) ESTIMATED π USING METHODS THAT HINT AT CALCULUS.

CALCULUS 3–5

A BRANCH OF MATHEMATICS THAT DEALS WITH DERIVATIVES AND INTEGRALS.

HISTORY OF CALCULUS 8

CAVALIERI, BONAVENTURA 33, 37, 169–70

AN ITALIAN MATHEMATICIAN, CAVALIERI (1598–1647) DEVELOPED IDEAS THAT HINT AT CALCULUS.

CHEKHOV'S GUN 157, 185

A PIECE OF WRITING ADVICE FROM THE RUSSIAN DOCTOR AND PLAYWRIGHT ANTON CHEKHOV (1860–1904): "IF YOU SAY IN THE FIRST CHAPTER THAT THERE IS A RIFLE HANGING ON THE WALL, IN THE SECOND OR THIRD CHAPTER IT ABSOLUTELY MUST GO OFF. IF IT'S NOT GOING TO BE FIRED, IT SHOULDN'T BE HANGING THERE."

CHOICE VARIABLES 110

IN AN OPTIMIZATION PROBLEM, A CHOICE VARIABLE IS A VARIABLE THAT IS UNDER YOUR CONTROL. FOR EXAMPLE, GIVEN MARKET PRICE p, A FIRM MIGHT CHOOSE HOW MUCH OUTPUT q TO PRODUCE IN ORDER TO MAXIMIZE PROFIT; IN THIS EXAMPLE, q IS A CHOICE VARIABLE (AND p IS NOT).

CLIMATE CHANGE / GLOBAL WARMING 128

AN INCREASE IN GLOBAL AVERAGE TEMPERATURE, ESPECIALLY AS A RESULT OF BURNING FOSSIL FUELS AND OTHER HUMAN ACTIVITIES; SEE *THE CARTOON INTRODUCTION TO CLIMATE CHANGE.*

CONSTANT FUNCTION 78

A FUNCTION THAT ALWAYS OUTPUTS THE SAME CONSTANT, $f(x)=c$.

CONSTANT MULTIPLE RULE 88–89

THE DERIVATIVE OF $c \cdot f(x)$ is $\frac{d}{dx}[c \cdot f(x)] = c \cdot \frac{d}{dx} f(x)$.

CONSTRAINTS 110

IN AN OPTIMIZATION PROBLEM, A CONSTRAINT IS A LIMITATION—FOR EXAMPLE, THAT THE CHOICE VARIABLE x HAS TO BE NON-NEGATIVE.

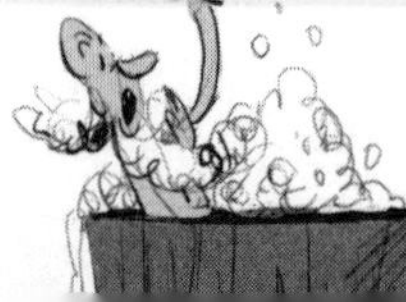

CONTINUOUS 66–67

A FUNCTION $f(x)$ IS **CONTINUOUS AT POINT** a IF $\lim_{x\to a} f(x) = f(a)$; IF IT IS CONTINUOUS AT EVERY RELEVANT POINT a THEN THE ENTIRE FUNCTION IS SAID TO BE CONTINUOUS. IN NON-MATHEMATICAL TERMS, CONTINUITY IS RELATED TO THE IDEA THAT YOU CAN DRAW PART OR ALL OF A FUNCTION WITHOUT LIFTING YOUR PENCIL FROM THE PAPER.

CORNER SOLUTION 101, 105–106, 114, 116, 129

A SOLUTION TO AN OPTIMIZATION PROBLEM THAT OCCURS NOT IN THE MIDDLE OF A SMOOTH PATH BUT AT A CORNER OR OTHER EDGE. FOR EXAMPLE, THE FUNCTION $f(x) = x$ WITH THE CONSTRAINT $x \geq 0$ HAS A MINIMUM VALUE AT $x = 0$. NOTE THAT $x = \infty$ CAN ALSO BE THOUGHT OF AS A CORNER SOLUTION; FOR EXAMPLE, THE MAXIMUM VALUE OF $f(x) = x$ IS THE "CORNER AT INFINITY" WHERE $x = \infty$.

CRITICAL POINT 105, 115

A POINT WHERE THE DERIVATIVE DOES NOT EXIST OR THE DERIVATIVE IS ZERO.

DERIVATIVE 4, 44, 71–82

IF IT EXISTS, THE DERIVATIVE OF A FUNCTION $f(x)$ AT POINT x MEASURES THE INSTANTANEOUS RATE OF CHANGE OF THE FUNCTION AT POINT x; EQUIVALENTLY, IT MEASURES THE SLOPE. IT IS DEFINED AS $\frac{d}{dx} f(x) = \lim_{h\to 0} \frac{f(x+h)-f(x)}{h}$.

INTUITION 80–81

THE STUDY OF DERIVATIVES IS CALLED **DIFFERENTIAL CALCULUS.** 4, 28, 76

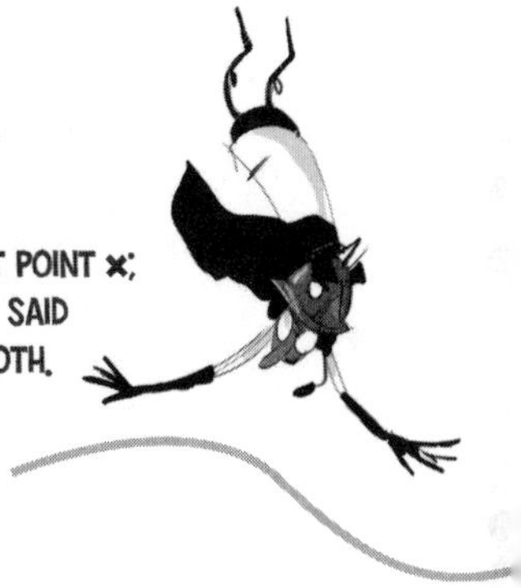

DIFFERENTIABLE 66–67, 103, 105

A FUNCTION $f(x)$ IS **DIFFERENTIABLE AT POINT** x IF THE DERIVATIVE EXISTS AT POINT x; IF IT IS DIFFERENTIABLE AT EVERY RELEVANT POINT x, THEN THE ENTIRE FUNCTION IS SAID TO BE DIFFERENTIABLE. IN NON-MATHEMATICAL TERMS, DIFFERENTIABLE MEANS SMOOTH.

DIFFERENTIAL EQUATION 197

AN EQUATION THAT INCLUDES BOTH A **FUNCTION** AND ITS **DERIVATIVE.**

DISCONTINUITY 100–101, 104–106

A POINT x WHERE THE FUNCTION $f(x)$ IS NOT CONTINUOUS; FOR EXAMPLE, THE FUNCTION THAT EQUALS 1 AT $x = 0$ AND 2 AT $x \neq 0$ HAS A DISCONTINUITY AT $x = 0$.

ECONOMICS 81, 108, 119–30

THE STUDY OF THE ACTIONS AND INTERACTIONS OF OPTIMIZING INDIVIDUALS; SEE *THE CARTOON INTRODUCTION TO ECONOMICS.*

EUCLID 192

A GREEK MATHEMATICIAN WHOSE BOOK *ELEMENTS* WAS USED AS A GEOMETRY TEXTBOOK UNTIL THE 20TH CENTURY, EUCLID (BORN ≈325 B.C.E.) ALSO MADE CONTRIBUTIONS TO **NUMBER THEORY**, THE STUDY OF PRIME NUMBERS AND OTHER INTEGERS.

EXTREME VALUES 95–106

FOR A FUNCTION $f(x)$, POINT a IS AN EXTREME VALUE IF $f(a)$ IS THE MAXIMUM OR MINIMUM VALUE OF $f(x)$ IN SOME NEIGHBORHOOD AROUND a; THESE POINTS ARE CALLED LOCAL MAXIMUMS OR MINIMUMS. GLOBAL MAXIMUMS OR MINIMUMS ARE POINTS a WHERE $f(a)$ TAKES A MAXIMUM OR MINIMUM VALUE, PERIOD. (THE TOP OF ANY HILL IS A LOCAL MAXIMUM; ONLY THE TOP OF MOUNT EVEREST IS A GLOBAL MAXIMUM.)

FERMAT'S LAST THEOREM 171

FIRST SUGGESTED BY THE FRENCH MATHEMATICIAN PIERRE DE FERMAT (1606–1665), THIS THEOREM STATES THAT THE EQUATION $a^n + b^n = c^n$ HAS NO POSITIVE INTEGER SOLUTIONS FOR $n > 2$. A PROOF WAS FINALLY PUBLISHED IN 1995, MORE THAN 350 YEARS AFTER FERMAT LEFT A TANTALIZING NOTE ABOUT IT ("I HAVE DISCOVERED A TRULY MARVELOUS PROOF OF THIS, WHICH THIS MARGIN IS TOO NARROW TO CONTAIN") IN ONE OF HIS MATH BOOKS.

FLUXIONS AND FLUENTS 8, 21

NOW HISTORICAL CURIOSITIES, THESE TERMS WERE USED BY ISAAC NEWTON IN HIS DESCRIPTION OF CALCULUS: A FLUENT WAS HIS TERM FOR A FUNCTION $f(x)$, AND A FLUXION WAS HIS TERM FOR THE DERIVATIVE $\frac{d}{dx}f(x)$.

FRACTAL GEOMETRY 190

THE STUDY OF GEOMETRIC SHAPES (CALLED **FRACTALS**) THAT HAVE THE PROPERTY THAT THEIR PARTS ARE IN SOME WAY SIMILAR TO THE WHOLE (AS WITH THE **BARNSLEY FERN**).

FUNCTION 51

A FUNCTION $f(x)$ IS A RELATIONSHIP BETWEEN INPUT VALUES x AND OUTPUT VALUES $f(x)$; THERE IS AT MOST ONE OUTPUT VALUE $f(x)$ FOR EACH INPUT VALUE x. A FUNCTION IS OFTEN DESCRIBED WITH A GRAPH THAT SHOWS x ON THE HORIZONTAL AXIS AND $y=f(x)$ ON THE VERTICAL AXIS.

FUNDAMENTAL THEOREM OF CALCULUS 6, 43–54

A THEOREM THAT RELATES DERIVATIVES AND INTEGRALS AND HENCE IS FUNDAMENTAL TO CALCULUS.

PART ONE OF THE FUNDAMENTAL THEOREM SAYS THAT IF $f(x)$ IS A CONTINUOUS FUNCTION, THEN $\frac{d}{dx}\left[\int_a^x f(t)dt\right]=f(x)$. 50–53, 172

PART TWO OF THE FUNDAMENTAL THEOREM SAYS THAT IF $f(x)$ IS A CONTINUOUS FUNCTION AND $F(x)$ IS AN ANTI-DERIVATIVE OF $f(x)$, THEN $\int_a^b f(x)dx=F(b)-F(a)$. 50, 152–53, 159–71

NOTE THAT SOME TEXTBOOKS CALL THESE THE FIRST AND SECOND FUNDAMENTAL THEOREMS INSTEAD OF PARTS ONE AND TWO OF THE FUNDAMENTAL THEOREM. AND TO CONFUSE MATTERS EVEN MORE, THEY DON'T AGREE ON WHICH PART COMES FIRST!

GHOSTS OF DEPARTED QUANTITIES 21

A SATIRICAL PHRASE USED BY GEORGE BERKELEY (1685–1753) TO CRITIQUE THE IDEA OF DERIVATIVES; IN FAIRNESS TO BERKELEY, NOTE THAT A RIGOROUS DEFINITION OF LIMITS WAS NOT DEVELOPED UNTIL THE 1800S.

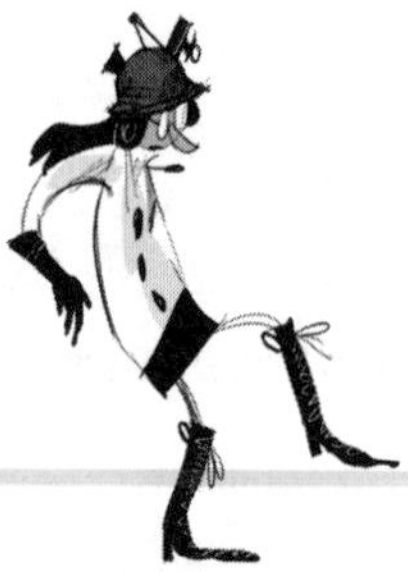

GLOBAL MAXIMUM/MINIMUM

SEE **EXTREME VALUES.**

INDUCTION 92–93, 141

A TWO-STEP PROOF THAT TYPICALLY SHOWS THAT SOME STATEMENT IS TRUE FOR ALL POSITIVE INTEGERS—FOR EXAMPLE, THAT THE SUM OF THE FIRST n INTEGERS IS $\frac{n(n+1)}{2}$. THE FIRST STEP IS TO SHOW THAT THE STATEMENT IS TRUE FOR $n=1$; THE SECOND STEP IS TO SHOW THAT IF THE STATEMENT IS TRUE FOR ANY n, THEN IT IS ALSO TRUE FOR $n+1$. COMPLETING BOTH OF THESE STEPS PROVES THAT THE STATEMENT IS TRUE FOR ALL POSITIVE INTEGERS.

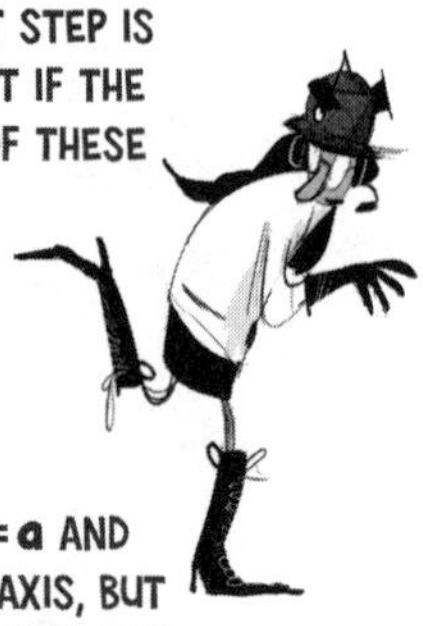

INFINITE SEQUENCE 62–63

AN INFINITELY LONG ORDERED LIST OF NUMBERS, SUCH AS $\frac{1}{2}, \frac{3}{4}, \frac{7}{8}\ldots$

INTEGRAL 5, 30, 45, 133–46, 149–53, 157

THE INTEGRAL $\int_a^b f(x)dx$ MEASURES THE AREA UNDER THE FUNCTION $f(x)$ BETWEEN $x=a$ AND $x=b$; FOR SIMPLICITY, ALL THE INTEGRALS IN THIS BOOK ARE OF AREAS ABOVE THE x-AXIS, BUT IN GENERAL NOTE THAT INTEGRALS MEASURE **SIGNED AREA**, MEANING THAT AREAS UNDER THE x-AXIS COUNT AS NEGATIVES.

IRRATIONAL NUMBER 34, 139

A NUMBER LIKE π OR $\sqrt{2}$ THAT CANNOT BE EXPRESSED AS A FRACTION OF TWO INTEGERS; COMPARE WITH **RATIONAL NUMBERS** LIKE **4** OR $\frac{1}{2}$.

KOVALEVSKAYA, SOFIA 42

CALLED THE "GREATEST KNOWN WOMAN SCIENTIST BEFORE THE TWENTIETH CENTURY," KOVALEVSKAYA (1850–1891) WAS A RUSSIAN MATHEMATICIAN WHO ADVANCED OUR UNDERSTANDING OF CALCULUS.

LAST DOLLAR RULE 199

AN OPTIMIZATION RULE FOR ECONOMICS BASED ON MARGINAL ANALYSIS.

LEIBNIZ, GOTTFRIED WILHELM 7, 21, 174

A GERMAN MATHEMATICIAN AND PHILOSOPHER, LEIBNIZ (1646–1716) INVENTED CALCULUS INDEPENDENTLY OF NEWTON. (FOR DETAILS ON THEIR FEUD, READ *THE CALCULUS WARS*.)

LEIBNIZ'S RULE IS ANOTHER NAME FOR THE **PRODUCT RULE**. 90

LIMIT 19, 55–68, 137

THE IDEA OF A MATHEMATICAL SEQUENCE OR PATH THAT IS HEADING ARBITRARILY CLOSE TO A GIVEN NUMBER EVEN IF IT NEVER ACTUALLY GETS THERE.

AND **ACCELERATION** 177

AND **DERIVATIVES** 76

AND **INTEGRALS** 137

AND **SLOPE OF TANGENT LINE** 74–75

AND **SPEED/VELOCITY** 72–73

LIU HUI 8, 34–42

A THIRD–CENTURY CHINESE MATHEMATICIAN, LIU HUI ($\approx$225–295) ESTIMATED π USING METHODS THAT HINT AT CALCULUS.

MARGINAL ANALYSIS 81, 123–30

AN ECONOMICS TERM FOR USING DERIVATIVES TO SEARCH FOR OPTIMAL VALUES.

MAXIMUM/MINIMUM

SEE **EXTREME VALUES**.

MULTI–VARIABLE CALCULUS 198–99

THE STUDY OF DERIVATIVES AND INTEGRALS IN FUNCTIONS WITH MORE THAN ONE OUTPUT VARIABLE AND/OR MORE THAN ONE INPUT VARIABLE. AN EXAMPLE OF THE FORMER IS POSITION IN THREE DIMENSIONS AS A FUNCTION OF TIME: $(x,y,z) = F(t)$. AN EXAMPLE OF THE LATTER IS OUTPUT AS A FUNCTION OF LABOR AND CAPITAL: $Q = F(K,L)$; IN THIS CASE, **PARTIAL DERIVATIVES** MEASURE THE RATE OF CHANGE IF YOU VARY ONE INPUT VARIABLE AND FIX THE OTHERS.

NEWTON, ISAAC 7, 21, 23, 42, 174–76, 180, 186

AN ENGLISH MATHEMATICIAN AND SCIENTIST, NEWTON (1642–1726) INVENTED CALCULUS INDEPENDENTLY OF LEIBNIZ. (FOR DETAILS ON THEIR FEUD, READ *THE CALCULUS WARS*.)

NEWTON IS FAMOUS FOR HIS **LAWS OF MOTION** AND HIS **LAW OF GRAVITY**, WHICH WAS INSPIRED BY A FALLING APPLE. 23, 175–76, 180

HE IS LESS FAMOUS FOR HAVING DEVOTED CONSIDERABLE TIME AND ENERGY TO NUMEROLOGY AND ALCHEMY. 186

OBJECTIVE 110

IN AN OPTIMIZATION PROBLEM, THE OBJECTIVE IS THE GOAL: FOR EXAMPLE, TO FIND THE MAXIMUM VALUE OF $F(x)$.

OPTIMIZATION 107–18

A TYPE OF MATHEMATICAL PROBLEM THAT INVOLVES FINDING THE BIGGEST, SMALLEST, MOST-EST, OR OTHER EXTREME VALUES.

AND **LAST DOLLAR RULE** 199

PARTIAL DERIVATIVES

SEE **MULTI-VARIABLE CALCULUS**.

PI

SEE π ON PAGE 201.

POWER RULE 92–93

FOR ANY INTEGER n (AND IN FACT FOR ANY REAL NUMBER n), THE DERIVATIVE OF $f(x)=x^n$ IS $\frac{d}{dx}x^n = nx^{n-1}$.

PRODUCT RULE 90–91

FOR DIFFERENTIABLE FUNCTIONS $f(x)$ AND $g(x)$, THE DERIVATIVE OF THEIR PRODUCT IS $\frac{d}{dx}[f(x) \cdot g(x)] = f(x) \cdot \frac{d}{dx}g(x) + g(x) \cdot \frac{d}{dx}f(x)$.

PYTHAGOREAN THEOREM 39

A THEOREM RELATING THE LENGTHS OF THE SIDES OF A RIGHT TRIANGLE: $a^2 + b^2 = c^2$. DESPITE THE GREAT QUOTE FROM THE BASKETBALL LEGEND SHAQUILLE O'NEAL (WHO SAID THAT HIS GAME WAS "LIKE THE PYTHAGOREAN THEOREM: HARD TO FIGURE OUT"), THERE ARE IN FACT HUNDREDS OF PROOFS OF THE PYTHAGOREAN THEOREM GOING BACK THOUSANDS OF YEARS.

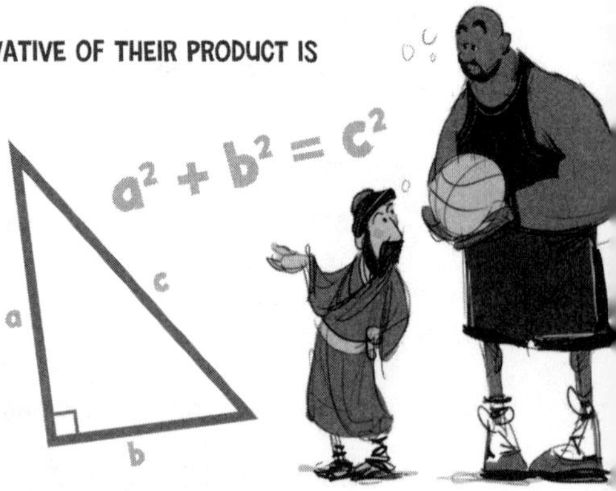
a² + b² = c²
a
b
c

RAMANUJAN 191

AN INDIAN MATHEMATICAL GENIUS, RAMANUJAN (1887–1920) DIED AT THE AGE OF 32 BUT LIVES ON IN HIS MATHEMATICAL WORKS AND IN BOOKS AND MOVIES, INCLUDING *THE MAN WHO KNEW INFINITY*.

RATIONAL NUMBER 139

A NUMBER LIKE **4** OR $\frac{1}{2}$ THAT CAN BE EXPRESSED AS A FRACTION OF TWO INTEGERS (CONTRAST WITH **IRRATIONAL NUMBERS**).

RIEMANN SUM 136–42

A WAY OF APPROXIMATING THE AREA UNDER A CURVE BY SUMMING UP THE AREA OF RECTANGLES. THIS APPROACH WAS FORMALIZED BY THE GERMAN MATHEMATICIAN **BERNHARD RIEMANN** (1826–1866), WHO ALSO PROPOSED ONE OF THE MOST FAMOUS UNSOLVED PROBLEMS IN MATHEMATICS, KNOWN AS THE **RIEMANN HYPOTHESIS**.

SANDWICH PROOF 143–45

A METHOD OF PROVING THAT (FOR EXAMPLE) THE NUMBERS IN A SET ARE ALL EQUAL TO x BY SHOWING THAT THE LARGEST AND SMALLEST NUMBERS IN THE SET ARE BOTH EQUAL TO x.

SECANT LINE 26–27, 58, 75

A LINE CONNECTING TWO POINTS ON A CURVE; THE SLOPE OF A SECANT LINE CAN BE USED TO APPROXIMATE THE SLOPE OF A TANGENT LINE.

SECOND DERIVATIVE 178

    THE DERIVATIVE OF A DERIVATIVE.

SET THEORY 11

    A BRANCH OF MATHEMATICS THAT STUDIES COLLECTIONS
    OF OBJECTS (THAT IS, SETS).

SLOPE 22–28, 58, 74–75

    THE INSTANTANEOUS RATE OF CHANGE OF A FUNCTION
    AT A POINT; FOR A STRAIGHT LINE, THE SLOPE IS RISE
    OVER RUN.

SPEED 15–28, 58, 72–73

    THE INSTANTANEOUS RATE OF CHANGE OF POSITION; SPEED IS THE
    ABSOLUTE VALUE OF **VELOCITY**.

SUM RULE 86–87

    FOR DIFFERENTIABLE FUNCTIONS $f(x)$ AND $g(x)$, THE DERIVATIVE OF THEIR SUM IS
    $\frac{d}{dx}[f(x)+g(x)] = \frac{d}{dx}f(x) + \frac{d}{dx}g(x)$.

TANGENT LINE 25–28, 74–75

    BEST DESCRIBED GRAPHICALLY, THE TANGENT LINE TO A CURVE AT A POINT IS THE LINE THAT "JUST
    TOUCHES" THE CURVE AT THAT POINT.

VELOCITY 72–73, 81

    THE INSTANTANEOUS RATE OF CHANGE OF POSITION *IN A GIVEN DIRECTION*.

    AND **ACCELERATION** 177

    AND **CONSTANT MULTIPLE RULE** 88

    AND **FUNDAMENTAL THEOREM OF CALCULUS** 161–65

    AND **SUM RULE** 86

ZENO 56–57, 62

    A GREEK PHILOSOPHER, ZENO ($\approx 490-430$ B.C.E.) IS BEST KNOWN FOR HIS PARADOXES.

THE **SLOPE** AT ANY POINT IS MEASURED WITH THE **TANGENT LINE**.

RISE

RUN

A NOTE ABOUT THE AUTHORS

GRADY KLEIN IS A CARTOONIST, AN ANIMATOR, AND A GRAPHIC DESIGNER WHO SPECIALIZES IN SIMPLIFYING COMPLEX SUBJECTS. HIS MOST RECENT BOOK IS *PSYCHOLOGY: THE COMIC BOOK INTRODUCTION*. SAMPLES OF HIS WORK CAN BE FOUND AT WWW.GRADYKLEIN.COM.

YORAM BAUMAN, Ph.D., IS AN ECONOMIST WHO PERFORMS AT UNIVERSITIES AND CORPORATE EVENTS AROUND THE WORLD AS "THE WORLD'S FIRST AND ONLY STAND-UP ECONOMIST." HIS WEBSITE IS WWW.STANDUPECONOMIST.COM.

KLEIN AND BAUMAN HAVE PREVIOUSLY COLLABORATED ON *THE CARTOON INTRODUCTION TO CLIMATE CHANGE*, THE TWO-VOLUME *CARTOON INTRODUCTION TO ECONOMICS*, AND, MOST RECENTLY, *THE CARTOON INTRODUCTION TO DIGITAL ETHICS*.